Industrial Safety
and Health Management

Industrial Safety and Health Management

C. Ray Asfahl, Ph.D., P.E., C.S.P.

Professor of Industrial Engineering and Chairman of Studies
Operations Management
University of Arkansas, Fayetteville

Prentice-Hall, Inc.   Englewood Cliffs, New Jersey 07632

Library of Congress Cataloging in Publication Data

Asfahl, C. Ray, 1938—
  Industrial safety and health management.

  Bibliography: p.
  Includes index.
  1. Industrial safety. 2. Industrial hygiene.
I. Title. [DNLM: 1. Accidents, Occupational—Prevention
and control—United States. 2. Occupational Diseases—
Prevention and control—United States. 3. Occupational
Health Services—Organization and Administration—
United States. 4. United States. Occupational Safety
and Health Administration. WA 400 A817i]
T55.A83 1984   613.6'2   83-11023
ISBN 0-13-463141-2

Editorial/production supervision and interior design: *Shari Ingerman*
Cover design: *Jeannette Jacobs*
Manufacturing buyer: *Tony Caruso*

Printed in the United States of America

10 9 8 7 6 5 4 3 2 1

ISBN 0-13-463141-2

Prentice-Hall International, Inc., *London*
Prentice-Hall of Australia Pty. Limited, *Sydney*
Editora Prentice-Hall do Brasil, Ltda., *Rio de Janeiro*
Prentice-Hall Canada Inc., *Toronto*
Prentice-Hall of India Private Limited, *New Delhi*
Prentice-Hall of Japan, Inc., *Tokyo*
Prentice-Hall of Southeast Asia Pte. Ltd., *Singapore*
Whitehall Books Limited, *Wellington, New Zealand*

To Vicki and to Eloise Crowley
of Republic Safety Institute

Contents

3 CONCEPTS OF HAZARD AVOIDANCE   35

4 IMPACT OF FEDERAL REGULATION   59

5 BUILDINGS AND FACILITIES   74

6 HEALTH AND ENVIRONMENTAL CONTROL   93

Preface

If your desire is to read a book that carries on the crusade for occupational safety and health, put this one aside. Rather than attempt to persuade you, this book tries to help you to *understand* risks and perceive alternatives to eliminate them or reduce their impact. What you decide to do with your knowledge is up to you, but if you are like me, learning about safety and health indirectly makes you become more careful and at least aware of risks.

This book is intended for managers—for those who are already managers and for others, including college students, who would like to become managers. Managers plan, coordinate, and control the work of other people. Some Safety and Health Managers manage safety and health alone. But even more of them are general business executives who have a lot of other things to manage, too. Either way, Safety and Health Managers need all the help they can get.

Safety and Health Managers formerly dealt principally with physical risks. This is still true, but a new dimension has been added—the legal risk of violation of applicable standards. This book is about both kinds of risks.

The federal Occupational Safety and Health Administration (OSHA) made an indelible mark on the field of industrial safety and health during the decade of the 1970s. While making this mark, OSHA maligned many industries and became a very controversial agency. From the change in political climate of the early 1980s emerged a powerful movement to either reshape OSHA or abolish it. Regardless of OSHA's fate, its legacy remains: The field of industrial safety and health will never be the same, and this book reflects the impact OSHA has had.

I have no particular authority to write this book. Actually, the thoughts are not mine but those of the many professionals in the field to whom I have had the privilege to listen. I can think of the names of many of these people but have decided not to name any of them here for two reasons. The first is that it is difficult to draw the line. As soon as I think of someone I should credit, I can think of others who would be conspicuous in their absence from the list. But there is another, more important reason for not mentioning any names—read on.

I have many friends who think OSHA is *great,* and as many other friends who *despise* OSHA. My friends who love OSHA are going to think I am biased *against* OSHA. My friends who hate OSHA are going to think I am biased *for* OSHA. I suppose that this predicament is inevitable for anyone who writes about a controversial subject. I would ask this one favor of readers of this book: If you find a particular part of it that is offensive to you, please do not label me as biased until you have read all of the book. Then after you have finished, draw your conclusion about my biases—in either direction. Remember to ask yourself whether the bias is yours or mine.

If I had decided to mention names, the first ones on the list would be employees of the Arkansas Department of Labor, especially the consultants of the Department who go out into industry, when requested, to assist in safety and health problems and in compliance with federal standards. For several years the Department contracted for me to provide in-house training for their consultants. But in truth they trained me more than I trained them. Their contribution to my knowledge and to this book is greatly appreciated. Many of the guest speakers in these in-house training sessions have also contributed to my knowledge.

The federal OSHA agency itself has been very helpful in providing information. I received courteous service from the local office here in Arkansas, from the Dallas Regional Office, from the OSHA training school in Des Plaines, Illinois, and from the national office whenever I inquired. They openly spread their facts and figures and policies before me. Of course, this is what Americans should expect of their democracy.

Many of my students in industrial engineering at the University of Arkansas have done library research and computer analyses which have contributed to the content of this book. I am proud that after graduation some of these students have chosen careers in the field. At least one has become an OSHA officer; others are Safety and Health Managers in corporations around the country.

Finally, I would like to credit Safety and Health Managers from industry who have given me ideas. Most of this contribution has come from seminars, sponsored by University of Arkansas here in my home state and by other public and private institutions nationwide.

Ray Asfahl

1

The Safety and Health Manager

Everyone wants a safe and healthful workplace, but what each person is willing to do to achieve this worthwhile objective can vary a great deal. As a result, the management of each firm must decide at what level, along a broad spectrum, the safety and health effort will be aimed. Some managers deny this responsibility and attempt to leave the decision to employees. This strategy seems to square with hallowed principles of personal freedom and individual responsibility. But such a denial of responsibility by management results in a decision by default, and usually the result is a relatively low level of safety and health in the workplace.

Is the foregoing an indictment on the judgment of the individual worker? Not really, because without a commitment on the part of management the worker usually is unable singlehandedly to build safety into his or her job station. It has been proven that the attitude of the worker is the most important determinant for his or her safety, but attitude alone cannot make a dangerous job safe. Furthermore, even if a given worker has a strong inclination to be careful and to guard his or her health, there are plenty of production incentives and other quite natural incentives to undermine safe attitudes when management has not made a commitment to safety and health.

One person, usually designated as "Safety Director" or "Industrial Hygienist," sets the tone of the safety and health program within a firm. In fact, right at the start it says something about the commitment of management when a firm decides to designate a person by title to the responsibility of safety and health. But naming someone "Safety Director" or "Manager of Safety and Health" is just a beginning step. Many such persons have little authority and have been largely ignored by management and worker alike, especially in the past. It was not unusual for a

Safety Director's work to be typified by public relations activities, such as posting motivational signs and compiling statistics. These are still important functions, but much more responsibility for this function is now recognized.

Something happened in the 1970s to change dramatically the role of the typical Safety Director in industrial firms throughout the country. The passage of the Occupational Safety and Health Act of 1970 created the Occupational Safety and Health Administration (OSHA), a federal agency whose regulations would have a large impact on the role of the typical Safety Director. Chapter 4 discusses this impact in detail, but the balance of this chapter discusses the enlarged role of the person charged with industrial safety and health. Prior to OSHA the field of occupational health was largely ignored by everyone in the plant except the plant nurse. Interfering with production schedules to alleviate a safety or health problem was virtually never attempted.

In describing the functions of today's executive charged with the safety and health responsibility, this text will use the designation Safety and Health Manager, recognizing the dual nature of the job. Also, the term "manager" envisions the enlarged scope of responsibility which includes analysis of hazards, compliance with standards, and capital investment planning, in addition to the conventional functions described earlier. The purpose of this text is to provide tools and guidelines to Safety and Health Managers to help them execute their enlarged duties.

Dealing with applicable standards is the greatest challenge facing today's Safety and Health Manager, and to meet this challenge is the primary purpose of this book. Since only 10% of the standards generate 90% of the activity, Safety and Health Managers need guides to the important parts of the standards. Frequently cited standards should receive prime attention because they indicate areas in which industries are having difficulty complying or areas in which enforcement agencies are giving a great deal of attention. In either event, Safety and Health Managers have a need to know these frequently cited standards so that they can bring their facilities within compliance. Besides the frequency of citation, Safety and Health Managers need to know the "why" behind the standards. Until the Safety and Health Manager learns what hazards a particular standard is intended to prevent, he or she will have a difficult time persuading either management or employees that a given situation needs correction.

A REASONABLE OBJECTIVE

Top management sometimes turns a deaf ear to the pleas of the Safety and Health Manager for plant improvements. But Safety and Health Managers sometimes are crusaders with a one-track mind. Any Safety and Health Manager who feels that elimination of workplace hazards is an indisputable goal is naive. In the real world we must choose between:

1. Hazards that are physically feasible to correct

Real world

2. Hazards that are physically feasible, but are economically infeasible, to correct
3. Hazards that are economically feasible to correct

Until the Safety and Health Manager comes to grips with this reality, he or she cannot expect to enjoy the approval of top management. Some Safety and Health Managers have faced this reality on the surface, but within their hearts they resent the attitudes of top executives who are unwilling to support their efforts to eliminate all workplace hazards. But this resentment is unjustified because it is an unrealistic and naive strategy to attempt to eliminate all hazards.

It may be surprising to some readers to discover that this book, which is supposed to be about safety and health, does not really advocate elimination of all workplace hazards. Such a goal is unattainable, and to reach for it is poor strategy because it ignores the need for discriminating among hazards to be corrected. To see how such a naive strategy is not even in the interest of safety or health, consider the following case study.

> **Case Study 1.1.** A Safety and Health Manager receives three suggestions from three different operating personnel as follows:
>
> 1. Install a drain to remove water that occasionally collects around the die-casting area.
> 2. Post a warning sign that advises forklift truck drivers to slow down.
> 3. Improve sanitation by cleaning the rest rooms more frequently.

There is a safety and/or health rationale for the correction of all three of the problems listed above. Should they be corrected?

Some managers would accept the safety and health rationale as *all they need* to begin action to correct the problems listed in Case Study 1.1. But this would be a naive response. More data are needed to decide what to do. While busying the plant maintenance department to correct the foregoing three problems, which may or may not be consequential, a serious electrocution or respiratory hazard may be going unchecked or maybe even unnoticed. By reacting to every hazard that happens to show up, the Safety and Health Manager may be missing opportunities to have a really significant impact on worker safety and health. At the same time such overreaction may also be deteriorating the Safety and Health Manager's credibility with top management. Even the law does not call for the elimination of all hazards, just the "recognized" ones. Therefore, let it be clearly understood that our objective is to eliminate unreasonable risks, not all risks in the workplace. The goal of this book, then, is to assist the Safety and Health Manager in detecting hazards and in deciding which ones are worthwhile to correct. The goal is an ambitious one, and this book certainly claims no breakthroughs for solving this difficult problem. But any light that can be shed on the mystery of which hazards are more significant and which standards, among the thousands, are most important is sorely needed by Safety and Health Managers everywhere.

SAFETY VERSUS HEALTH

This chapter has already implied that early "safety directors" did not emphasize health problems. It is essential that today's Safety and Health Manager give sufficient attention not only to safety hazards but also to health hazards which are steadily gaining in importance as new data about industrial disease are being uncovered.

What really is the difference between safety and health? The words are so common that almost everyone has a firm image of the concept of safety versus the concept of health. There is no question that machine guarding is a safety consideration, and airborne asbestos is a health hazard, but some hazards, such as those associated with paint spray areas and welding operations, are not so easy to classify. Some situations may be both a health and a safety hazard. This text will draw the following line between safety and health:

Safety deals with acute hazards, whereas health deals with chronic hazards.

An acute effect is a sudden reaction to a severe condition; a chronic effect is a long-term deterioration due to a prolonged exposure to a milder adverse condition. Everyday concepts of health and safety fit this definition which separates the two. Industrial noise, for instance, is usually a health hazard because it is usually the long-term exposure to noise levels in the range 90 to 100 decibels that do the permanent damage. But noise can also be a safety hazard because a sudden acute exposure to impact noise can *injure* the hearing system. Many chemical exposures have both an acute and a chronic effect and are thus both safety and health hazards.

Industrial hygienists, those who concentrate on health hazards, are known by their sophisticated instruments and scientific expertise. These tools are necessary to the industrial hygienist because of the tiny effects they must measure to determine whether a chronic hazard exists. By contrast, the safety specialist, instead of being an expert with precise scientific instruments, usually has more industrial process experience and practical on-the-job knowledge. This difference in backgrounds generates some confrontation between safety and health professionals, and though they should be partners, they often compete.

The bases of competition between safety and health professionals are classics: young versus old, new versus old, and education versus experience. Safety professionals are usually older and have more industrial experience; their career field is more traditional and more entrenched in industrial organizations. Health professionals are typically younger, have more college education, and occupy newer job positions.

Degree of hazard is another point of contention between safety and health: both sides think their hazards are more grave. Safety professionals can point to fatalities on the job and feel an urgency in protecting the worker from imminent danger from accidents. To them the industrial hygienists with their meters, pumps, and test tubes are checking for microscopic hazards which are not really pressing. But to the industrial hygienist the safety professional's work sometimes seems superficial. The industrial hygienist is fighting those insidious unseen hazards on the job which can be just as lethal as a falling crane. There are probably more occupational health fatalities than safety fatalities, but the statistics will not reflect this difference because the health fatalities are delayed and are often never diagnosed.

Because of age and experience a safety professional will often have managerial control over health functions, and this can rankle the industrial hygienist. The Safety and Health Manager should be aware of this sensitivity and be careful to recognize the importance of both fields of interest.

ROLE IN THE CORPORATE STRUCTURE

Most Safety and Health Managers wear several hats, especially in smaller firms. Often they are responsible also for security, although the security function has little in common with workplace safety and health. Some Safety and Health Managers are also personnel managers, and even more frequently they report to the personnel manager. This is a fairly natural arrangement in that it emphasizes the importance of worker training, statistics, job placement, and the industrial relations aspect of safety and health. The growing importance of engineering to workplace safety and health, however, strains the placement of the Safety and Health Manager within the personnel department, which traditionally has little interaction with engineering.

The Safety and Health Manager is virtually never associated with the purchasing function, but one of the first goals of the Safety and Health Manager should be to obtain some input to the purchasing process. Used equipment dealers, even new equipment dealers, often have bargain-priced machines, compressors, tractors, forklifts, and other equipment which fail in some way or another to meet standards. The purchasing agent is usually not knowledgeable in safety and health standards and is easy prey for these dealers because the price is right. What is needed is a knowledgeable person to check specifications and prevent the costly purchasing error of buying equipment that does not meet current safety and health standards. When standards change, another category of equipment sometimes becomes obsolete, and the Safety and Health Manager should warn Purchasing when these changes occur.

A recent concept of the Safety and Health Manager is as a liaison with government agencies, a condition brought about by the arrival of OSHA. Some Safety and Health Managers have a dual responsibility for environmental protection activities. Sometimes the Safety and Health Manager is considered a staff member of the legal department. This arrangement stresses the adversarial role and is not recommended because it tends to detract both from workplace safety and health and also from constructive relations with enforcement agencies.

A related field is consumer product safety. The Consumer Product Safety Commission (CPSC) is a federal agency whose organization is obviously patterned after OSHA's. The Product Safety and Liability Act was passed the year after the passage of the OSHA Act, and the wording of the two laws is remarkably similar. Although both fields consider the safety of machines and equipment, the CPSC concentrates on the responsibility of the manufacturers of the machines and equipment, whereas OSHA concentrates on the responsibility of the employer who places the equipment into use in the workplace.

RESOURCES AT HAND

The demands for training materials, the need for ideas for hazard correction, and the importance of the latest interpretations of standards guarantees that the successful Safety and Health Manager will not try to work alone, insulated from the career field. A variety of resources have arisen to meet these needs of the Safety and Health Professional.

Professional Certification

Safety and Health Managers can establish themselves with their peers as well as with their employers by achieving professional certification. This requires relevant work experience, letters of recommendation, and an examination. Education is also required, but there is partial trade-off between education and experience. Safety professionals should apply to

> Board of Certified Safety Professionals of America
> 501 South 6th Street
> Champaign IL 61820

Health professionals should apply to

> American Board of Industrial Hygiene
> 475 Wolf Ledges Parkway
> Akron OH 44311

The examinations are quite difficult, and the background screening process is stringent. Very few individuals qualify for certification in both fields.

Professional Societies

Two professional societies dominate the career field of occupational safety and health:

> American Society of Safety Engineers (ASSE)
> 850 Busse Highway
> Park Ridge IL 60068

American Industrial Hygiene Association (AIHA)
66 South Miller Road
Akron OH 44313

A somewhat narrower but highly significant organization is

American Conference of Government Industrial Hygienists (ACGIH)
PO Box 1937
Cincinnati OH 45201

A very important committee of this organization is

Committee on Industrial Ventilation
PO Box 16153
Lansing MI 48901

This committee's publication *Industrial Ventilation* (Ref. 41) is probably the best recognized manual of recommended practice in the field of ventilation.

National Safety Council

Although not a professional society in the strict sense of the term, the National Safety Council (NSC) is the most important organization to the field. The Council is headquartered at

National Safety Council
444 North Michigan Avenue
Chicago IL 60611

The NSC is broad in scope and encompasses all kinds of safety, not just occupational safety. The membership of the NSC is principally organizations and businesses. It has been in existence for over 70 years.

Corporate membership in the National Safety Council affords many benefits for the Safety and Health Manager. The Council is the principal focal point for information about safety hazards. The library at the Council's national headquarters contains a wealth of information, and although it is open to the public, companies who are members of the Council have special privileges for copying services and research. Each year the Council publishes comprehensive summaries of accident statistics in its book *Accident Facts* (Ref. 1). The NSC's *Accident Prevention Manual for Industrial Operations* (Ref. 47a), now in its eighth edition and in two volumes, is the most authoritative and comprehensive reference in the field.

The National Safety Council makes use of its current statistical data to continually recognize and award member firms with excellent safety awards. The annual edition of *Work Injury and Illness Rates* (Ref. 78) lists company names by Standard Industrial Classification (SIC) Code which have the "best records known in industry."

Standards Institutes

The age of OSHA enforcement has brought increased recognition of the national standards-producing organizations. Most prominent among these are

> American National Standards Institute (ANSI)
> 1430 Broadway
> New York NY 10018
>
> National Fire Protection Association (NFPA)
> 470 Atlantic Avenue
> Boston MA 02210
>
> American Society of Mechanical Engineers (ASME)
> 345 East 47th Street
> New York NY 10017
>
> American Society for Testing and Materials (ASTM)
> 1916 Race Street
> Philadelphia PA 19103

These organizations have standing committees that invite public comment and prepare voluntary standards for occupational safety and health. At the outset OSHA issued many existing standards produced by these organizations, designating such existing standards as representing the "national consensus." Most of OSHA's national consensus standards are derived from either ANSI or NFPA standards.

Trade Associations

If a problem pertains to a specific industry or type of equipment, an association of manufacturers may be called upon to furnish safety and health data. Some people decry such use of trade association data as biased, but many careful studies of safety and health problems have been performed by trade associations. Useful associations in this regard are

1. American Foundrymen's Society (AFS)
2. American Iron and Steel Institute (AISI)
3. American Metal Stamping Association (AMSA)
4. American Petroleum Institute (API)
5. American Welding Society (AWS)
6. Associated General Contractors of America (AGCA)

7. Compressed Gas Association (CGA)
8. Industrial Safety Equipment Association (ISEA)
9. Institute of Makers of Explosives (IME)
10. National Electrical Manufacturers Association (NEMA)
11. National LP-Gas Association (NLPGA)
12. National Machine Tool Builders Association (NMTBA)
13. Scaffolding, Shoring, and Forming Institute (SSFI)

National trade associations are especially useful in providing audiovisual training materials if one is willing to accept the fact that such materials also serve to promote the industry's products.

Government Agencies

A program of free consultation is usually provided by state agencies and in some states by private consulting firms. There is an understandable reluctance to call a government agency for help with a safety standards problem, but the purpose of these consultation agencies is to assist, not to write citations. In most states the consultation function is performed by a completely separate agency from the enforcement agency. In states that have comprehensive state plans, with one agency responsible for both enforcement and consultation, care is taken to maintain confidentiality of records.

The National Institute for Occupational Safety and Health (NIOSH) has a wealth of research data on the hazards of specific materials and processes. NIOSH uses these data to write criteria for recommended new standards. In recent years, however, NIOSH has suffered from severe cutbacks in funding.

OSHA itself can be of value to the Safety and Health Manager who seeks information. Some Safety and Health Managers would never consider calling OSHA to discuss a problem for fear of precipitating an inspection. However, problems can be posed in a hypothetical setting, and OSHA personnel can understand the need for keeping such questions hypothetical. Most OSHA personnel will be glad to furnish whatever answers to questions are available in the interest of encouraging employers to keep their facilities safe and healthful. OSHA has also opened the doors to its national training institute in Des Plaines, Illinois, for training the general public in "voluntary compliance."

SUMMARY

Safety and Health Management is a career field that has been both enhanced and made more challenging by the institution of the federal OSHA agency. This chapter has provided a brief introduction to the career field and to some of the organizations and boards that have given the field identity and assistance in fulfilling its mission.

Chapter 2 proceeds to describe how Safety and Health Managers can go about performing their responsibilities within their organizations. Chapter 3 addresses the principal goal of the Safety and Health Manager—the reduction of workplace hazards—describing four basic approaches to this problem. OSHA has had such an impact on the field that it deserves a chapter of its own; Chapter 4 describes OSHA in general, including both the positive and the negative aspects of this controversial agency. The remaining chapters address specific hazard categories, advising Safety and Health Managers what to do to eliminate hazards while complying with established standards. Any factual data or approach strategies that can assist the Manager in understanding the hazard mechanisms and in tackling the most significant problems first will be helpful, even if many questions are left unanswered.

EXERCISES AND STUDY QUESTIONS

1.1. Why are some safety and health standards cited more frequently than others?

1.2. What else, besides frequency of citation, does a Safety and Health Manager need to know about safety and health standards?

1.3. Should Safety and Health Managers attempt to eliminate all workplace hazards? Why or why not?

1.4. Identify three categories of hazards with respect to feasibility of correction.

1.5. Describe at least two disadvantages of overreacting to minor hazards in the workplace.

1.6. How does a safety hazard differ from a health hazard?

1.7. Name three safety hazards and three health hazards.

1.8. Name some chemicals that are both health and safety hazards.

1.9. Name some physical agents that can be both health and safety hazards.

1.10. Why does an industrial hygienist need more scientific instruments to evaluate hazards than does a safety specialist?

1.11. Which type of hazard appears to be more grave: safety or health?

1.12. What aspects of the Safety and Health Manager's job are related to the personnel department?

1.13. What disadvantage is associated with placing the Safety and Health Manager within the personnel department?

1.14. What disadvantage can be seen in placing the Safety and Health Manager within the legal department of a firm?

1.15. Compare the missions of the two federal agencies: OSHA and CPSC.

1.16. What national organization is of the greatest importance to the Safety and Health Manager?

1.17. What is ANSI, and what relationship does it have to the field of safety and health?

2

Development of the Safety and Health Function

The safety and health function has both line and staff characteristics, and the Safety and Health Manager needs to recognize which elements of the function belong to which. The physical accomplishment of workplace safety and health is a line function. For example, operator work practices are the responsibility of the workers themselves as directed by their line supervision. In industries where maintenance departments are recognized as another line function, the correction of facilities problems is again the direct responsibility of the maintenance operators and their line supervision. The Safety and Health Manager then performs a staff function by acting as a "facilitator" in assisting, motivating, and advising the line function in achieving worker safety and health.

The interest of line personnel in receiving this advice and assistance from the Safety and Health Manager is going to depend on how important the goal of safety and health is to top management. The successful Safety and Health Manager will be keenly aware of this need for top management support. Respect and approval of top management must be won by responsible decisions and actions by the Safety and Health Manager. A necessary ingredient of these decisions and actions is a recognition of the important principle stated in Chapter 1, that the goal is to eliminate unreasonable hazards, not all hazards. The respect and approval of top management is difficult to win because Safety and Health Managers too often are such emotional crusaders for the cause that they lose their credibility and with it lose their eligibility to be considered a "manager." On the other hand, federal regulation has added a measure of urgency and credibility to efforts to make industries safe and healthful.

Similarities can be drawn between the safety function and other staff functions, such as quality control and production control. Like safety and health, quality and production goals must be achieved by line personnel facilitated by the staff function. This principle is recognized in such clichés as "You can't inspect quality into a product" and "Safety is everybody's business."

Once the approval of top management has been won, the Safety and Health Manager is advised to document this approval in a written safety and health policy statement issued by top management. This written policy becomes the documented authority to line personnel that top management does have safety and health goals and wants these goals met. The everyday actions of management then reinforce the written policy, as discussed in Chapter 1. If, however, top management fails to practice what it preaches in the policy statement, it is the duty of the Safety and Health Manager to go back to management and redetermine its level of commitment to safety and health.

Having established by word and deed management's commitment to safety and health, the Safety and Health Manager is ready to proceed with the staff function of facilitating the safety and health program throughout the plant. To do this, plant operating personnel will have needs that the Safety and Health Manager can satisfy. To make workers aware of hazards, supervisors and the workers themselves need regular training in hazard recognition and correction. Statistics and accident records are needed to keep management and operating personnel advised of how well the company and its departments are doing in achieving their safety and health goals. Sometimes the Safety and Health Manager can provide a plant-wide stimulus for worker safety and health by means of contests and awards for safety performance. Finally, of increasing importance is the Safety and Health Manager's role in dealing with safety and health standards and in assisting operating personnel in achieving compliance with these standards. The remainder of this chapter itemizes staff functions of the Safety and Health Manager's office with guidance for the development of each.

WORKERS' COMPENSATION

"Workers' Compensation" laws, called "Workmen's Compensation" prior to the sexism issue, provided an initial structure to industrial safety. The first such laws were introduced in state legislatures in 1909, and now all states have comparable legislation. It is invariably the duty of the Safety and Health Manager to implement the Workers' Compensation system within the plant, so it is considered here as the first staff function to be discussed.

Workers' Compensation legislation has the ostensible purpose of protecting the worker by providing statutory compensation levels to be paid by the employer for various injuries that may be incurred by the worker. There is an ulterior feature, however, which provokes labor to be dissatisfied with the Workers' Compensation system. This feature is the immunity from additional liability that the Workers' Compensation system grants to the employer, except in cases in which "gross negligence" can be proved.

Table 2.1 itemizes example statutory compensation levels for various permanent injury types. To most people the various levels seem too low to compensate adequately for the permanent injury to the worker. Historical evolution of the rates is slow, and it can be seen that public sensitivity to worker injuries has increased over the years. This sensitivity has resulted in outcries for reform of the Workers' Compensation system. On the other side of the issue is management's position that industry can never fully compensate monetarily for everything that may happen to workers in the course of their duties. Since some risk is inescapable in any line of work, management's general position is that in consideration of the salary and wages that workers receive, part of the normal risk of injury must be borne by the worker.

Typically, the firm does not pay the Workers' Compensation payments

TABLE 2.1 SAMPLE STATUTORY COMPENSATION LEVELS FOR PERMANENT INJURIES (COMPENSATION AT 66⅔% OF AVERAGE WEEKLY PAY)

Type of permanent injury	Compensation level[a] (weeks)
Arm amputated	
At or above the elbow	200
Below the elbow	150
Leg amputated	
At or above the knee	175
Below the knee	125
Hand amputated	150
Thumb amputated	60
Finger(s) amputated	
First	35
Second	30
Third	20
Fourth	15
Foot amputated	125
Toe amputated	
Great toe	30
Other toes, per toe	10
Loss of sight in one eye	100
Loss of hearing in one ear	40
Loss of hearing in both ears	150
Loss of testicle	50
Loss of both testicles	150

[a]These compensation levels are in addition to any compensation paid for the healing period. The sample compensation levels were obtained from Workers' Compensation levels for Arkansas and are intended only as an approximate guide. Exceptions and special cases exist, and tables vary somewhat from state to state.

Source: Arkansas Workers' Compensation Commission, Ref. 3.

directly; rather, it carries insurance against compensation claims. The insurance company is vitally interested in the safety and health within the plant, and this provides a major impetus to the development of the safety and health program. The accident experience of the firm is reflected in the levels of the insurance premium, which can be adjusted up or down depending on plant safety experience. A good insurance company will make regular inspections of facilities to be sure that installations and practices are safe. This is a direct and measurable monetary stimulus to the safety program.

Some companies choose to self-insure against Workers' Compensation claims. This may make economic sense when claims experience and premium levels are compared. But the intangible benefits provided by the insurance company must also be considered in order to make a rational decision. Besides the regular inspections mentioned earlier, the insurance companies are valuable sources of technical advice to their clients. Many insurance companies provide training films and other valuable aids to the conduct of the safety and health program. If the Safety and Health Manager is not receiving these aids and services from the insurance company, he or she should request them and also perhaps consider alternative vendors when policy renewal time presents itself.

The number of companies who have elected to self-insure has led to a new type of consultant called a "loss control representative." This consultant's objective is to keep Workers' Compensation claims low by supplying the type of services normally provided by the insurance carrier. A significant part of these services is maintaining close relationships with employees who do file claims. This serves the purpose of showing an interest in the well-being and encouragement of the honest claimant and at the same time uncovering evidence of fraudulent claims in the case of the dishonest worker who is either malingering or is truly injured but whose injury occurred off the job.

RECORDKEEPING

The National Safety Council established the first national system of industrial safety recordkeeping standardized and designated the Z16.1 system by the American National Standards Institute. In the 1970s, the federal OSHA agency set mandatory recordkeeping requirements very similar to the Z16.1 system, which was voluntary. There were some differences, however, which make year-by-year comparisons of safety and health records infeasible when one year is based on traditional Z16.1 records and another on the federal system. This is particularly unfortunate in that it confounds attempts to determine from statistical records whether the federal agency has had any beneficial impact on worker safety and health. Some specific industries and hazard categories, such as trenching and excavation cave-ins in construction, have shown visible improvements since OSHA's inception, but other gains have been obscured by the change in the statistical records system. Other variations in conditions, such as employment levels and recession cycles, have also acted to blur statistical comparisons.

The recording of worker fatalities is more consistent than that of injuries and illnesses; thus fatality statistics can be used to observe trends both before and after federal regulation. Figure 2.1 shows that the long-range trend of industrial fatalities is downward. Note that the implementation of the OSHA law in the early 1970s has had very little visible impact on this trend when all industries are considered together.

Traditional Indexes

Familiar statistical measures are "frequency" and "severity," which were defined by the old Z16.1 system. Frequency measured the numbers of cases per standard quantity of workhours, and severity measured the total impact of these cases in terms of "lost workdays" per standard quantity of workhours.

Some injuries, such as amputations, are quite severe but might result in few or no lost workdays. To avoid a distortion in severity rates in such cases, standard lost-workday charges were arbitrarily set for permanent injuries such as amputations or loss of eyesight. The greatest need for arbitrary severity charges was for fatalities, because when you think about it, a fatality is not really a lost-workday case in the literal sense of the term; neither is a permanent total disability because the worker never works again.

Another obsolete term is "seriousness," which was the ratio of severity to frequency. This produced a measure of the relative average importance of injuries and illnesses without regard to the number of hours worked during the study period.

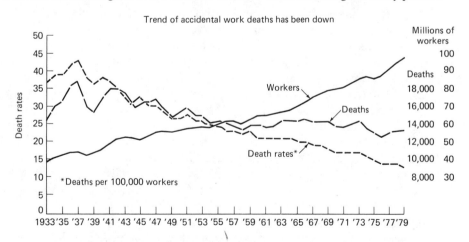

Figure 2.1. Accidental work deaths (1980 rates). (Courtesy of the National Safety Council, Chicago; used with permission.)

Incidence Rates

The current system of recordkeeping represents an enlargement of the old Z16.1 system. The total injury/illness treatment incidence rate includes all injuries or illnesses which require medical treatment. Compare this with the traditional "frequency" rate, which included only those cases in which the worker missed at least a day of work.[1] "Medical treatment" does not include simple first aid, preventive medicine (such as tetanus shots), or medical diagnostic procedures. As in the old system, the number of injuries is divided by the number of hours worked during the period covered by the study and then is multiplied by a standard factor to make the rate more understandable. Specifically,

$$\frac{\text{total injury/illness}}{\text{incident rate}} = \frac{\text{number of injuries and illnesses} \times 200,000}{\text{total hours worked by all employees during the period covered}} \qquad (2.1)$$

Without the factor of 200,000, the incidence rate would be a very small fraction indeed, as it should be. One should expect a very small number of recordable injuries and illnesses per single hour worked! The choice of the number 200,000 is not entirely arbitrary. A full-time worker typically works approximately 50 weeks per year at 40 hours per week. Thus the number of hours worked per year per worker is approximately

$$40 \text{ hours/week} \times 50 \text{ weeks/year} = 2000 \text{ hours/year}$$

So 200,000 hours represents the number of workhours spent by 100 workers in a year:

$$100 \text{ workers} \times 2000 \text{ hours/year/worker} = 200,000 \text{ hours/year}$$

Thus the total injury/illness incidence rate represents the number of injuries expected by a 100-employee firm in a full year, if injuries and illnesses during the year follow the same frequency as observed during the study period. Note from equation (2.1) that the actual period for gathering the incidence rate data need not be a year or any other specific time period. A fairly long period is needed, however, to obtain a representative number of cases, especially when the incidence is low. A typical data collection period is one year.

Sometimes the Safety and Health Manager will want to relate current total injury/illness incidence rates to the traditional "frequency" rate. The old "frequency" rate used a factor of 1,000,000 hours instead of 200,000. Thus rates were standard as "per million manhours," as they were called in those days. Note that such a factor related to a standard year for a firm employing 500 employees, not 100 employees. Thus the old "frequency" rates should be higher than the current total injury/illness incidence rates—but they generally are not because one must remember that the current rate includes all cases involving medical treatment, not

---

[1] The ANSI Z16.1 system labeled such lost-workday cases as "disabling injuries" whether the disability was temporary or permanent.

just lost-workday cases. Also, days in which the worker was still on the job but was unable to perform his or her regular job due to a injury or illness are now taken into account. These days are called "restricted work activity" days and may be lumped together with lost workdays or considered separately, depending on the statistic desired.

The term "incidence rate" is really a general term and in addition to the total injury/illness incidence rate includes the following:

1. Injury incidence rate
2. Illness incidence rate
3. Fatality incidence rate
4. Lost-workday-cases incidence rate
5. Number-of-lost-workdays rate
6. Specific hazard incidence rate

$$\frac{X\ 200,000}{\text{Total hrs worked during specific period}}$$

All of the foregoing rates use the standard 200,000 factor. The lost-workday-cases incidence rate compares approximately to the old frequency rate except for the factor of 200,000 instead of 1,000,000.

In counting the number of lost workdays the date of the injury or onset of illness should not be counted even though the employee may leave work for most of that day. Thus if the employee returns to his or her *regular* job and is able to perform all regular duties full time on the day after the injury or illness, no lost workdays are counted. Also, when counting lost workdays, weekends, or other normal days off should not be counted if the worker would not have worked those days anyway. The number-of-lost-workdays rate compares to the old severity rate except that no arbitrary charges are assessed for permanent partial disabilities, and except for the 200,000 factor.

The specific hazard incidence rate is useful in observing a narrow slice of the total hazards picture. For specific hazards, injury incidence, illness incidence, fatality incidence, and all of the other rates can be computed. Care must be taken in selecting the corresponding total hours worked to be used in the denominator in calculating specific hazard incidence rates. Since specific hazards are more narrow and fewer workers are exposed, data should be collected over several years to achieve meaningful results for specific hazards incidence rates.

The National Safety Council has computed overall incidence rates by Standard Industrial Classification (SIC) code, and a chart of these computed rates is shown in Figure 2.2.

Recordkeeping Forms

The format for keeping injury and illness records has now been standardized nationwide. The basic form is the "Log of Occupational Injuries and Illnesses," displayed in Figure 2.3 (see pp. 20-21). The right half of the form is used as a summary to post

1980 incidence rates* of principal industries

Reporters to the National Safety Council

Incidence Rates*
CASES INVOLVING DAYS AWAY FROM WORK & DEATHS

Incidence Rates*
DAYS AWAY FROM WORK

Industry (Cases Involving Days Away from Work & Deaths)	Rate		Industry (Days Away from Work)	Rate
CHEMICAL	0.69 (3.23) △		COMMUNICATION	(26) †14
COMMUNICATION	0.86 (2.00)		CHEMICAL	(23) 14
AIRCRAFT	0.86 (3.32)		AIRCRAFT	(17) 16
AGRICULTURAL CHEMICALS	1.04 (3.95)		AGRICULTURAL CHEMICALS	(25) 20
TEXTILE	1.11 (5.75)		ELECTRIC, ELECTRONIC EQUIP.	(36) 23
ELECTRIC, ELECTRONIC EQUIP.	1.13 (4.12)		PETROLEUM & COAL PRODS.	(41) 27
PETROLEUM & COAL PRODS.	1.23 (6.05)		TEXTILE	(37) 29
PIPELINE TRANSPORTATION	1.46 (5.08)		PIPELINE TRANSPORTATION	(36) 35
OIL & GAS EXTRACTION	1.49 (5.36)		MOTOR VEHICLES	(59) 37
FURNITURE & FIXTURES	1.82 (13.78)		GAS	(48) 38
PRIMARY NONFERROUS METALS	1.89 (10.53)		FURNITURE & FIXTURES	(89) 40
MOTOR VEHICLES	1.97 (7.27)		ELECTRIC SERVICE	(63) 45
ELECTRIC SERVICE	2.15 (6.72)		RUBBER & PLASTICS	(69) 46
STEEL	2.19 (10.89)		LEATHER	(59) 47
WATER TRANSPORTATION	2.28 (5.04)		ALL INDUSTRIES	(61) 49
METAL MINING	2.41 (13.41)		OIL & GAS EXTRACTION	(50) 49
NONFERROUS ROLLING, DRAWING	2.54 (8.60)		SERVICES	(60) 50
RUBBER & PLASTICS	2.66 (6.93)		FABRICATED METAL PRODS.	(62) 52
MACHINERY, EXC. ELECTRICAL	2.71 (9.09)		PRIMARY NONFERROUS METALS	(90) 53
LEATHER	2.72 (9.46)		WHOLESALE & RETAIL TRADE	(58) 54
CEMENT	2.77 (16.50)		MACHINERY, EXC. ELECTRICAL	(74) 58
ALL INDUSTRIES	2.79 (7.81)		STEEL	(71) 58
BITUMINOUS COAL	2.81 (7.25)		NONFERROUS ROLLING, DRAWING	(79) 62
WHOLESALE & RETAIL TRADE	2.82 (6.87)		BITUMINOUS COAL	(66) 65
GAS	2.85 (7.96)		TRANSIT	(101) 66
FABRICATED METAL PRODS.	2.97 (10.16)		CONSTRUCTION	(74) 68
PAPER	3.52 (10.48)		CEMENT	(70) 70
SERVICES	3.78 (8.46)		SHIP & BOAT BUILDING	(111) 72
NONMETALLIC MINERALS	3.89 (15.20)		IRON & STEEL FOUNDRIES	(106) 76
FOOD	4.19 (11.76)		FOOD	(89) 78
CONSTRUCTION	4.29 (12.03)		METAL MINING	(92) 81
PRINTING & PUBLISHING	4.41 (9.23)		AIR TRANSPORTATION	(98) 83
TRANSIT	4.45 (8.22)		PRINTING & PUBLISHING	(89) 85
IRON & STEEL FOUNDRIES	4.83 (13.67)		PAPER	(102) 93
STONE, CLAY & GLASS	4.94 (12.62)		STONE, CLAY & GLASS	(114) 96
SHIP & BOAT BUILDING	5.40 (30.41)		WATER TRANSPORTATION	(102) 100
LUMBER & WOOD PRODUCTS	6.06 (15.32)		GENERAL GOVERNMENT	(118) 108
RAILROAD TRANSPORTATION	6.50 (11.66)		RAILROAD TRANSPORTATION	(126) 110
AIR TRANSPORTATION	7.09 (10.85)		NONMETALLIC MINERALS	(117) 113
GENERAL GOVERNMENT	7.20 (15.07)		MEAT PRODUCTS	(167) 131
RAILROAD EQUIPMENT	7.67 (17.83)		LUMBER & WOOD PRODUCTS	(185) 161
MEAT PRODUCTS	8.47 (24.57)		RAILROAD EQUIPMENT	(260) 218
TRUCKING	17.65 (36.83)		TRUCKING	(270) 247

Data compiled in accordance with OSHA Recordkeeping definitions.

* Incidence rates per 100 full-time employees, using 200,000 employee hours as the equivalent.

△ Rates in parentheses are for Total Recordable Cases.

† Rates in parentheses are for Total Lost Workdays.

ACCIDENT FACTS 1981 Edition   NATIONAL SAFETY COUNCIL

Figure 2.2. Comparison of incidence rates for various industries by Standard Industrial Classification (SIC) code. (Courtesy of the National Safety Council, Chicago; used with permission.)

annually, for employees to see what injuries and illnesses have been recorded for the year. The summary is required to be posted in a prominent position in the workplace on February 1 each year and to remain posted for 30 days. It is the employer's responsibility to enter data correctly into the log/summary. All records are required to be saved for a period of at least five years.

Besides the log/summary, there is a "Supplementary Record of Occupational Injuries and Illnesses" (See Fig. 2.4 on p. 22). Each page of the supplementary record corresponds to a single-line entry in the log.

To illustrate the calculation of the various incidence rates and to demonstrate the use of the standard forms, a case study will now be analyzed.

Case Study 2.1   A metal products fabrication and assembly plant employs 650 workers and has the following injury/illness experience for the year (workers are employed on a regular 40-hour-workweek basis):

Files 1, 9, 14	Workers sprain ankles in slips and falls; treatment received; workers remain on job
File 2	Worker cuts hand on metal strapping in receiving department; first aid received; no treatment; worker stays on job
Files 3, 11	Workers' skin irritated by solvents; treatment received; workers remain on job
File 4	Worker falls from platform and breaks leg; misses 1 week of work; spends 6 additional weeks on a secondary activity in the plant because of inability to walk
File 5	Worker electrocuted by portable electric drill
File 6	Lifting injury to back; 1 week off work; 2 weeks in a different job not requiring lifting
File 7	Severe dermatitis from solvent; 2 days away from work; 3 days of restricted work in areas not using solvents
File 8	Carbon tetrachloride poisoning effects observed; worker became ill; 1 workday in restricted activity
File 10	Fingertip amputated on box stitching machine; worker loses 2 days' work
File 12	Worker strikes head on overhead conveyor structure; 20 stitches taken; 1 workday lost
File 13	Worker burns hand (second-degree burn) on hot casting; treated, released; no lost workdays
File 15	Wrist sprain; worked 2 days as maintenance dispatcher before returning to regular job
File 16	Worker injured in weekend water-skiing accident; loses 1 week of work

Analysis. The table on p. 23 displays entries in appropriate numbered columns of the Log of Occupational Injuries and Illnesses (refer to Figure 2.3 for column definitions).

Bureau of Labor Statistics
Log and Summary of Occupational
Injuries and Illnesses

NOTE:	This form is required by Public Law 91-596 and must be kept in the establishment for 5 years. Failure to maintain and post can result in the issuance of citations and assessment of penalties. *(See posting requirements on the other side of form.)*	**RECORDABLE CASES:** You are required to record information about every occupational **death**; every nonfatal occupational **illness**; and those nonfatal occupational **injuries** which involve one or more of the following: loss of consciousness, restriction of work or motion, transfer to another job, or medical treatment (other than first aid). *(See definitions on the other side of form.)*

Case or File Number	Date of Injury or Onset of Illness	Employee's Name	Occupation	Department	Description of Injury or Illness
Enter a nondupli-cating number which will facilitate com-perisons with supple-mentary records.	Enter Mo./day.	Enter first name or initial, middle initial, last name.	Enter regular job title, not activity employee was per-forming when injured or at onset of illness. In the absence of a formal title, enter a brief description of the employee's duties.	Enter department in which the employee is regularly employed or a description of normal workplace to which employee is assigned, even though temporarily working in another depart-ment at the time of injury or illness.	Enter a brief description of the injury or illness and indicate the part or parts of body affected.
					Typical entries for this column might be: Amputation of 1st joint right forefinger; Strain of lower back; Contact dermatitis on both hands; Electrocution—body.
(A)	(B)	(C)	(D)	(E)	(F)
					PREVIOUS PAGE TOTALS ➡
					TOTALS (Instructions on other side of form.) ➡

OSHA No. 200

Figure 2.3. OSHA Form No. 200: Log of Occupational Injuries and Illnesses.

For Calendar Year 19 _____   Page _____ of_____

Company Name

Establishment Name

Establishment Address

Form Approved
O.M.B. No. 44R 1453

Extent of and Outcome of INJURY						Type, Extent of, and Outcome of ILLNESS												
Fatalities	Nonfatal Injuries					Type of Illness							Fatalities	Nonfatal Illnesses				
Injury Related	Injuries With Lost Workdays				Injuries Without Lost Workdays	CHECK Only One Column for Each Illness *(See other side of form for terminations or permanent transfers.)*							Illness Related	Illnesses With Lost Workdays				Illnesses Without Lost Workdays
Enter DATE of death.	Enter a CHECK if injury involves days away from work, or days of restricted work activity, or both.	Enter a CHECK if injury involves days away from work.	Enter number of DAYS *away from work.*	Enter number of DAYS of *restricted work activity.*	Enter a CHECK if no entry was made in columns 1 or 2 but the injury is recordable as defined above.	Occupational skin diseases or disorders	Dust diseases of the lungs	Respiratory conditions due to toxic agents	Poisoning (systemic effects of toxic materials)	Disorders due to physical agents	Disorders associated with repeated trauma	All other occupational illnesses	Enter DATE of death.	Enter a CHECK if illness involves days away from work, or days of restricted work activity, or both.	Enter a CHECK if illness involves days away from work.	Enter number of DAYS *away from work.*	Enter number of DAYS of *restricted work activity.*	Enter a CHECK if no entry was made in columns 8 or 9.
Mo./day/yr.													Mo./day/yr.					
(1)	(2)	(3)	(4)	(5)	(6)	(a)	(b)	(c)	(d)	(e)	(f)	(g)	(8)	(9)	(10)	(11)	(12)	(13)
												(7)						

(watermarks: INJURIES ... ILLNESSES)

Certification of Annual Summary Totals By _____ Title _____ Date _____

FOLD

OSHA No. 200   **POST ONLY THIS PORTION OF THE LAST PAGE NO LATER THAN FEBRUARY 1.**

OSHA No. 101
Case or File No. _____

Form approved
OMB No. 44R 1453

Supplementary Record of Occupational Injuries and Illnesses

EMPLOYER

1. Name _____

2. Mail address _____
                   (No. and street)   (City or town)   (State)

3. Location, if different from mail address _____

INJURED OR ILL EMPLOYEE

4. Name _____ Social Security No. _____
        (First name)   (Middle name)   (Last name)

5. Home address _____
             (No. and street)   (City or town)   (State)

6. Age _____   7. Sex: Male_____ Female_____ (Check one)

8. Occupation _____
            (Enter regular job title, *not* the specific activity he was performing at time of injury.)

9. Department _____
            (Enter name of department or division in which the injured person is regularly employed, even
            though he may have been temporarily working in another department at the time of injury.)

THE ACCIDENT OR EXPOSURE TO OCCUPATIONAL ILLNESS

10. Place of accident or exposure _____
               (No. and street)   (City or town)   (State)
    If accident or exposure occurred on employer's premises, give address of plant or establishment in which
    it occurred. Do not indicate department or division within the plant or establishment. If accident oc-
    curred outside employer's premises at an identifiable address, give that address. If it occurred on a pub-
    lic highway or at any other place which cannot be identified by number and street, please provide place
    references locating the place of injury as accurately as possible.

11. Was place of accident or exposure on employer's premises? _____ (Yes or No)

12. What was the employee doing when injured? _____
                          (Be specific. If he was using tools or equipment or handling material,
    _____
              name them and tell what he was doing with them.)

13. How did the accident occur? _____
                    (Describe fully the events which resulted in the injury or occupational illness. Tell what
    _____
    happened and how it happened. Name any objects or substances involved and tell how they were involved. Give
    _____
    full details on all factors which led or contributed to the accident. Use separate sheet for additional space.)

OCCUPATIONAL INJURY OR OCCUPATIONAL ILLNESS

14. Describe the injury or illness in detail and indicate the part of body affected. _____
    _____
            at second joint; fracture of ribs; lead poisoning; dermatitis of left hand, etc.)

15. Name the object or substance which directly injured the employee. (For example, the machine or thing
    he struck against or which struck him; the vapor or poison he inhaled or swallowed; the chemical or ra-
    diation which irritated his skin; or in cases of strains, hernias, etc., the thing he was lifting, pulling, etc.)
    _____

16. Date of injury or initial diagnosis of occupational illness _____
                         (Date)

17. Did employee die? _____ (Yes or No)

OTHER

18. Name and address of physician _____

19. If hospitalized, name and address of hospital _____
    _____
       Date of report _____ Prepared by _____
       Official position _____

Figure 2.4. OSHA Form No. 101: Supplementary Record of Occupational Injuries and Illnesses.

22

File Number	2	3	4	5	6	a	b	c	d	e	f	g	8	9	10	11	12	13
1						✔												
2	(not recordable)																	
3						✔												✔
4	✔	✔	5	30														
5	3/9/83																	
6	✔	✔	5	10														
7						✔								✔	✔	2	3	
8								✔						✔			1	
9							✔											
10	✔	✔	2															
11						✔												✔
12	✔	✔	1															
13						✔ (colum 6)												≡
14							✔											
15	✔		2															
16	(not recordable)																	
Totals	1	5	4	13	42	4	3			1				2	1	2	4	2

Calculation of Incidence Rate

Injury and illness incidence rate $= \dfrac{(5 + 3 + 2 + 3) \times 200{,}000}{650 \times 2000}$

$= 2.00$

Injury incidence rate $= \dfrac{(5 + 3) \times 200{,}000}{650 \times 2000} = 1.23$

colum 2 + 6

Illness incidence rate $= \dfrac{(2 + 3) \times 200{,}000}{650 \times 2000} = 0.77 \quad 0.62$

Fatality incidence rate $= \dfrac{1 \times 200{,}000}{650 \times 2000} = 0.15$

Lost-workday-cases incidence rate $= \dfrac{(5 + 2) \times 200{,}000}{650 \times 2000} = 1.08$

$$\text{Number-of-lost-workdays rate} = \frac{(13 + 42 + 2 + 4) \times 200,000}{650 \times 2000}$$

$$= 9.38$$

$$\text{Specific hazard incidence rate (dermatitis and skin disease)} = \frac{3 \times 200,000}{650 \times 2000} = 0.46$$

The 650-employee firm in Case Study 2.1 provides ample data to show meaningful calculations for the various incidence rates. But most firms are much smaller. For very small firms the calculations are obviously inappropriate. It is not uncommon for small businesses to operate for several years without a single injury or illness. Recognizing that injury/illness records can be ridiculous for very small firms, Congress has exempted small firms with 10 or fewer employees from general recordkeeping requirements.

ACCIDENT CAUSE ANALYSIS

So far this chapter has discussed the more visible busy functions of the job of the Safety and Health Manager, many of which are required by state or federal agencies. But even more important to the health and safety of workers are some of the jobs that the Safety and Health Manager is not required to, but should, do. One of these voluntary but important tasks is a thorough analysis of the potential causes of injuries and illnesses that have already occurred in the plant. Even accidents or incidents that may not actually have caused injuries or illnesses, but which could have, should be studied to prevent their recurrence. Any occurrence of an unplanned, unwanted event is a piece of data to consider in the prevention of future illnesses and injuries. Accident cause analysis and subsequent dissemination of this information to personnel who will be exposed to the hazards in the future is believed to be the most effective way of preventing injuries and illnesses. The literature of injury case histories is filled with accounts of cases where workers are killed by conditions that previously had caused accidents or injuries to others. One case will be used to illustrate this point.

> Case Study 2.2.   A worker was struck in the head and killed by the sudden movement of a large wrench for releasing gates on the bottom of railroad hopper cars. The worker used a powerful, 3- to 4-foot-long wrench to trip a mechanical latch on the bottom gate of the car. The wrench was supposed to be of a ratchet type so that when the gate was tripped, the tremendous weight of the bulk material in the hopper car would not suddenly force the wrench back on the worker. But the ratchet wrench for some reason was not available, and workers had been using an ordinary rigid wrench to release the latch. Only a week before the fatality occurred another worker had narrowly escaped the same injury when he lost control of the same wrench in the same operation.

Sometimes the accident analysis leads to a design change in a product or process. In other cases work procedures are changed to prevent future occurrences, or to minimize the adverse effects of these occurrences. Even when nothing can be changed to prevent a future occurrence, at least workers can be informed of what happened, what caused the accident, under what conditions the accident might occur again, and how to protect themselves in such an event. Informing workers of the facts and causes of accidents that have already happened to their co-workers is the single most effective method of training workers to avoid injury and illness. Thus accident cause analysis is the foundation upon which safety and health engineering, capital investment planning, training, motivation, and other functions are constructed. There are other types of accident analysis: statistical frequency analysis was discussed earlier, and cost analysis will be discussed later in this chapter. But no type of analysis is as important as the determination of causes of accidents that have already occurred and might occur again.

ORGANIZATION OF COMMITTEES

The value of using safety and health committees has long been recognized. Committees are appointed from the ranks of the operating personnel of the regular line organization. The appointments are temporary so that workers throughout the organization rotate on and off a committee periodically. The committees then make facilities inspections, evaluate safety and health suggestions, analyze accident causes, and make recommendations.

Several natural advantages of the committee approach make it a winning strategy. In general, operating personnel know a lot more about their processes and machines than does the Safety and Health Manager. Many valuable and practical ideas can come from operating personnel if staff persons will listen. Also, operating personnel may more readily accept new policies and procedures if these procedures arise from other operating personnel like themselves. Then there is the advantage of exposure. Sooner or later nearly everyone has his or her turn on a safety committee, which means that the direct activity of the safety and health program is a product of plantwide participation. Some workers have no appreciation for or sensitivity to safety and health hazards until they take their turns on the committee. Indirectly, then, the committee becomes a vehicle for safety and health training.

Despite its advantages there are pitfalls to the committee approach. The Safety and Health Manager should provide resources and guidance to the committee so that it will have the necessary tools and knowledge to function effectively. Otherwise, the committee may make ridiculous suggestions and be disappointed when management does not approve or does not follow through with capital support. Also, committees must be conditioned not to expect miracles. Some orientation or training is necessary so that committee members will comprehend the goal of eliminating recognized and *unreasonable* hazards, but not all hazards. Finally, committees should not be allowed to degenerate into spy parties with the objective of discrediting the processes or procedures of other departments.

SAFETY AND HEALTH ECONOMICS

Safety and Health Managers are sometimes dismayed to discover that top management bases safety and health decisions on dollars and cents.[2] But the cold reality is that business exists to make profits, and everything a business does is either directly or indirectly related to economics. Safety and Health Managers who are naive enough to think that the humanitarian objective of worker safety and health transcends the more crude isues of profit and loss should ask themselves the following question: How *much* safety and health staff activity is justified by the humanitarian objective?

The prevention of employee injuries and illnesses can be formulated as an economic objective which is more meaningful to management than vague humanitarian aspirations. Accidents, injuries, and illnesses have undeniable costs that contribute nothing to the value of products manufactured by the firm or services performed. Occupational injuries alone have been estimated to total over $20 billion annually. The annual cost of injuries and illnesses in many industries dwarfs the total profits picture. This is a reality that almost any top manager will want to consider. While it is true that many of these costs are subtle and difficult to estimate, the existence of these costs is in no way diminished by this fact.

One obvious and direct category of costs from injuries and illnesses is the payment of Workers' Compensation insurance premiums, which are based on the firm's injury and illness experience. Self-insured firms have the actual claims data on which to calculate these direct costs. In addition to these claims are medical costs which may be covered by insurance. Since these costs are directly identified to injuries and illnesses in accounting records, they are sometimes called "direct costs" of injuries and illnesses. Workers' Compensation costs typically range from 1 to 2% of the total payroll of the firm. Certain hazardous industries may have Workers' Compensation costs of around 3% of payroll, according to a 1980 survey by the United States Chamber of Commerce.

The figure 1 or 2% of payroll, which may or may not seem significant to the reader, has been referred to by some analysts as the "tip of the iceberg." The intangible costs of accidents, although hidden, appear to be much greater than the so-called "direct costs." It is the job of the Safety and Health Manager to attempt to estimate these costs and to keep management apprised so that rational investment decisions can be made.

The National Safety Council in its *Accident Prevention Manual for Industrial Operations*[3] lists the following categories of hidden costs of accidents:

---

[2]"Dollars and sense" has been suggested by the Arkansas Department of Labor in its pamphlet *A Common Goal* published in 1975 (Ref. 10).

[3]*Accident Prevention Manual for Industrial Operations: Administration and Programs Volume*, 8th ed. Chicago: National Safety Council, 1981, pp. 214–215 (used with permission).

1. **Cost of wages paid for time lost by workers who were not injured.** These are employees who stopped work to watch or assist after the accident or to talk about it, or who lost time because they needed equipment damaged in the accident or because they needed the output or the aid of the injured worker.

2. **Cost of damage to material or equipment.** The validity of property damage as a cost can scarcely be questioned. Occasionally, there is no property damage, but a substantial cost is incurred in putting back in order material or equipment which has been thrown into a state of disorder. The charge should, however, be confined to the net cost of repairing or putting in order material or equipment that has been damaged or displaced, or to the current worth of the equipment less salvage value if it is damaged beyond repair.

   An estimate of property damage should have the approval of the cost accountant, particularly if the current worth of the damaged property used in the cost estimate differs from the depreciated value established by the accounting department.

3. **Cost of wages paid for time lost by the injured worker,** other than workers' compensation payments. Payments made under workers' compensation laws for time lost after the waiting period are not included in this element of cost.

4. **Extra cost of overtime work necessitated by the accident.** The charge against an accident for overtime work necessitated by the accident is the difference between normal wages and overtime wages for the time needed to make up lost production, and the cost of extra supervision, heat, light, cleaning, and other extra services.

5. **Cost of wages paid supervisors for time** required for activities necessitated by the accident. The most satisfactory way of estimating this cost is to charge the wages paid to the foreman for the time spent away from normal activities as a result of the accident.

6. **Wage cost caused by decreased output of injured worker after return to work.** If the injured worker's previous wage payments are continued despite a 40 percent reduction in his output, the accident should be charged with 40 percent of his wages during the period of such low output.

7. **Cost of learning period of new worker.** If a replacement worker produces only half as much in his first two weeks as the injured worker would have produced for the same pay, then half of the new worker's wages for the two weeks' period should be considered part of the cost of the accident that made it necessary to hire him.

   A wage cost for time spent by supervisors or others in training the new worker also should be attributed to the accident.

8. **Uninsured medical cost borne by the company.** This cost is usually that of medical services provided at the plant dispensary. There is no great difficulty in estimating an average cost per visit for this medical attention.

   The question may be raised, however, whether this expense may properly be considered a variable cost. That is, would a reduction in accidents result in lower expenses for operating the dispensary?

9. **Cost of time spent by higher supervision and clerical workers** on investigations or in the processing of compensation application forms. Time spent by supervision (other than the foreman or supervisor covered in Item 5) and by clerical employees in investi-

gating an accident, or settling claims arising from it, is chargeable to the accident.

10. **Miscellaneous usual costs.** This category includes the less typical costs, the validity of which must be clearly shown by the investigator on individual accident reports. Among such possible costs are public liability claims, cost of renting equipment, loss of profit on contracts canceled or orders lost if the accident causes a net long-run reduction in total sales, loss of bonuses by the company, cost of hiring new employees if the additional hiring expense is significant, cost of *excess* spoilage (above normal) by new employees, and demurrage. These cost factors and any others not suggested above would need to be well substantiated.

Every firm is different, and if time and staff resources permit, the best way to estimate hidden costs of accidents is to survey and analyze the individual company's recent accident data. When performing such an analysis it must be remembered that noninjury accidents can also be costly and are generally caused by the same types of conditions and practices that result in injury accidents. Therefore, noninjury accidents should also be included when one attempts to assess the total costs of accidents.

Most firms cannot afford the luxury of a comprehensive, statistically reliable, in-house study of hidden accident costs. An alternative is to turn to national studies of average costs of various accident categories, and apply these survey averages as estimates of in-house costs. Two well-known studies of uninsured costs of accidents were reported by Grimaldi and Simonds (Ref. 30) and Imre (Ref. 38) in 1975. Although the data were gathered over a span of several years, when the dollar figures were adjusted for inflation to a common representative year, the results of the two studies were shown to corroborate each other, recognizing that rough approximations are all that can be hoped for in such studies.

To make use of the Grimaldi and Simonds and Imre data, the Safety and Health Manager must determine or estimate the numbers of accident cases that fall into the following four general categories as standardized by the National Safety Council:[4]

- *Class 1:* Cases involving lost workdays, if records are kept under OSHAct, or permanent partial disabilities and temporary total disabilities, if records are based on ANSI Z16.1.
- *Class 2:* Medical treatment cases requiring the attention of a physician outside the plant.
- *Class 3:* Medical treatment cases requiring only first aid or local dispensary treatment and resulting in property damage of less than $20.00 or loss of less than 8 hours' working time.

[4]*Accident Prevention Manual for Industrial Operations: Administration and Programs Volume,* 8th ed. Chicago: National Safety Council, 1981, p. 209 (used with permission).

- *Class 4:* Accidents which either cause no injury or cause minor injury not requiring the attention of a physician, and which result in property damage of $20.00 or more, or loss of 8 or more employee-hours.

Class 1 and 2 cases are taken almost directly from required records kept in-house. Class 3 and 4 cases require additional recordkeeping. Class 3 cases are minor accidents, and although their individual costs are usually small, the total cost generated by this category is significant because of the large numbers of accidents in this category. The total number of cases in Class 3 is typically larger than for any of the other categories, and if such is not found to be true of any given study, the analyst should probe deeper for more Class 3 cases that may have been overlooked. Class 4 cases may be even more difficult to track. Class 3 cases at least could be counted by examining infirmary records. Since Class 4 cases are difficult to recognize, the more major cases are the ones that are counted most often. Since this was no doubt also true of the cost studies that generated survey averages, the average cost figures for Class 4 cases turn out to be surprisingly large.

The following figures are suggested for general use for estimates of uninsured costs, as recommended by Grimaldi and Simonds[5] in terms of 1974 dollars:

Lost-time cases	$220
Doctors' cases	55
First-aid cases	12
No-injury cases	40

The National Safety Council suggests the Consumer Price Index for adjusting accident cost estimates for inflation. Figure 2.5 uses this method to extrapolate the survey data through 1986. The extrapolation follows the 25-year trend of the Consumer Price Index.

Having gathered the numbers of accident cases in each of the four classes, the Safety and Health Manager can then compute total uninsured costs by multiplying by the cost factors obtained from Figure 2.5.

Case Study 2.3. Estimate the total annual uninsured cost of accidents in a firm that has records of the following numbers of accidents in the first six months of 1981.

Lost-time cases	12
Doctors' cases	10
First-aid cases	46
No-injury cases	18

[5]John V. Grimaldi, and Rollin H. Simonds, *Safety Management,* 3rd ed., Homewood, IL: Richard D. Irwin, Inc., © copyright 1975, p. 614 (used with permission).

Figure 2.5. Estimate of uninsured costs per industrial accident adjusted by the Consumer Price Index.

Calculation

ESTIMATE OF UNINSURED COSTS PER ACCIDENT BY TYPE, ADJUSTED BY CONSUMER PRICE INDEX (see Figure 2.5)

	1974	CPI ratio	1981
Lost-time cases	220	1.91	420
Doctors' cases	55	1.91	105
First-aid cases	12	1.91	23
No-injury cases	400	1.91	764

The annual costs are therefore:

Lost-time: 2 (12 cases × $420/case) = $10,080/year

Doctors': 2 (10 cases × $105/case) = $ 2,100/year

First-aid: 2 (46 cases × $ 23/case) = $ 2,116/year

No-injury: 2 (18 cases × $764/case) = $27,504/year

Total $41,668/year

Costs of Workers' Compensation insurance and other direct costs should be added to the uninsured cost estimates. Cost estimates of potential fines and legal costs can also be added, although these costs are usually dwarfed by other costs which are sure to be incurred whether or not an enforcement inspection takes place.

TRAINING

Training or training support may be the most important staff function to be performed by the Safety and Health Manager. Despite a recent trend toward concentration on unsafe conditions, experts still attribute most worker injuries and illnesses to unsafe acts. Unsafe work habits are deeply rooted, even in new, young workers. Our society and its standards of status, as influenced by the media, especially television, places a premium on high-risk activity. From an early age children learn that heroes are people who are daring, lucky, and risk their lives, especially in their life work. In some jobs, such as space exploration, the military, law enforcement, and firefighting, it is occasionally both necessary and rational to take big risks. And those who take these risks indeed deserve to be called heroes. Unfortunately, though, the desire for recognition, status, and the esteem of their peers causes people to take unncessary risks in activities that do not warrant such risk. A good example of this phenomenon is exhibited in the automobile driving habits of people of all ages. Deep-rooted unsafe habits and lack of knowledge about specific job hazards are major barriers to worker safety and health. It is upon these two problems that the training program should be focused, and training is perhaps the most challenging and important function of the Safety and Health Manager.

One of the biggest mistakes a Safety and Health Manager can make is to assume that he or she is the principal trainer in safety and health. The principal trainers in safety or health or in any other aspect of the job are the first-line supervisors. Their direct contact with the workers will determine how the job will be done. A corollary to this principle is that most training in safety and health is informal and is conducted on the job. In fact, training by example is a very important delivery mode, and new workers are more influenced by what the supervisor and experienced workers *do* than by what they *say*.

Recognizing that most training takes place between supervisor and worker, there is still a need for classroom training in safety and health principles, standards, and hazards recognition, especially for supervisors. The Safety and Health Manager can provide this training directly or can act as a facilitator in bringing useful information and training aids into the plant in subjects needed.

A word of caution is in order, however: Safety and Health Managers should not "try to reinvent the wheel" in their development of training materials. Audio-visual packages and outlines are available, and when the Safety and Health Manager's time, including overhead, is considered, it is usually much more reasonable to purchase or rent training materials than to attempt to create original material in-house. Chapter 1 enumerated some sources to assist Safety and Health

Managers in developing this aspect of their function. Chapter 3 will also address the subject of training and how it becomes a part of the overall objective of hazard avoidance.

SUMMARY

Safety and health on the job, like production quality or any other desirable factory characteristic, is achieved by the workers themselves. Thus the actual achievement of safety and health is a line function. The Safety and Health Manager then has a staff function in facilitating the line organization, especially first line supervisors, in achieving the goal of safety and health.

It cannot be assumed that safety and health is really a "goal" of line management. Everyone wants safety and health in the workplace, more or less, but the degree of management's commitment to this goal must be sized up and documented by the Safety and Health Manager.

Once management's commitment to the goal of safety and health is ascertained, the Safety and Health Manager can get to the important functions of dealing with Workers' Compensation, collecting and analyzing statistical records, economic analysis, safety and health training, and dealing with both hazards and violations of safety and health standards.

EXERCISES AND STUDY QUESTIONS

2.1. Is worker safety a line or staff function?

2.2. What is the Safety and Health Manager's relationship to the line organization?

2.3. Why is it often difficult for Safety and Health Managers to win the respect and approval of top management?

2.4. What principle is embodied in the statement "Safety is everybody's business"?

2.5. What should the Safety and Health Manager do if top management does not "practice what it preaches" in safety and health policies?

2.6. What is the difference between the terms "Workmen's Compensation" and "Workers' Compensation"?

2.7. What are the ostensible and ulterior purposes of the Workers' Compensation system?

2.8. Should employers bear all the risk of injury to their workers?

2.9. Who pays the Workers' Compensation claims of injured workers?

2.10. What is a "loss control representative"?

2.11. What is the Z16.1 system?

2.12. Why is it difficult to prove by statistics whether OSHA has been of any benefit?

2.13. What is the disadvantage of using "lost workdays" as a measure of injury or illness severity?

2.14. What constitutes a "recordable injury or illness"?

2.15. A firm employing 300 workers has 25 recordable injuries/illnesses in a year. What is its total injury/illness incidence rate?

2.16. How does the injury/illness incidence rate compare with the traditional "frequency" rate?

2.17. Compare the terms "frequency," "severity," and "seriousness."

2.18. What is important about February 1 in the life of a Safety and Health Manager?

2.19. How long are files of records required to be kept?

2.20. Are there advantages and pitfalls to the use of safety and health committees?

2.21. Compare the importance of direct and hidden costs of accidents.

2.22. Name some categories of hidden costs of accidents.

2.23. How can records of noninjury accidents be important to the Safety and Health Manager?

2.24. Who are the principal trainers in safety and health in industry?

2.25. During a six-month period a firm employing 50 employees has 18 injuries and illnesses requiring medical treatment, in four of which cases the employee lost at least 1 day from work.

    (a) What is the general injury/illness incidence rate?

    (b) What is the traditional "frequency" rate?

    (c) Is this a very dangerous industry?

2.26. **Case Study.** A chemical plant employing 900 employees (employees work a regular 40-hour workweek) has the following safety/health record for the year 1983:

    File 1: Forklift truck drops pallet load of packaged raw material; no injuries; some material wasted; pallet destroyed; extensive cleanup required

    File 2: Worker suffers heat cramps (illness) from continuous exposure to hot process; sent to hospital for emergency treatment; back on the job the next day

    File 3: Worker burns hand on steam pipe; first aid received and worker returns to workstation

    File 4: Worker suffers dermatitis from contact with solvent; 1 week of work lost (regular workweek); another 4 weeks of work restricted to an assembly job

    File 5: Worker fractures finger in packaging machine; worker sent to hospital for treatment; back on the job the next day

    File 6: Maintenance worker lacerates hand when screwdriver slips; five stitches taken; worker back on the job the next day

    File 7: Pressure vessel explodes; extensive damage to processing area; miraculously no one is injured

    File 8: Worker gets poison ivy from exposure a week earlier while removing weeds around the plant perimeter fence; worker receives medication but no workdays are lost

    File 9: Worker becomes ill from continuous exposure to hydrogen sulfide leaks from furnace area; misses 2 weeks' work; leaks are repaired

    File 10: Worker gets severe poison ivy from weekend outing with Boy Scout troop; misses 2 days of work

File 11: Maintenance worker falls from fractionating tower and is killed

File 12: Worker fractures arm in power transmission system powering pulverizer mill; loses 3 days of work and an additional 6 weeks of work is in the production scheduling office before returning to regular job

(a) Calculate the following incidence rates:
1. Total injury/illness incidence rate
2. Injury incidence rate
3. Illness incidence rate
4. Fatality incidence rate
5. Lost-workday-cases incidence rate
6. Number-of-lost-workdays rate
7. Specific hazard incidence rate (fractures)

(b) Estimate the hidden costs of accidents using survey average cost data.

(c) Does the use of survey average cost data seem appropriate in this case? Why or why not?

(d) How does the safety/health record of this firm compare with that of other chemical companies and with industries in general?

3

Concepts
of Hazard Avoidance

Hazards involve risk or chance, and these words deal with the unknown. As soon as the unknown element is eliminated, the problem is no longer one of safety or health. For example, everyone knows what would happen if he or she jumped off a 10-story building. Immediate death would be virtually a certainty, and such an act is not properly described as "unsafe"; it is described as "suicidal." But to work on the roof of a 10-story building with no intention of falling off becomes a matter of safety. Workers without fall protection on the unguarded rooftop of a building are exposed to a recognized hazard. This is not to say that such workers will be killed or even to say that they will come to any harm whatsoever, but there is that chance, that unknown element.

Dealing with the unknown makes the job of the Safety and Health Manager a difficult one. If the Safety and Health Manager pushes for a capital investment to enhance safety or health, who is able to *prove* later that the investment was worthwhile? Improved injury and illness statistics help and may look impressive, but they do not actually prove that the capital investment was worthwhile because no one really knows what the statistics would have shown had the investment not been made. It is in the realm of the unknown.

Since safety and health deal with the unknown, there is no step-by-step recipe for eliminating hazards within the workplace. Instead, there are merely concepts or approaches to take to whittle away at the problem. All the approaches have merit, but none is a panacea. Drawing on their own strengths, different Safety and Health Managers tend to concentrate on certain favorite approaches familiar to them. The objective of this chapter is to present various approaches so that the Safety and Health Manager will have a variety of tools, not just one or two, to deal with the unknown elements of worker safety and health. Both the good and the bad will be discussed for each approach. The good is often obvious or taken for granted. But the drawbacks of each approach must be squarely faced, too, so that Safety and Health Managers can see the limitations and draw on the strengths of each approach to accomplish their missions.

THE ENFORCEMENT APPROACH   *OSHA approach*

This is the approach initially taken by OSHA, but OSHA was certainly not the first to employ it. Safety rules with penalties for breaking them have existed almost since people first began to deal with risks. The pure enforcement approach says that since people do not properly assess hazards and take prudent precautions, they should be told rules to follow and be subject to penalties for breaking these rules.

The enforcement approach is simple and direct; there is no question that it has an impact. The enforcement must be swift and sure and the penalties sufficiently severe, but if these conditions are met, people will follow rules to some extent. Using the enforcement approach, OSHA has without doubt forced thousands of industries to comply with regulations that have changed workplaces and made millions of jobs safer and more healthful. The preceding statement sounds like a glowing success story for OSHA, but the reader knows that the enforcement approach has failed to do the whole job. It is difficult to see any general improvement in injury and illness statistics as a result of enforcement, although some categories, such as trenching and excavation cave-ins, have shown marked improvement. Despite its advantages there are some basic weaknesses in the enforcement approach, as the statistics suggest. These weaknesses will now be examined.

At the foundation of any enforcement approach lies a set of mandatory standards. Mandatory standards must be worded in absolutes such as "always do this" or "never do that." The wording of complicated exceptions can alleviate the problem somewhat but requires the anticipation of every circumstance to be encountered. Within the framework of the stated scope of the standard, recognizing all exempt situations, each rule must be absolutely mandatory to be enforceable. But mandatory language employing the words "always" and "never" is really inappropriate when dealing with the uncertainties of safety and health hazards. To see how this is true, consider the following example.

Example 3.1

Suppose that a properly grounded electrical appliance for the resuscitation of injured employees is equipped with a three-prong plug. But in the midst of an emergency it is discovered that the wall receptacle is the old, ungrounded, two-hole variety. With no adapter in sight and an employee in desperate need of the appliance, who would not bend back or cut off the grounding plug and proceed to save the employee's life?

Of course the example above states an extreme case, and "we must be reasonable" and "use our professional judgment," but in the arena of enforcement and mandatory standards, who is going to say what is "reasonable?" Everyone knows what is reasonable in such an extreme case, but countless borderline cases occur every day in which it is not certain whether the proper course of action is to violate or not to violate the rule. Consider the following actual case.

Example 3.2

A dangerous fire was in progress as flammable liquids were burning in tanks. To shut off the source of fuel, a thinking employee quickly turned off the adjacent tank valves to avert a more dangerous fire which could have cost many lives, not to speak of property damage. Did the employee receive a medal for his meritorious act? No, the company received an OSHA citation because the employee was not wearing gloves! The valves were hot, and because the employee went ahead and closed the valves, burning his hands, the company was issued a citation.

If a government agency will issue a citation for failure to wear gloves while closing a valve in an emergency, who will have the courage to "be reasonable" and to go ahead and act even if a violation is the consequence? In the face of a pure enforcement approach, industry employees and employers alike will gradually retreat to a defensive position, failing to produce, and blaming their lack of productivity on the government.

As stated earlier, OSHA did not invent the enforcement approach for dealing with hazards. Other mandatory safety rules and laws are familiar to everyone. Sometimes overzealous and oppressive rules can destroy themselves by alienating the very persons they are intended to protect. A notorious example is the mandatory helmet law for motorcycle riders. The helmet manufacturers can present impressive statistics which show that helmets save lives, at least in some accidents. Such statistics are a strong motivation to motorcyclists to wear helmets. But in certain situations, the use of a helmet has disadvantages which cause motorcyclists to hate the law that requires them *always* to wear a helmet. Thus it is illegal to give a friend a trial ride around the block without obtaining an extra helmet for the passenger for that one ride. If a temporary skin rash or scalp treatment prevents an operator from wearing the helmet for a day or two, he or she must not ride even if the motorcycle is the sole means of transportation. Where to put the helmet during a brief stop can also become extremely awkward in some situations. One motorcyclist in Houston became so frustrated that he complied with the letter of the law while he defied its spirit by riding his motorcycle around and about Houston wearing his helmet on his elbow!

THE PSYCHOLOGICAL APPROACH

Contrasted with the enforcement approach is an approach that attempts to reward safe attitudes. This is the approach employed by many Safety and Health Managers and may be identified as the psychological approach. The familiar elements of the psychological approach are posters and signs reminding employees to work safely. A large sign may be at the front gate of the plant displaying the number of days since a lost-time injury. Safety meetings, departmental awards, drawings, prizes, and picnics can be used to recognize and reward safe attitudes.

Religion versus Science

The psychological approach emphasizes the religion of safety and health versus science. Safety meetings in which the psychological approach is used are typified by attempts at persuasion, sometimes called "pep talks." The idea is that employees can be rewarded into wanting to have safe work habits. Peer pressure can be brought to bear on the individual when the whole department may suffer when one person has an injury or illness.

Top Management Support

The psychological approach is very sensitive to the support of top management, and if such support is absent, the approach is very vulnerable. Recognition pins, certificates, and even monetary prizes are small reward if workers feel that to win these rewards they are not pursuing the real goals of top management.

Workers are able to sense the extent of management's commitment to safety by the day-to-day decisions it makes, not by written proclamations that everyone should "be safe." A rule requiring safety glasses to be worn in the production area is undermined when top management does not wear safety glasses when visiting the production floor. If safe practices are ordered to be cast aside when production must be expedited to fill an order on time, workers find out just how much worker safety and health mean to top management. Most Safety and Health Managers will want to get a written endorsement of the plant safety program from top management, but unless top management really understands and believes in the safety and health program, the written endorsement is not very valuable. The true orientation of top management will soon show through. Safety and Health Managers should beware of this pitfall when seeking such a written endorsement.

Young Workers

New workers, especially new young workers, are particularly influenced by the psychological approach to safety and health. Workers in their late teens or early twenties enter the workplace having recently emerged from a social structure which places a great deal of importance on daring and risk taking. These new workers are watching supervisors and more experienced co-workers to determine what kind of behavior or work habits earn respect in the industrial setting. If their older, more experienced peers wear respirators or ear protection, the young workers may also adopt safe habits. If highly respected co-workers laugh at or ignore safety principles, young workers may get off to a very bad start, never taking safety and health seriously.

Accident reports confirm that a large percentage of injuries are caused by unsafe acts by workers. This fact underlines the importance of the psychological approach in developing good worker attitudes toward safety and health. This approach can be bolstered by training in the hazards of specific operations. Once the subtle hazards are made known to workers who would not know about these hazards from their general experience, the development of safe attitudes becomes less difficult.

THE ENGINEERING APPROACH

For decades safety engineers have attributed most workplace injuries to unsafe worker acts, not unsafe conditions. The origin of this thinking has been traced to the great pioneering work in the field by the late H. W. Heinrich (Ref. 35), the first safety engineer so recognized. Heinrich's studies resulted in the widely known ratio 88:10:2:

Unsafe acts	88%
Unsafe conditions	10%
Unsafe causes	2%
Total causes of workplace accidents	100%

Recently, Heinrich's ratios have been questioned, and efforts to recover Heinrich's original research data have produced sketchy results. The current trend is to give increasing emphasis to the workplace machinery, environment, guards, protective systems (i.e., the *conditions* of the workplace). Accident analyses are probing more deeply to determine whether incidents which at first appear to be caused by "worker carelessness" could have been prevented by a process redesign. This development has greatly enhanced the importance of the "engineering approach" to dealing with workplace hazards.

Three Lines of Defense

From the profession has emerged a definite preference for the engineering approach when dealing with health hazards. When the process is noisy or presents airborne exposures to toxic materials, the firm should first try to redesign or revise the process to "engineer out" the hazard. Thus engineering controls receive first preference in what might be called "three lines of defense" against health hazards. These three lines of defense are identified as follows:

1. Engineering controls
2. Administrative or work practice controls
3. Personal protective equipment

The advantages of the engineering approach are obvious. Engineering controls deal directly with the hazard by removing it, ventilating it, suppressing it, or otherwise rendering the workplace safe and healthful. This removes the necessity of living with the hazard and minimizing its effects as contrasted with the strategies of administrative controls and the use of personal protective equipment. This preference of strategies will be considered again in Chapter 8.

Safety Factors

Engineers have long recognized the chance element in safety and know that margins for error must be provided. This basic principle of engineering design appears at various places in safety standards. For example, the safety factor for the design of scaffold components is 4:1. For overhead crane hoists the factor is 5:1, and for scaffold *ropes*, the factor is 6:1 (i.e., scaffold ropes are designed to withstand six times the intended load).

The selection of safety factors is an important responsibility. It would be nice if all safety factors could be 10:1, but there are trade-offs which make such large safety factors unreasonable, even infeasible, in some situations. Cost is the obvious, but not the only trade-off. Weight, supporting structure, speed, horsepower, and size are all factors that may be affected by selection of too large a safety factor. The drawbacks of large safety factors must be weighed against the consequences of system failure, in order to arrive at a rational decision. There are many degrees of difference between situations when evaluating the consequences of system failures. Compare the importance of safety factor in the Kansas City hotel disaster[1] of 1981 to a failure in which the only loss is some material or damaged equipment. Obviously, the former situation should employ a larger safety factor than the latter. Selection of safety factors depends on the evaluation or classification of degree of hazard, a subject that will be treated in depth later in this chapter.

Fail-Safe Principles

Besides the engineering principle of safety factors, there are additional principles of engineering design which consider the consequences of component failure within the system. These principles are labeled here as "fail-safe principles," and three are identified as follows:

1. General fail-safe principle
2. Fail-safe principle of redundancy
3. Principle of worst case

Each of these principles is explained below, and their applications will appear again and again in subsequent chapters dealing with specific hazards.

---

[1] Two skywalks collapsed in the crowded multistory lobby of the Hyatt Regency Hotel in Kansas City, Missouri, on July 17, 1981, killing 113 persons.

GENERAL FAIL-SAFE PRINCIPLE

The resulting status of a system, in event of failure of one of its components, shall be in a safe mode.

Systems or subsystems generally have two modes: active and inert. With most machines the inert mode is the safer of the two. Thus product safety engineering is usually quite simple: If you "pull the plug" on the machine, it cannot hurt you. But the inert mode is not always the safer mode. Suppose that the system is a complicated one with subsystems built in to protect the operator and others in the area in the event of a failure within the system. In this case, "pulling the plug" to disconnect the machine might deactivate the safety subsystems so essential to protect the operator and others in the area. With such a system, disconnecting the power might render the system more unsafe than when the power was on. Design engineers need to consider the General Fail-Safe Principle so as to assure that a failure within the system will result in a safe mode. Thus it may be necessary to provide backup power for the proper functioning of safety subsystems. Some examples will now be used to illustrate the concept.

Example 3.3

An electric drill has a trigger switch which might be continuously depressed to operate the drill. The trigger switch is loaded with a spring so that if some failure (on the part of the operator in this case) results in the release of the trigger, the machine will return to a safe mode (off, in this case). Such a switch is often called a "deadman control." This example illustrates the common situation in which the inert state of the system is the safer one.

Example 3.4

This example illustrates the more uncommon situation in which the inert state of the system is the more dangerous. Consider an automobile with power steering and power brakes. When the engine dies, both steering and braking become very difficult; so at least as far as these subsystems are concerned, the inert state is more dangerous than the active one.

As will be seen in later chapters, industrial systems have characteristics similar to those described in the examples above. Sometimes safety standards specify that system failures be accommodated in such a way that subsystems for safety will continue to be operative.

The General Fail-Safe Principle stated above is the one that embodies the literal meaning of the words "fail-safe." However, industry and technology often associate another concept with the term "fail-safe," and that is the concept of *redundancy*, stated as follows:

FAIL-SAFE PRINCIPLE OF REDUNDANCY

A critically important function of a system, subsystem, or components can be preserved by alternate parallel or standby units.

The design principle of redundancy has been widely utilized in the aerospace industry. When systems are so complicated and of such critical importance as a large aircraft or a space vehicle, the function is too important to allow the failure of a tiny component to bring down the entire system. Therefore, engineers back up primary subsystems with standby units. Sometimes, dual units can be specified right down at the component level. For extremely critical functions, three or four backup systems can be specified. In the field of occupational safety and health, some systems are seen as so vital as to require design redundancy. Mechanical power presses are an example.

Still another fail-safe design principle is the principle of "worst case":

PRINCIPLE OF WORST CASE

The design of a system should consider the worst situation to which it may be subjected in use.

This principle is really a recognition of "Murphy's Law," which states that "if anything can go wrong, it will." Murphy's Law is no joke; it is a simple observation of the result of chance occurrences over a long period of time. Random events that have a constant hazard of occurrence are called "Poisson processes." The design of a system must consider the possibility of the occurrence of some chance event that can have an adverse effect on safety and health.

An application of the Principle of Worst Case is seen in the specification of explosion-proof motors in ventilation systems for rooms in which flammable liquids are handled. Explosion-proof motors are much more expensive than ordinary motors, and industries may resist the requirement to install explosion-proof motors, especially in those processes where the vapor levels of the substances mixed never even get close to the explosive range. But consider the scenario presented by a hot summer day in which a spill happens to occur. The hot weather raises the vapor level of the flammable liquid being handled. A spill at such an unfortunate time dramatically increases the liquid surface exposure, which makes the problem many times worse. At no other time would the ventilation system be more important. But if the motor is not explosion-proof and is exposed to the critical concentration of vapors, a catastrophic explosion would occur as soon as the ventilation system switches on.

The concept of "defensive driving" is well known to all drivers and serves to explain the Principle of Worst Case. Defensive drivers control their vehicles to the extent that they are prepared for the worst random event they can reasonably imagine.

Engineering Pitfalls

It is easy to get caught up in the idea that technology can solve our problems, including elimination of workplace hazards. Certainly, an inventor of a new gadget to prevent injuries or illnesses can quickly become enamored with it and present a convincing argument that the new invention should be installed in workplaces everywhere. When standards writers are persuaded by these arguments, they often require all appropriate industries to install the new device. Several things can go wrong, however.

Recalling the case against the enforcement approach, certain unusual circumstances can make the engineering solution inappropriate or even unsafe. A good example is the use of spring-loaded cutoff valves in air lines for compressed air tools. The purpose of the cutoff valve is to prevent hose whip action by stopping the flow of air if the tool accidentally separates from the hose. The sudden flow of air overcomes the spring-loaded valve and closes it, stopping the flow. The problem comes when several tools are operated from the same main hose and flow reaches maximum even during normal use. The cutoff then becomes a nuisance and impedes production.

A second problem with the engineering approach is related to the first: workers remove or defeat the purpose of engineering controls or safety devices. The most obvious example is the removal of guards from machines. Before faulting the worker for such behavior, take a close look at the guard design. Some guards are so awkward that they make the work nearly impossible. Some machine guards are so impractical that they conjure up doubts about the motives of the equipment manufacturer. There is a legal motivation to install an impractical guard on a new machine so that users will take the guard off before putting the machine in service. When the user modifies a machine by removing a guard, the manufacturer is absolved of guilt for any accident which later occurs but which might reasonably have been prevented by the guard.

An irony of the engineering approach is that if the engineered system does not do the job for which it was intended, it can do more harm than good by engendering a false sense of security. A printing press operator once lost the end of his finger when he attempted to reach into the danger zone, which was protected by a photoelectric sensing device. Believing the system to be infallible, he reached as fast as he could, thinking that the control system would stop the machine in time; but it did not.

Such false sense of security can even lead to new operator procedures which depend on the safety device to control the operation so that the work can be speeded up. The best example that comes to mind is the hoist limit switch on an overhead crane. The switch is tripped, shutting off the hoist motor, if the hoist load block approaches too close to the bridge. The idea sounds good, but the operator can take advantage of the device by *depending* on the switch to stop the load during normal operation. The hoist limit switch is not intended as an operating control, but workers can and do use it that way. The only defense against such use appears to be proper training and safe attitudes on the part of the operator: that is, the psychological approach.

Finally, the engineered system can sometimes cause a hazard, as illustrated in the following example in which a pneumatic press ram pinned an operator's hand on the *upstroke* (see Figure 3.1). The press was equipped with a two-hand control that

Two-hand control
palm button

Two-hand control
palm button

Figure 3.1. Pneumatic press ram pins operator's hands on the upstroke; two-hand control safety device prevents the operator from reactivating the press to release her hand.

was designed for safety's sake not to allow the press to be actuated except by both hands of the operator. Ironically, the two-hand control created a hazard. This press was later redesigned to place a shield in front of the ram so that the operator would be unable to reach into the area above the ram.

As another example, robots are being used to work in hot, noisy environments, to lift heavy objects, and to otherwise serve where humans might be injured or suffer health hazards. Most industrial robots are simply mechanical arms computer programmed to feed material to machines or to do welding. But the mindless swinging of these mechanical arms can cause injury to workers who get in the path of the robot. The irony is that a hazard is created by the robot, the very purpose of which was to reduce hazards. One solution is to make the robot more sophisticated, giving it sensors to detect when a foreign object or person gets into its path. Another solution is simply to install guardrails around the robot or otherwise keep personnel out of the danger zone.

In summary, the engineering approach is a good one and deserves the emphasis it is receiving in the current decade. However, there are pitfalls, and the Safety and Health Manager needs a certain sophistication to see both the advantages and disadvantages in the proposed capital equipment investments in safety and health systems. Upon review of the foregoing examples of engineering pitfalls it can be seen that almost every problem can be dealt with if some additional thought is given to the design of the equipment or its intended operation. The conclusion to be reached is that engineering can solve safety and health problems, but the Safety and Health Manager should not naively assume that the solutions will be simple.

THE ANALYTICAL APPROACH

The analytical approach deals with hazards by studying their mechanisms, analyzing statistical histories, computing probabilities of accidents, conducting epidemiological and toxicological studies, and weighing costs and benefits of hazards elimination. Many but not all analytical methods involve computations.

Accident Analysis

Accident and incident (near miss) analysis is so important that it has already been discussed extensively in Chapter 2. No safety and health program within an industrial plant is complete without some form of review of mishaps that actually have occurred. The subject is mentioned again here only to classify it as within the analytical approach and to show its relationship to other methods of hazard avoidance. Its only drawback is that after an accident has occurred it is too late to prevent the consequences of that accident. But the value of the analysis for future accident prevention is critical.

Accident analysis is not used nearly enough to assist in the other approaches to hazard avoidance. The enforcement approach would be much more palatable to the public if the enforcing agency would spend more time analyzing accident histories so that citations are written only for the most important violations. The psychological approach could also be strengthened a great deal by substantiating persuasive appeals with actual results of accidents. The engineering approach needs accident analysis to know where the problems are and to design the solution to deal with all the accident mechanisms.

Failure Modes and Effects Analysis   FMEA

Sometimes a hazard has several origins, and a detailed analysis of potential causes must be made. Reliability engineers use a method called "Failure Modes and Effects Analysis" (FMEA) to trace the effects of individual component failures on the overall or "catastrophic" failure of equipment. Such an analysis is equipment oriented instead of hazard oriented. In their own interest, manufacturers of equipment sometimes perform an FMEA before a new product is released. Users of these products can sometimes benefit from an examination of the manufacturer's FMEA in determining what caused a particular piece of equipment to fail in use.

The FMEA can become important to the Safety and Health Manager when the failure of a piece of equipment can result in an industrial injury or illness. If a piece of equipment is critical to the health or safety of employees, the Safety and Health Manager may desire to request a report of an FMEA performed by the manufacturer or potential bidder for the equipment. In practice, however, FMEA is usually ignored by Safety and Health Managers until *after* an accident has occurred. Certainly, Safety and Health Managers should at least know what the initials FMEA represent so that they are not dazzled by the term in the courtroom should equipment manufacturers use it to defend the safety of their products.

One beneficial way of using FMEA *before* an accident occurs is in preventive maintenance. Every component of equipment has some feasible mechanism for eventual failure. To simply use equipment until it eventually fails can sometimes have tragic consequences. Consider, for example, the wire rope on a crane, or the chain links in a sling, or the brakes on a forklift truck. The FMEA can direct attention to critical components that should be set up on a preventive maintenance policy which permits parts to be inspected and replaced *before* failure.

Fault Tree Analysis

A very similar but more general method of analysis than FMEA is "fault tree analysis." Whereas FMEA focuses upon component reliability, fault tree analysis concentrates on the end result, which is usually an accident or some other adverse consequence. Accidents are caused at least as often by procedural errors as by equipment failures, and fault tree analysis considers all causes—procedural and/or equipment. The method was developed by Bell Laboratories under contract with the U.S. Air Force in the early 1960s. The objective was to avoid a potential missile system disaster.

The term "fault tree" arises from the appearance of the logic diagram which is used to analyze the probabilities associated with the various causes and their effects. The leaves and branches of the fault tree are the myriad individual circumstances or events which can contribute to an accident. The base or trunk of the tree is the catastrophic accident or other undesirable result being studied. Figure 3.2 shows an example fault tree diagram of the network of causal relationships that contribute to electrocution of a worker using a portable electric drill.

The diagram in Figure 3.2 reveals the use of two symbols in coding causal relationships. Figure 3.3 deciphers this code. It is essential that the analyst be able to distinguish between the "AND" versus the "OR" relationship for event conditions. All of the causal event conditions are required to be present to cause a result to occur when these conditions are connected by an AND gate. However, a single condition is

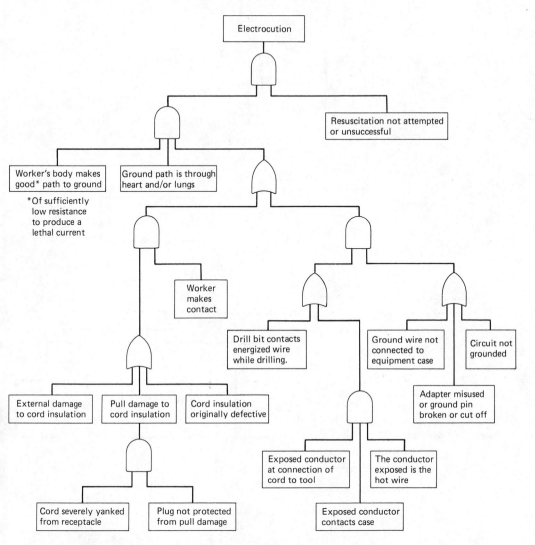

Figure. 3.2. Fault tree analysis of hazard origins in the electrocution of workers using portable electric drills (not double-insulated).

sufficient to cause a result to occur when conditions are connected by an OR gate. For example, oxygen, heat, and fuel are all required to produce fire, so they are connected by an AND gate, as shown in Figure 3.4. But *either* an open flame or a static spark may be sufficient to produce ignition heat for a given substance, so these conditions are connected by an OR gate, as shown in Figure 3.5. Note that Figures 3.4 and 3.5 could be combined to start the buildup of a branch of a fault tree.

One difficulty with fault tree analysis is that it requires each condition to be stated in absolute "yes/no" or "go/nogo" language. The analysis breaks down if a condition as stated may or may not cause a specified result. When the analyst is con-

(a)   (b)

Figure 3.3. Logic code for fault tree diagrams: (a) AND-gate symbol; (b) OR-gate symbol.

Figure 3.4. Example use of the AND gate in fault tree diagrams.

Figure 3.5. Example use of the OR gate in fault tree analysis.

fronted with a "maybe" situation, it usually means that the cause has not been sufficiently analyzed to determine what additional conditions are also necessary to effect
the result. Therefore, the difficult of dealing with a "maybe" situation forces the
analyst to look more deeply into the fault relationships, so the "difficulty" may be a
benefit after all.

Fault tree analysis permits the analyst to compute quantitative measures of the
probabilities of accident occurrence. The computation is tricky, however, and at
best is only as good as the estimates of the probabilities of occurrence of the causal
conditions. It seems intuitive that one should add the probabilities of events leading
into an OR gate and multiply the probabilities of events leading to an AND gate.
This intuition is wrong in both cases and will be shown for the OR gate as follows.
Suppose in the example shown in Figure 3.5 that the probability of occurrence of the
two event causes were as follows:

Event cause	Probability of occurrence (%)
Open flame	50
Static spark	50

To add these two probabilities would result in a 100% chance of reaching "ignition
heat," an obviously erroneous result. A 50–50 chance of either of two possible
causes is insufficient basis to find a 100% certainty of any resulting event. To make
the logic even more convincing, the reader should rework the example using the
following probabilities:

Event cause	Probability of occurrence (%)
Open flame	60
Static spark	70

Obviously, no result can have a probability of 130%! The correct probability computation is to subtract the "intersection" or probability that both independent
causes would occur as follows:

$$
\begin{aligned}
P[\text{ignition heat}] &= P[\text{open flame}] + P[\text{static spark}] \\
&\quad - P[\text{open flame}] \times P[\text{static spark}] \\
&= 0.6 + 0.7 - 0.6 \times 0.7 \\
&= 1.3 - 0.42 = 0.88 = 88\%
\end{aligned}
$$

For OR gates in which there are several event causes, the computation becomes very
complex. Adding to the complexity is the question of whether the event causes are

"independent"; that is, the occurrence of one event does not affect the likelihood of the occurrence of the other event cause(s). A special case of "dependent" events are events which are "mutually exclusive"; that is, the occurrence of one event *precludes* the occurrence of the other(s). The "mutually exclusive" condition simplifies the computation, but event causes in fault tree diagrams are typically not mutually exclusive.

Really to resolve the problem of fault tree computations would require a treatise on probability theory which is beyond the scope of this book. Suffice it to say here that fault tree probability computations are more complex than most people think and are usually done incorrectly.

Despite the computational problems associated with fault tree computations, the fault tree diagram itself is a useful analytical tool. The diagramming process itself forces the analyst to think about various event causes and their relationship to the overall problem. The completed diagram permits certain logical conclusions to be reached without computation. For instance, in Figure 3.2 the event "worker makes contact" is a key event because from the diagram it can be seen that prevention of this event will preclude any of the five events to the lower left in the diagram from causing electrocution. Even more important is the event "worker's body makes good path to ground." Prevention of this one single event is sufficient to prevent electrocution, according to the diagram. The reader can no doubt draw other interesting conclusions from the diagram in Figure 3.2. These conclusions result in a more sophisticated understanding of the hazard and in turn may lead to revisions to make the diagram more realistic. Such a developmental process leads to the overall goal of hazard avoidance.

Toxicology

Toxicology is the study of the nature and effects of poisons. Industrial toxicology is especially concerned with identifying what industrial materials or contaminants can harm workers and what should be done to control these materials. This is really a broad statement because virtually every material is harmful to living organisms if the exposure rate or quantity is great enough.

Many toxicological studies are performed on animals to provide the basis for conclusion about the hazards to humans. These animal studies are essential because most toxicological experiments would cause death or serious harm to human subjects. The disadvantage is that animal defenses to various toxic substances vary between species. The more the field of toxicology advances, however, the more various toxic materials can be classified and their effects predicted even before experiment. Rabbit/man, monkey/man, mouse/man, and guinea pig/man comparisons of species defenses to various agents are becoming fairly well known.

Epidemiological Studies

Epidemiology is contrasted with toxicology in that epidemiology studies are strictly of people, not animals. The word obviously derives from the word "epidemic," and in the literal sense epidemiology is the study of epidemics. The epidemiological approach is to examine populations of people to associate various patterns of possible disease causes with the occurrence of the disease. It draws heavily on the analytical tools of mathematical statistics.

A classic epidemiological study was the association of the disease rubella (German measles) in pregnant women to birth defects in the infants born of those pregnancies. The study began with a curiosity observed by N. McAlister Gregg, an Australian opthalmologist, in 1941. He observed eye cataracts among infants born of mothers who had had German measles during pregnancy in 1939 and 1940. The phenomenon might have been unnoticed except for the Australian epidemic of that disease during the World War II buildup. Years of statistical epidemiological studies later confirmed a strong relationship between rubella during pregnancy and a wide variety of birth defects in the infants born of those pregnancies.

Epidemics are usually thought of as attacking a general population at a specific time in a specific geographical area. Examples are the bubonic plague in Europe in the mid-fourteenth century and the rubella epidemic in Australia in 1939–1940. But more subtle epidemics attack a specific subgroup of people who may be spread out over time and place. In other words, the victims of the epidemic may not live in one place or at one time but have some other common characteristic, such as what they *do*. This aspect of epidemiology is what makes it important to occupational safety and health. Thus lung fibrosis may not be a very common disease in any location or at any time, but when the population of only those persons who have worked with asbestos is examined, it can be seen that after a long latency period lung fibrosis can be considered an epidemic. Epidemiological studies linking lung fibrosis to asbestos have led to the identification of a type of lung fibrosis known as "asbestosis." Other epidemiological links are "brown lung" to textile workers, "black lung" to coal miners, and angiosarcoma to vinyl chloride workers.

Both epidemiology and toxicology are important elements in the analytical approach to avoiding hazards, but the Safety and Health Manager does not typically perform such studies. The studies provide the basis for mandatory standards which are subsequently used in the enforcement approach. Safety and Health Managers may also use the results of such analytical studies to substantiate the psychological approach or as a justification for an engineering approach to a health problem.

Cost/Benefit Analysis

Chapter 1 set the record straight on the importance of hazard costs. Like it or not, people do make cost judgments on occupational safety and health—not just management, but workers, too. In the world of reality, funds do have limitations, and cost/benefit analyses must be used to compare capital investment alternatives. Safety and Health Managers who feel that they can justify at any cost capital investment proposals which can be shown to have the possibility of preventing injuries and illnesses can easily appear naive. There is always more than one opportunity to improve safety and health, and cost/benefit analyses provide the basis for deciding which ones to undertake first.

The biggest difficulty with cost/benefit analysis is the estimation of the benefit side of the picture. Benefits to safety and health consist of hazard reduction, and to make the cost/benefit analysis computation, some quantitative assessment of hazard must be made. Such probabilities of injury or illness are very difficult to determine. Statistical data are being compiled at the state level, as discussed in Chapter 2. But these data are still usually of insufficient detail to permit a quantitative determination of existing risk. Besides this determination an estimate of expected risk *after* the improvement must also be made, since it cannot be assumed in general that every safety and health improvement will completely eliminate hazards. An example will be used to illustrate the method of cost/benefit analysis.

EXAMPLE 3.5   Cost/Benefit Analysis of Installing a Particular Machine Guard

---

Cost
Amortization of initial investment:
Initial cost	$4000
Expected useful life	8 years
Salvage value	0
Interest cost on invested capital	20%

Annual cost = $4000 × (20% interest factor for 8 years)
= $4000 × 0.26061 = **$1042**

Expected cost of annual maintenance (if any)	0
Expected annual cost due to reduction in production rate (if any)	800
Total expected annual cost	$1842

Benefit
Estimated tangible costs per injury of this type	$ 350
Estimated intangible costs per injury of this type	2400
Total costs per injury	$2750
Average number of injuries per year on this machine due to this hazard	1.2
Expected number of injuries of this type after guarding	0.1
Expected reduction in injuries per year	1.1

Expected benefit = $2750 × 1.1 = **$3025**

Since the total expected benefit of $3025 is more than the total expected cost of $1842, the conclusion is that it would be worthwhile to install this machine guard.

In Example 3.5 the reader will perceive the large degree of speculation in the estimates of hazard costs. Such speculation casts doubt upon the entire analysis, but it is better to speculate and calculate than to completely ignore the cost/benefit trade-off. This is an opportune point to lead into the next section of this chapter—hazards classification. There is a great deal to be gained from a subjective analysis of hazard costs, without resorting to overly sophisticated quantitative analysis, the credibility of which "hard" data may not support.

HAZARDS CLASSIFICATION SCALE

The absence of hard data to support quantitative cost/benefit analyses leaves a void of tools or benchmarks for use by the Safety and Health Manager, safety committee, or other party on whom the decision responsibility for safety and health improvement may be thrust. Some sort of ranking or scale is needed to distinguish between serious hazards and minor ones so that rational decisions can be made to eliminate hazards on a worst-first basis.

OSHA does recognize four categories of hazards or standards violations as follows:

- Imminent danger
- Serious violations
- Nonserious violations
- De minimus violations

Chapter 4 explains these categories in more detail. However, the categories are rather loosely defined and are distinguished chiefly by the extent of the penalty authorized for each type. The "imminent danger" category qualifies OSHA to seek a U.S. district court injunction to force the employer to remove the hazard or face a court-ordered shutdown of the operation. The "de minimus" violations by contrast are merely technical violations which bear little relationship to safety or health and typically do not carry a monetary penalty. But the designations of which violations will be recognized as falling into which category of hazard is especially unclear.

It is perhaps impossible to define clear-cut categories in every instance, but much is to be gained by some type of subjective ranking of workplace hazards. The thesis of this text is that a scale of 1 to 10 should be attempted, crude as that scale might be. Until people start talking about degrees of hazards on such a quantitative scale, little progress can be made toward establishing an effective and orderly strategy for hazard elimination. On a 10 point scale, a "10" is characterized as the worst hazard imaginable and a "1" the least significant or mildest of hazards.

The 10-point scale is recommended because such a scale has become very popular in everyday speech. Facilitated by the media, especially television, the public has come to understand a statement such as "on a scale of 1 to 10 that item (tennis match, ski slope, kiss, etc.) was at least a 9." The familiarity of this popular jargon can be employed at work to characterize workplace hazards.

Table 3.1 is a first attempt to describe subjectively each of 10 levels of hazards. The definitions address basically four types of hazards: fatalities, health hazards, industrial noise hazards, and safety (injury) hazards. A clear-cut delineation is obviously difficult, and some readers will no doubt disagree with the wording of the definitions. Criticisms of the scale will reflect both the shortcomings of the definitions and the biases of the critics themselves. Acoustical experts, for instance, may want to place a high degree of emphasis on excessive noise hazards. Other specialists will want to emphasize other areas.

TABLE 3.1 CATEGORY DESCRIPTIONS FOR A 10-POINT SCALE FOR WORKPLACE HAZARDS

1	2	3	4	5	6	7	8	9	10
"Technical" violations; OSHA standards may be violated but no real occupational health or safety hazard exists	No real fatality hazard Health hazards minor or unverified Even minor injuries unlikely	Fatality hazard not of real concern Health hazards have exceeded designated action levels *or* Sound exposure action levels exceeded (e.g., continuous exposure in the range 85–90 dBA) *or* Minor injury risk exists but major injury hazard is very unlikely	Fatality hazard either remote or nonexistent Health hazards characterized by illnesses which are usually temporary, controls or personal protective equipment may or may not be required *or* Temporary hearing damage usually will result without controls or protection and a few workers may incur partial permanent damage *or* Minor injuries likely, such as cuts or abrasions, but major injury risk is low	Fatality hazard either remote or not applicable Long-range health *may* be at risk; controls or personal protective equipment *advisable* or *required* by OSHA *or* Hearing damage may be permanent without controls or protection (e.g., continuous 8-hr exposure in the range 95–100 dBA) *or* Major injuries such as amputation not very likely	Fatality hazard unlikely Long-range health definitely at risk; controls or personal protective equipment *required* by OSHA *or* Hearing damage likely to be permanent without controls or protection (e.g., continuous 8-hr exposure in the range 100–105 dBA) *or* Major injury such as amputation not very likely but definitely *could* occur	Fatality not very likely, but still a consideration Serious long-range health hazards are proven; controls or personal protective equipment *required* to prevent *serious* occupational illness *or* Hearing damage obviously would be *severe* and *permanent* without protection (e.g., continuous 8-hr exposure in excess of 105 dBA) *or* Major injury such as amputation could easily occur	Fatality possible; this operation has never produced a fatality, but a fatality could easily occur at any time *or* Severe long-range health hazards are *obvious*; controls or personal protective equipment essential to prevent *fatal* occupational illness *or* Major injury is likely; amputations or other major injuries have *already* occurred in this operation in the past	Fatality likely; similar conditions have produced fatalities in the past; conditions too risky for normal operation; rescue operations are undertaken for injured workers with rescuers using personal protective equipment	Fatality imminent; risks are grave; some employees earlier in the day have died or are dying; conditions are too risky even for daring rescue operations except perhaps with exotic rescue protection

One critical test is met by the proposed scale. Within each hazard type each successive level of the scale describes a progressively more severe hazard. Individual industries may devise more suitable definitions, but the idea is to start talking and thinking about hazards in terms of a 10-point scale.

One criterion omitted from the scale definitions is cost of compliance or cost of correcting a given hazard. Cost is a completely different criterion and is almost independent of level of hazard. That is, it can easily cost just as much or more to correct a Category 2 hazard as it does to correct a Category 9 hazard. Cost is an important criterion in the decision-making algorithm but is omitted from the scale definitions to permit a clear ranking of hazard priority first. Once the hazards have been sorted out, costs of hazard correction can be estimated and capital allocated for correction according to a rational capital investment policy.

Another missing criterion, this one perhaps conspicuous, is any mention of OSHA legal definitions of "imminent danger," "serious violation," and so on. To place these legal designations into fixed positions in the hazard scale would undermine the objectivity of the classification. Many persons have a preconceived notion of what legal penalty should be imposed or not imposed for a given hazardous situation. This bias is found in OSHA officers and their industrial counterparts as well. For instance, if a plant Safety and Health Manager should happen to believe that a given situation is not serious enough to warrant a plant shutdown, he or she would tend to prohibit selection of any designation that would bear the legal OSHA designation "imminent danger." This is despite the fact that reason would place the Category 10 definition into the imminent danger category.

Three credible profiles of the OSHA legal classification superimposed on the proposed 10-point scale are displayed in Figure 3.6. Profile A has an industry flavor and represents the viewpoints of some business executives. Profile A cannot be considered an extreme position because thousands of American businesses are headed by persons who believe that no government official should have the right to step in and close their businesses with or without a court order, regardless of hazards. Therefore, some business executives do not recognize the legal category "imminent

Figure 3.6. Three profiles of legal designations for hazards superimposed on the 10-point hazard classification scale.

danger.'' Profile A does at least recognize the imminent danger category, although the profile is skewed to the right. Profile C shows a contrasting position, being skewed to the left. Some OSHA officers have demonstrated that their positions are very close to profile C. Profile C likewise is not the ultimate in left extreme because some people believe that any fatality or amputation hazard should be classified as serious regardless of the remoteness of the hazard. Profile B represents a sort of middle-of-the-road position.

A final consideration in the classification of hazards would be the industry environment in which the hazard is portrayed. What seems hazardous to a structural steelworker working 100 feet above the ground on a narrow beam is in an entirely different category from what seems hazardous to an accountant. A similar comparison could be drawn between a coal miner and a computer analyst. It is only hoped that as a first step at least, coal miners should know what other coal miners mean when they talk about Category 4 hazards. Similarly, accountants should be uniform in their conception of Category 9 hazards, for example.

Except for some very broad categories, such as ''serious'' and ''nonserious,'' federal law has little to say about degree of hazard. No consistent criterion is available to Safety and Health Managers, who must decide which problems must be tackled first. The 10-point scale recommended by this text for ranking safety and health hazards offers an opportunity to bring some order to this perplexing problem. The scale is a vehicle for encouraging all parties to focus on the degree of each hazard so that rational decisions can be made to correct those problems which are indeed significant.

SUMMARY

The sophisticated Safety and Health Manager is not content with one approach to deal with workplace hazards. There is too much uncertainty to solve the enormous problems neatly with a simple approach such as ''awards for no lost-time accidents'' or ''fines for anyone who breaks the rules.'' These two approaches, together with all the others, have their place, but an integrated program utilizing the strengths of all approaches has the greatest potential for success.

While becoming more sophisticated, it is easy for Safety and Health Managers to fall into the trap of becoming more enamored with their impressive-looking analyses, scientific formulas, and statistics. Some of the best analyses may be subjective, not quantitative. The 10-point hazard classification scale suggested in this chapter represents an opportunity for Safety and Health Managers to talk and think about workplace hazards in terms of *degree*. Corporate top management has been looking for years for that breed of Safety and Health Manager to emerge—the kind who can discern between the really significant problems, the ordinary problems, and the trivial ones.

EXERCISES AND STUDY QUESTIONS

3.1. A well-known safety rule is not to pull an electrical plug out of a wall socket by means of the cord. Have you broken this rule? Do you regularly do so? What is the reason for this rule? Do you think the rule is an effective one?

3.2. Do you believe that safety or health rules are "made to be broken?" Why or why not?

3.3. Name four basic approaches to hazard avoidance.

3.4. Recall your first supervisor in your first full-time job. Did he or she ever mention safety or health on the job? Was your supervisor's influence on your safety habits positive, negative, or neutral?

3.5. What is the Heinrich ratio?

3.6. What are the three lines of defense against health hazards?

3.7. What is the standard safety factor for crane hoists? for scaffolds? for scaffold ropes?

3.8. Name three general fail-safe principles. Can you think of a real-world example of any of them?

3.9. What has Murphy's Law to do with occupational safety and health?

3.10. What is the purpose of FMEA?

3.11. Construct a fault tree diagram describing potential causes of fire in paint spray areas. Compare your analysis with others in the class.

3.12. Construct a fault tree diagram that describes the ways a pair of dice can be thrown to "roll 7." Calculate the "risk" or chances of rolling 7.

3.13. Explain the difference between toxicology and epidemiology.

3.14. Alter the fault tree diagram of Figure 3.2 to consider the possibility that the portable electric drill might be "double-insulated" (i.e., sheathed in an approved plastic case to prevent electrical contact with the metal tool case).

3.15. A certain ventilation system would cost a firm approximately $60,000, which can be amortized over its useful life at a cost of $15,000 per year. Annual maintenance costs are expected to be $600 per year and monthly operating costs (utilities) at $150 per month. The system is expected to facilitate production by reducing the amount of machine cleaning for an expected annual savings of $1200 per year. The primary benefit of the proposed ventilation system is expected to be the elimination of the need for respirators, which cost the company $4000 per year in equipment, maintenance, employee training, and respiratory system management. The ventilation system is expected to reduce short-term illness complaints by an average of 6 per year and long-term illnesses by an average of 0.2 per year. The short-term illnesses result in a total cost of $600 per occurrence, including intangibles. The long-term illnesses are expected to result in a total cost of $30,000 per occurrence, including intangibles. Use a cost/benefit analysis to determine whether the ventilation system should be installed. What is the primary benefit of the ventilation system?

3.16. On a scale of 1 to 10 (10 is worst), how would you rate each of the following hazards?
   (a) Ten-foot balcony not provided with a guardrail. Workers regularly work close to the edge every day without fall protection.
   (b) Same as part (a) except that the working surface is outdoors and is very slippery in rainy weather.

(c) No guardrail on a 10-foot balcony accessed only twice per year by a maintenance worker to service an air conditioner.

(d) An unguarded flat rooftop accessed only by air conditioner maintenance personnel. The closest necessary approach to the edge of the roof is 25 feet.

(e) Broken ladder rung (middle rung of a 12-foot ladder).

(f) Overfilled waste receptacles in lunchroom.

(g) Two-ton hoist rope with dangerously frayed and broken wires in several strands.

(h) Mushroomed head on a cold chisel.

3.17. Consider the following accident case history:

On July 4, 1980, three workers, ages 14, 16, and 17, were installing a sign at a bait shop along a state highway. They were using a truck-mounted extensible metal ladder to unload and position an upright steel support for the sign. Two of the workers were holding and guiding the steel support while the third stood on the flatbed truck operating the extensible ladder controls. The metal ladder came into contact with a 13,200-volt overhead power line. The two workers guiding the steel support were standing on the ground and were subjected to immediate electrocution. The third worker attempted to break contact with the power line by operating the controls, but the controls had become useless, probably because the control wiring had become burned by the high-voltage contact. The worker then leaped from the truck and ran around to the front to try to drive the truck away to break contact. As he grasped for the door handle to the truck cab, he was still standing on the ground, which provided a path for the current through his body. The utility company had equipped the line with a "recloser" which normally would have broken the circuit under these conditions, but for a variety of reasons failed to do so in this case. Therefore, the power remained on for a rather lengthy period. The high voltage and burning current finally broke down the dielectric strength of the rubber tires and the tires blew out. This shifted the truck's position and contact with the power line was broken, but not before one worker had been burned in half, and another's legs were both burned off; all three workers died.

How could future accidents of this type be prevented? Compare the four basic approaches to hazard avoidance in preventing accidents of this type.

4

Impact of Federal Regulation

OSHA - 1970
Dep of Labor
Consumer Product Saftey Commission
Investments contact list

The field of industrial safety and health took a giant leap in the early 1970s. There is some question, however, whether the leap was forward or backward. On December 29, 1970, Congress passed the Williams-Steiger Occupational Safety and Health Act of 1970, creating OSHA, the Occupational Safety and Health Administration of the U.S. Department of Labor. OSHA got off to a bad start and immediately became the target of sharp public criticism. But at the same time the agency created an immediate focus of attention on the field of industrial safety and health. Changes in federal administrations have changed OSHA's methods, and these methods need to be examined here, as well as the basic approaches upon which OSHA was founded. Regardless of how OSHA may change, its impact on the field is a permanent one, and this book would not be complete without examining this impact.

Most people think of OSHA when the subject of federal safety and health regulations is mentioned. But there is also MSHA, the Mine Safety and Health Administration; TOSCA, the Toxic Substances Control Act; and CPSC, the Consumer Product Safety Commission. Most of these other agencies and legislation governing safety and health are patterned after the OSHA agency and legislation. Accordingly, this chapter examines the basic approach of OSHA and the impact it has had on the field.

STANDARDS

The most significant change that OSHA brought to industry was a book of federal standards. Never before had virtually all general industries been subjected to a book of federally prescribed, mandatory rules for worker safety and health. This set of

59

rules formed the basis for inspection, citations, penalties, and virtually every activity in which OSHA is engaged. One rule, however, was not in the book; it is the most important rule of all and will be discussed first.

General Duty Clause

Congress decided to set up one general rule for all to follow, and this rule was included in its entirety right in the text of the law that created OSHA. This rule is called the "General Duty Clause," and might be called the First Commandment of OSHA. It is quoted as follows:

> Public Law 91-596
>
> Section 5(a)   Each employer. . .
>
>
> (1) shall furnish to each of his employees employment and a place of employment which are free from recognized hazards that are causing or are likely to cause death or serious physical harm to his employees. . . .

The General Duty Clause is cited by OSHA whenever a serious safety or health violation is alleged for which no specific rule seems to apply. This has drawn some criticism from industry because after a serious accident has happened, it is easy to conclude that a situation was unsafe. However, before the accident occurred, it might not be clear what should have been done, especially in the absence of any specific rule to provide guidance. Sometimes a specific rule will also apply somewhat, and OSHA has cited both the General Duty Clause and the specific standard. When a situation is clearly in violation of a specific standard, OSHA usually does not bother to cite the General Duty Clause.

If the General Duty Clause is the First Commandment of OSHA, the second is like unto it, and is quoted as follows:

> Public Law 91-596
>
> Section 5(b)   Each employee shall comply with occupational safety and health standards and all rules, regulations, and orders issued pursuant to this Act which are applicable to his own actions and conduct.

Note that Section 5(a)(1) describes a responsibility for employ*ers,* whereas Section 5(b) pertains to employ*ees.* Penalties are prescribed for employer violations, but no penalty is prescribed for employee violations. Section 5(a)(1) has been cited quite frequently, but as far as this author knows, there has never been an OSHA citation of Section 5(b).

Promulgation

In addition to the General Duty Clause, the law that created OSHA also set up machinery for OSHA to issue new standards too technical and detailed to be included one by one in the text of the law passed by Congress. The law provides safeguards intended to assure that OSHA will be fair and give all interested parties an opportunity to support or object to proposed new standards. In addition to promulgating new standards, OSHA is also permitted to revise old standards or even revoke them, following the same rulemaking procedure with prescribed safeguards.

"National Consensus"

The OSHA law recognized the existence of "national consensus standards" already in use prior to the enactment of the OSHA law. This is a very important part of the law because it gave OSHA the authority to *bypass the procedural safeguards* mentioned above and issue standards without consulting the public. The principle was that the standards by their prior existence had already been accepted by the public. OSHA's authority to issue national consensus standards expired in early 1973, two years after the effective date of the Act. Thus after two years OSHA was no longer permitted to "grandfather in" safety and health standards. The largest number of national consensus standards were developed by the two leading national standards-producing organizations, the American National Standards Institute (ANSI) and the National Fire Protection Association (NFPA).

Besides the national consensus standards, any "established Federal standard" was also permitted to be adopted as a general standard by OSHA. The established federal standards had formerly applied only to limited groups such as construction industries or government contracts. With the OSHA law Congress permitted OSHA to extend these standards to virtually all industries.

The biggest issue with the national consensus standards was whether they ever really represented the "national consensus." Virtually all of these standards had never been enforced, and the language used made it obvious that the original drafters of many of the standards had never intended for them to be mandatory rules to be enforced with monetary penalties for offending employers. This issue will be examined further in the section "Public Uproar."

Structure

When discussing standards, this text will sometimes use the term "horizontal standard" or "vertical standard." The Safety and Health Manager needs to understand these terms. Prior to OSHA, states enforced codes for safety and health by industry, issuing separate handbooks of rules for each industry to follow. These standards have come to be called vertical standards. OSHA's basic approach, by contrast, is to generalize and organize standards by hazard sources, regardless of industry. These are called "horizontal standards." There are certain standards that OSHA limits in scope to a particular industry, but these are the exceptions; the basic structure of the standards is horizontal.

OSHA almost had to use the horizontal approach. The nature of the OSHA law itself was to authorize the new agency to generalize on specific standards. Imagine for a moment the morass of books that would have been generated if OSHA had attempted to write a separate standards book for each industry, sifting through the national consensus standards to determine which standards should be included in each. Unusual industry categories would inevitably "fall through the cracks" under a vertical scheme, characterizing themselves as being under the scope of neither this nor that standard.

By using the horizontal approach, OSHA left the job of sifting through the standards to the public. Thus a general book of standards entitled *General Industry, Part 1910,* was published to cover virtually all industries. A manufacturer of teacups must read out of the same book as does an airline, and both must determine whether they need to be concerned with a provision entitled "horse scaffolds."

The construction industry is one for which OSHA published a vertical standard entitled *Construction Standards,* Part 1926, but even these special categories of industries are also covered by the more general Part 1910 with respect to any hazard for which no standard exists in the narrower vertical standard. OSHA did make the job of compliance a little easier by subsequently publishing a larger Construction Standard which included those parts of the General Industry standard which OSHA might cite at construction sites. However, they did not abdicate the right to turn to General Industry standards, which were not included in even the larger version of the Construction Standard.

The concept of vertical versus horizontal standards extends down within the subdivisions and individual paragraphs of the OSHA standard. For instance, OSHA standard 1910.106 pertains to flammable and combustible liquids and is basically a horizontal standard to be applied to general industry. However, within the standard there is paragraph 1910.106(i), which is entitled "Refiners, chemical plants, and distilleries." Therefore, that portion of 1910.106 should be considered a vertical standard applying only to those industries named.

One problem in determining the scope of all standards and of vertical standards in particular is the infeasibility of an exclusive classification of industries and equipment. Consider the diagram in Figure 4.1 (called a Venn diagram), which relates two different classes of equipment. The circles represent the two classes, and the shaded area is the overlap between the two groups. The circles could also be considered safety standards, and thus equipment represented by the shaded area would be subject to both standards.

Another classification for standards divides them into "specification" standards versus "performance" standards. The specification type is easier to enforce because it spells out in detail exactly what the employer must do and how to do

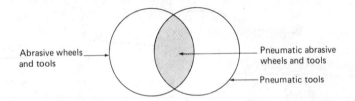

Figure 4.1. Venn diagram to represent the relationship of two different overlapping types of equipment.

it. The performance type has its advantages, however, in that it permits the employer latitude in devising innovative ways to eliminate or reduce hazards. In short, the specification standard emphasizes *methods* while the performance standard emphasizes *results*. The comparison is best shown by an example.

Example 4.1

    (a)   Example of a *specification* standard:

        OSHA standard 1910.110   Storage and handling of liquefied petroleum gases
        (c)   Cylinder systems
            (5)   Containers and equipment used inside buildings or structures
                (vi)   Containers are permitted to be used in industrial occupancies for processing, research, or experimental purposes as follows:
                    (a) ...
                    (b) Containers connected to a manifold shall have a total water capacity not greater than 735 pounds (nominal 300 pounds LP-gas capacity) and not more than one such manifold may be located in the same room unless separated at least 20 feet from a similar unit.

    (b)   Example of a *performance* standard:

        OSHA standard 1910.36   General requirements (under Subpart E—Means of Egress)
        (b) Fundamental requirements
           (2) Every building or structure shall be so constructed, arranged, equipped, maintained, and operated as to avoid undue danger to the lives and safety of its occupants from fire, smoke, fumes, or resulting panic during the period of time reasonably necessary for escape from the building or structure in case of fire or other emergency.

The standard in part (a) leaves no doubt in the reader's mind as to what must be done and specifies exactly the limitations of containers and manifolds within the room. There is no leeway for the employer to devise a better way—perhaps safer and cheaper—to minimize hazards. However, once the exact standards have been met, the employer can feel safe from citation.

In the performance standard stated in part (b), it can be seen that employers have all kinds of latitude to set up their buildings in ways to avoid "undue danger." If a new type of effective smoke alarm is developed and placed on the market, the employer has the choice of installing the new alarm or adopting a different strategy to enhance safety in the event of "fire or other emergency." But with the performance standard, a disagreement may develop between an enforcement inspector and the employer over the effectiveness of the methods selected by the employer to protect against "undue hazards."

Most standards do not exactly fit either type but do rather roughly fit one category or the other. Employers tend to prefer performance standards, while enforcement officials generally prefer specification standards. However, this issue does not clearly divide the two groups. Sometimes employers prefer specification standards because it is easier to determine whether a given facility or equipment meets a specification standard or not. Also, a few enforcement officials prefer the performance standard because then if an accident does occur, it tends to strengthen the case for a citation regardless of the specific steps the employer may have taken to prevent the hazard.

NIOSH

The National Institute for Occupational Safety and Health (NIOSH) was created by the OSHA law to carry on research and training. NIOSH was housed under the then Department of Health, Education and Welfare (HEW), which subsequently was changed to the Department of Health and Human Services (HHS). NIOSH is mentioned here because of its important role in recommending new standards to OSHA. OSHA has the sole authority to promulgate new standards, but NIOSH develops the criteria for new standards and conducts the research necessary to justify the need for new standards. NIOSH is often thought of as being concerned primarily with health and toxic materials, but it is important to remember that NIOSH also has responsibility for safety standards research and development.

ENFORCEMENT

Occupational safety and health ought to be topics of vital interest and importance regardless of OSHA, but most of us would not have realized this if OSHA had not been given the authority in the 1970s to inspect industries and issue citations with monetary penalties.

Inspections

The original OSHA law gave the OSHA official the right to enter a factory or other workplace without delay (at reasonable times) upon presenting credentials. Such credentials consisted of basic identification of the officer but no court-issued search warrant. The right of a government official to do this was later challenged by an Idaho businessman, and the U.S. Supreme Court ruled in his favor in the famous "Barlow decision" in 1978. Employers may now invoke the Fourth Amendment to the U.S. Constitution and require OSHA to obtain a search warrant to conduct an inspection.

Some company managers take the position that no time is a "reasonable time" to be visited by the OSHA inspector. They argue that their plant has such proprietary processes that a visit by the inspector would jeopardize their trade secrets. Congress anticipated this excuse and provided that any information gathered that might reveal a trade secret be kept confidential. Actually, this provision was essential to prevent OSHA from conflicting with existing laws to protect trade secrets.

OSHA inspections are generated according to the following priorities:

1. Imminent danger
2. Fatalities and major accidents
3. Employee complaints
4. High-hazard industries

"Imminent danger" would be a situation in which death or serious physical harm could be expected to occur immediately. Time is of the essence in these situations, and ordinary enforcement procedures may be too late to protect workers. OSHA may go to a U.S. District Court and get a temporary injunction to remove employees from the work area in an imminent danger situation. Such an action would be rare and must be reasonable in allowing some employees to remain to correct the condition and to permit a safe and orderly shutdown. If the process is continuous, a sudden shutdown might cripple the equipment and might even be unsafe. OSHA realizes this, and in the case of continuous process operations, the process does not have to be completely closed down; a pilot crew is permitted to remain and to maintain the capacity of the process to resume normal operations.

OSHA requires a telephone call or other notification within 48 hours upon occurrence of fatal accidents or accidents in which five or more persons are hospitalized. This notification will invariably trigger an OSHA inspection because this category of inspections is second in priority only to imminent danger inspections. In fact, at first fatalities were considered top priority; later it was decided that it would be better for OSHA to assign top priority to serious, perhaps fatal, hazards before the accident occurs rather than after the fact. OSHA has a policy of investigating both categories, however, within 24 hours of notification.

An employee can request that OSHA investigate a hazard by filing with OSHA a complaint describing a hazard believed to exist in the workplace. To be valid the complaint must be signed by the employee. Many workers fear recrimination or other reprisal if they are subsequently identified by their signatures. But OSHA is bound by law to keep the origins of an employee complaint confidential if the employee so requests. However, the employee has additional safeguards against discrimination, as will be seen later in this chapter.

Next in priority to complaints come the inspections of industries that are shown by statistical records to be particularly hazardous. Early in the implementation of OSHA's enforcement program a collection of industries were named "target industries." Subsequently, a "target health standards" group was named. Focus on the Target Industries Program (TIP) shifted later to a "special emphasis program," which named trenching and excavation cave-ins as the principal concern. Next came the "National Emphasis Program" (NEP), emphasizing foundries and metal stamping. The presidential administration of Ronald Reagan assigned a national priority to those firms whose lost-workday rate was above the national average. Thus the old target industries and national emphasis programs which emphasized whole *in-*

dustries were discarded in favor of a system that concentrates on *individual firms* that have poor safety and health records.

The shift in inspection strategy that occurred in the early 1980s permitted OSHA to redirect its efforts in the face of imperative budget cuts in agencies throughout the federal government. Since not enough OSHA field personnel were available to inspect firms frequently and thoroughly, the emphasis was shifted to those firms that need improvement the most. The revised strategy is a powerful signal to the Safety and Health Manager to achieve results in reducing lost-workday cases to avoid OSHA inspection. Although the total number of inspections has been reduced, the focus results in a greater chance of inspection if the firm's records show that it is really dangerous.

Citations

After the inspection OSHA may issue a citation for alleged violations of the standards or of the General Duty Clause. The statutory limit is six months, so if no citation has been received within six months, the employer can be assured that a citation will not be forthcoming. A growing percentage of firms do not receive a citation upon inspection. OSHA has on occasion pointed to this percentage and designated these firms as "in compliance." The designation may be a misnomer, however, because the depth of inspections varies considerably. A really thorough wall-to-wall inspection of a comprehensive manufacturing plant of any size except the very smallest is almost certain to reveal some OSHA violations. The OSHA Compliance Officer may be reluctant to issue a citation and may overlook some minor items if it is felt that the employer has shown a good-faith effort to comply with the standards.

OSHA recognizes that some conditions, even though they represent a violation of the letter of the law, have no direct or immediate relationship to safety or health. In these instances OSHA may issue a "de minimus notice" in lieu of a citation. De minimus notices do not carry a monetary penalty.

If a citation is issued, it is not a very private matter. To the chagrin of the Safety and Health Manager, the citation must be prominently posted near where the violation occurred. Employees and management alike have the opportunity to read the citation and note the alleged violation. This increases the overall awareness of OSHA's enforcement powers and may spawn complaints by employees for other potential violations. As far as management is concerned, the Safety and Health Manager can prepare top management by explaining that many companies do receive a citation when inspected by OSHA; there is no disgrace to being found out of compliance with a few detailed provisions of the standards.

OSHA penalties can be quite severe, as Table 4.1 indicates, but actual fines are usually quite inconsequential. Penalties for nonserious violations are usually less than $100. OSHA has a formula for computing a reduced penalty, taking into consideration size of the company (small businesses tend to be assessed smaller fines), the history of previous violations, and "good faith" shown by the employer. By the time a given citation is issued, the formula for computing a possible reduction in penalty will have already been applied.

Although most OSHA penalties are small, it is possible for fines to become quite severe. Note that the "failure-to-abate" penalty is assessed for *every day* a violation remains uncorrected. Some "willful and repeat" violations have resulted in fines of hundreds of thousands of dollars assessed on a single firm, although these cases are rare.

TABLE 4.1   OSHA PENALTIES (STATUTORY MAXIMUMS)

Offense	Maximum penalty	Offender
Nonserious violations	$1000	Employer
Serious violations	$1000	Employer
Willful violations	$10,000	Employer
Repeat violations	$10,000	Employer
Failure to abate a violation	$1000 per day	Employer
Willful violation resulting in death to employee	$10,000 and/or 6 months in prison	Employer
Second offense: willful violation resulting in death	$20,000 and/or 1 year in prison	Employer
Failure to post notices, citations, etc.	$1000	Employer
Giving advance notice of an inspection	$1000 and/or 6 months in prison	Anyone
Falsifying records or reports	$10,000 and/or 6 months in prison	Anyone
Killing an OSHA inspector	Life imprisonment	Anyone

Source: Ref. 65.

Two questions that are frequently asked in seminars about OSHA are:

1. Who in the organization goes to jail when the "employer" is found to be guilty of a willful violation resulting in death?
2. Where does the money collected in OSHA fines go—to OSHA's budget to pay inspectors?

In answer to the first question, "employer" in a complex organization can be construed to represent the entire chain of supervision from the supervisor of the employee victim to the chief executive officer of the firm. However, what few cases have been prosecuted indicate that OSHA will usually attempt to zero in on only one guilty person to go to prison. The rationale is that OSHA attempts to prosecute the manager with the highest authority who knew about the violation, had the authority to correct it, but failed to correct the situation that resulted in the employee's death.

The other question, "Where does the money go?", apparently has its origins in some state and local enforcement schemes. It would hardly seem reasonable to deposit fine money into the coffers of those agency officials who assess the fines. But the misconception that OSHA is permitted to keep the money from fines collected seems to persist in the minds of the public. The origins of this idea appear to be the tradition of state boiler inspection agency regulations which provide for money collected to be deposited into the boiler inspection division accounts. OSHA does not use this strategy, however, and all fines are deposited directly into the U.S. Treasury and not earmarked for OSHA's use. OSHA fines would be insufficient to operate OSHA anyway since the total amount collected annually is generally very low compared to OSHA's annual budget.

A big mistake made by some managers is to pay OSHA fines and to then consider the matter closed. Such a strategy ignores the far more important aspect of the citation—the prescribed abatement period. The OSHA fine itself may be inconsequential; the big impact of the OSHA citation, if it is accepted, is that each item listed in the citation must be *corrected*, regardless of cost. The cost of correcting violations is usually much greater than the amount of the OSHA penalty. Before accepting a citation the Safety and Health Manager should stop and think about whether it is really feasible to correct the violation. There is only a 15-day period (working days) after receipt of the citation during which a decision must be made whether or not to contest the citation. Appeals can even be taken through the judicial processes all the way to the Supreme Court.

Appeals of citations are not to be confused with variances. Appeals are for employers who have already been cited. Variances are for employers who need time to comply or have an alternative to compliance which is more practical and still protects employees.

Prescribed abatement periods can be seen to be very important. It is easy to overlook abatement periods and concentrate on the nature of the violations and the proposed penalties. The abatement period can seem to pass quickly, and then the firm is subject to a possible reinspection and potentially severe penalties. OSHA is known to be very reasonable in extending abatement periods if the employer contacts OSHA and gives reasons for a request to extend. But if the employer takes no action and the OSHA Compliance Officer returns, a severe penalty is likely. It is easy for OSHA to be too optimistic about timetables for changes in facilities, allowing insufficient time for administrative approvals, delayed equipment deliveries, installation schedules, and of course Murphy's Law. It is up to the Safety and Health Manager to bring up the subject of abatement and to assure that OSHA sets reasonable periods.

Employee Discrimination

One penalty which is sometimes costly does not appear in Table 4.1. This is the penalty paid by the employer who is found to have discriminated against an employee because that employee filed an OSHA complaint or answered the OSHA

Compliance Officer's questions during an inspection or exercised any other right afforded employees under the OSHA law. Company management should be very careful to document reasons for discharging any employee, especially if that employee has a history of complaining about safety and health hazards. This type of complaining is no basis for dismissal, and the company that dismisses such an employee may find itself later reinstating the employee with back pay. No fine must be paid to the government for such an infraction, but if the case has lain idle for many months, the cost of reinstatement with back pay can be substantial.

Besides dismissal of the employee, OSHA recognizes other, more subtle means of discrimination. Any of the following employer actions is considered by OSHA to be illegal discrimination if it is used as punishment for the employee's exercise of rights under OSHA.

- Termination
- Demotion
- Assignment to an undesirable job or shift
- Denial of promotion
- Threats or harassment
- Blacklisting the employee with other employers

OSHA has even told employees that the employer may be in violation of the law for suddenly punishing an employee for doing something else wrong after the employee has protested a hazardous condition. It would be especially incriminating to single out a complaining employee for punishment for some unrelated action that goes unpunished for other employees. As can be seen, the whole matter can be extremely delicate, and the Safety and Health Manager should assure that all persons in management, from first-line supervisors on up, be aware and careful not to discriminate either directly or indirectly against any employee for complaining about safety and health violations, either during the worker's employment with the company or in the future if and when the worker seeks employment elsewhere.

One questionable worker "right" is whether an employee can walk off the job because of unsafe or unhealthy conditions and still expect to be paid. The courts have ruled against employees who sue to be paid by their employer for not working when they have walked off the job due to unsafe or unhealthful conditions.

PUBLIC UPROAR

OSHA survived a very stormy first decade. Despite the validity of its purposes, OSHA quickly grew to be one of the most hated agencies the federal government had ever created. At times its demise seemed imminent, but it continued to survive.

At the root of public criticism of OSHA lie the OSHA standards. Much has been said about "nitpicking inspectors," "unjustified fines," and "gestapo techniques,'' but these criticisms would scarcely have arisen if the standards themselves had been set up differently.

The original standards had a few ancient and obsolete provisions which have subsequently been deleted. Another problem with the standards was that advisory provisions containing language such as "should" were incorporated as mandatory rules with language such as "shall." The courts frowned upon this, and so did the public. The government has since worked hard to eliminate advisory provisions from the standards. Federal standards have also been criticized for their level of detail, vagueness, redundancy, and irrelevance.

Some federal standards have seemed to do more for industries manufacturing safety equipment than they have done to protect the worker. A good example was the original standards for fire extinguishers. OSHA now permits alternatives to fire extinguishers for many applications.

A favorite weapon of OSHA's critics is the old question "Has it done any good?" They usually feel comfortable with the question without any examination of the record because they believe that OSHA could not have any measurable beneficial effect. It is difficult to assess the impact of federal regulation on worker safety and health because even the statistical recordkeeping has changed somewhat since OSHA was created. It is not even agreed whether the changes in recordkeeping have made injury and illness rates look better or worse. Some believe that the presence of OSHA enforcement with OSHA Compliance Officers examining injury and illness records has tempted management to hide injuries and illnesses, making the overall record of injuries and illnesses under OSHA look better than it really is. Others believe that because injuries and illnesses requiring medical treatment are now required by law to be recorded, statistical summaries will show more injuries and illnesses and thus make OSHA look worse. There is no question, however, that in some areas OSHA has had an impact on worker safety and health. The most dramatic results are probably with respect to reduction of fatalities from trench and excavation cave-ins.

The concept of cost effectiveness continues to gain in importance. Thus the old question "Does a federal regulation do any good?" is changing to "Does the regulation do *enough* good to warrant the *cost* of compliance?" The cost of compliance is a much greater consideration than is the cost of monetary penalties assessed by regulatory agencies.

ROLE OF THE STATES

Prior to OSHA, occupational safety and health were generally considered to be concerns of the states, not of the federal government. The general feeling among supporters of OSHA was that the states historically had not been doing a sufficient job in establishing and enforcing adequate standards for occupational safety and health. Recognizing, however, that some states might develop effective occupational safety and health standards and enforcement programs, the OSHA law provides for state plans to be submitted to OSHA for approval.

Enforcement

It must first be said about state enforcement that OSHA itself has been given no authority to regulate state agencies, counties, or municipalities. Even federal agencies are exempt from regular OSHA enforcement—a sore point with some private employers—but the OSHA law does provide for federal agency coverage as a part of the responsibilities of the heads of the various federal agencies. Obviously, it would be impractical for the federal government to cite and penalize itself, so OSHA does not make inspections and issue citations to its sister agencies. As far as states are concerned, if the federal government were to issue citations and assess penalties from state and local governments, there would be sovereignty problems, so that is prohibited.

But if a state submits a plan for occupational safety and health standards and enforcement, to be approved by OSHA the plan must contain a program applicable to employees of state agencies and political subdivisions of the state. In addition, the state standards and enforcement must be at least as effective as the corresponding federal standards and enforcement. About half of the states have achieved "as effective as" status and have had their state plans approved. Safety and Health Managers operating in any of the state-plan states should consult with appropriate state officials for copies of standards and enforcement procedures. In most state-plan states the standards are virtually identical to the federal OSHA standards.

Consultation

Besides standards and enforcement programs, OSHA also delegates to states authority and responsibility for consultative assistance in occupational safety and health to employers upon request of the employers seeking such assistance. OSHA has authority to make federal grants to the states to support enforcement, consultation, and other purposes of the OSHA law. Recent political tides have been in the direction of transferring federal agency authority and activity to the jurisdiction of the states. OSHA is pursuing this path with the state consultation programs. Cooperation with a state agency can even result in temporary immunity from OSHA citation in some cases. The possibility is worth checking into, and Safety and Health Managers should consider state consultation as a part of their overall strategies. There is no charge for state consultation, and as of this writing it was available in some form in every state.

FUTURE TRENDS

In light of the degree of controversy over OSHA and over federal regulation in general, a large segment of the American public has formed the opinion that OSHA is finished or at least on its way out. But despite its problems, OSHA appears to be quite durable and has survived repeated attempts to repeal or modify it.

The Barlow decision was an important one but really did not affect OSHA's enforcement operations as much as the public first thought it would. Those firms insisting on inspection warrants are usually inspected later; the procedure merely delays the process somewhat.

An attempt to exempt small businesses and family farms merely shifted enforcement emphasis. The same can be said of a strategy to inspect only those firms which have a lost-workday rate greater than the national average. An exemption of one category means emphasis on another, provided that the staffing and organization of the agency remain intact.

All federal agencies faced reductions in funding and manpower authorizations in the early 1980s with the change in administrations and approval of federal spending reductions. But OSHA fared no worse than other typical federal agencies, and in the 1970s OSHA fared much better than other federal agencies in expansion of its role and budget.

One indicator of OSHA's future are the congressional acts that have been patterned after the OSHA law and have been passed despite criticism directed at OSHA. The laws governing product safety and liability and mine safety were approved in the 1970s and were constructed and worded in a manner similar to the OSHA law. This illustrates that Congress has continued to endorse the concept of OSHA.

The evolution of OSHA standards is very slow. Once a standard has been adopted as the "national consensus," it becomes very difficult to revoke the standard at a later date by saying that the standard did not represent the "national consensus" after all. Such a step repudiates or at least reflects adversely on the thousands of citations enforcing the given standard in inspections prior to the revocation. Any revision of the standards that can be considered subjective or controversial can be challenged in the courts. All of this tends to preserve the status quo despite widespread controversy. OSHA has been able to overcome the inertia of existing standards by making some rather broad changes to some of the standards, the fire protection standard being the most notable example.

The overriding factor in consideration of the future of OSHA is a positive one. Despite its problems, OSHA has had a tremendous impact on employers, workers, and the public alike by raising everyone's consciousness of safety and health hazards. For this it will never be forgotten and probably will never be repealed, at least not in its entirety.

EXERCISES AND STUDY QUESTIONS

4.1. Suppose that you are a writer of new standards. Pick a familiar hazard and write a two- or three-sentence paragraph for a possible standard to protect against this hazard. Write the standard first in the style of a specification standard and then rewrite it in the language of a performance standard.

4.2. Compare advantages of "specification" standards versus "performance" standards from the viewpoint of the employer and then from the viewpoint of an enforcement agency.

4.3. List in order of priority four OSHA inspection categories.

4.4. What is the difference between a "repeat violation" and failure to correct a violation? How do the penalties differ?

4.5. Name some public criticisms of the OSHA standards.

5

Buildings and Facilities

13%

Percent of OSHA
general industry citations
addressing this subject

The text now turns to the business of examining hazards in various categories, high-lighting applicable standards and suggesting methods of bringing about change to eliminate or reduce hazards. Most enterprises begin with a building in which to conduct operations, and this is an appropriate topic to begin the examination of hazards. The enterprise is fortunate if its management considers hazards and applicable safety and health standards during its building design stages.

Safety standards for buildings, whether they be municipal, state, or federal, are usually called "codes." For the most part, building codes apply to the construction of new buildings or to their modification. Thus, although building codes change constantly, those buildings built or remodeled before a particular change in the code are not required to be torn down and rebuilt or remodeled according to the new code. Needless to say, most buildings in existence do not meet the latest provisions of current building codes.

Some federal standards for buildings have been applied to all buildings, regardless of age. The standards have included matters of relative permanence such as floors, aisles, doors, numbers and locations of exits, and stairway lengths, widths, riser design, angle, and vertical clearance. Industry's objection has been that such standards are unfair, not only because they apply to existing buildings but also because they are vague and generally worded. But despite these problems, industries have undertaken a large number of retrofit programs to update their buildings and facilities to satisfy federal standards.

In the defense of federal standards for buildings and facilities stands the fact that some of the most frequent categories of worker injuries and even fatalities arise from improper building design, lack of guardrails, and problems with exits.

Building equipment is designed and built without sufficient thought about the worker who must have access to this equipment to clean, maintain, repair, replace lightbulbs, or otherwise service the buildings or facilities. Some workers are even in locations where they would be unable to escape in event of fire. Aisle widths are often set up arbitrarily without giving thought to clearance between moving machinery and personnel.

Federal standards pertaining to buildings and facilities include the following categories:

- Walking–working surfaces
- Means of egress
- Powered platforms, manlifts, and vehicle-mounted work platforms
- General environmental controls

A few provisions of these standards generate almost all of the problems. This chapter will now single out those individual provisions and analyze each to determine what the Safety and Health Manager should do to alleviate hazards and conform to standards.

WALKING AND WORKING SURFACES

One does not normally call a floor a "walking or working surface," so why do the standards writers select such a complicated bit of jargon? This question is answered by reflecting on the hazardous locations in which people work. To be sure, many accidents, especially slips and falls, occur on floors, but consider the other "walking and working surfaces"—for instance, mezzanines and balconies. Then there are the even more hazardous platforms, catwalks, and scaffolds. Not to be forgotten are ramps, docks, stairways, and ladders.

Guarding Open Floors and Platforms

The most frequently cited standard in the Walking and Working Surfaces subpart is indeed one of the most frequently cited standards in all of the OSHA standards. It is repeated here in its entirety, due to its great importance.

> OSHA standard 1910.23—Guarding floor and wall openings and holes
> (c) Protection of open-sided floors, platforms, and runways
>   (1) Every open-sided floor or platform 4 feet or more above adjacent floor or ground level shall be guarded by a standard railing (or the equivalent as specified in paragraph (e) (3) of this section) on all open sides, except where there is entrance to a ramp, stairway, or fixed ladder. The railing shall be provided with a toeboard wherever, beneath the open sides,
>   (i) Persons can pass,
>   (ii) There is moving machinery, or
>   (iii) There is equipment with which falling materials could create a hazard.

To many people, 4 feet seems an innocuous height, but the reason for this illusion is that they are thinking of jumping, not falling. Almost everyone has *jumped* without injury from elevations of more than 4 feet, but few adults experience a *fall* from that height without injury. A surprising number of *fatalities* result from falls at heights of only 8 feet.

The illusion of safety of platforms in the range 4 to 14 feet has led to complacency on the part of building and facility designers as well as of Safety and Health Managers. To stand on the extreme point of a precipice overlooking the Grand Canyon without a guardrail would seem foolhardy to the average person. But many "average persons" would think nothing of standing on the unguarded top of a tank 10 feet high. An unexpected event in such a situation can easily result in the worker taking a reflex action which results in a fatal fall.

The top of a tank, in the example above, was considered a "working surface" to be reckoned with by the facility designer. This is an important thing to remember. Even if the surface is the top of a tank, or even the top of a piece of equipment that is in the process of being manufactured, it may act as a temporary walking and working surface. But in unusual temporary situations there are other ways to provide temporary fall protection for workers, as will be seen in Chapter 8.

Another problem area is the protection of personnel from falls from loading docks. A few loading docks are slightly less than 4 feet high, thereby avoiding the issue. Some Safety and Health Managers have had the area immediately adjacent to the dock built higher to comply with standards, but this is of dubious value in preventing injuries. Others have installed temporary, removable railings, and some have used chain-type gates. The chain-type gates do not qualify as "standard railings" but have sometimes been accepted as a practical substitute in an attempt to deal with the hazard in the face of a difficult situation.

A rooftop is, of course, a walking and working surface for roofing workers. This raises the question of whether guardrails, fall protection, or some other type of protection is needed for perimeters of rooftops where persons are working. A provision of the OSHA Construction Standard specifies "catch platforms," unless

1. The roof has a parapet.
2. The slope of the roof is flatter than 4 inches in 12 inches.
3. The workers are protected by a safety belt attached to a lifeline.
4. The roof is lower than 16 feet from the ground to the eaves.

The term "standard railings" mentioned in the platform guarding standard is further clarified by federal standards, and salient features are shown in Figure 5.1. Certain reasonable deviations from the specifications in Figure 5.1 are permitted. For instance, some other material in place of the midrail is permitted, provided that the alternative material provides protection equivalent to the midrail. Also, the height of the railing does not have to be exactly 42 inches high. Railings of 36 inches are acceptable if they do not present a hazard and are otherwise in compliance. This avoids the necessity of tearing down good railings and rebuilding to meet current

Figure 5.1. Standard railings. (From NIOSH.)

codes, as has actually been done by some employers, as shown in Figure 5.2.

One railing that is definitely not standard but still can afford reasonable protection in some cases is lines of parallel ductwork or pipes along the otherwise exposed sides. In many cases parallel ductwork or pipes completely eliminate the hazard of the otherwise open-sided floor, platform, or runway. In fact, one might argue that such a floor, platform, or runway is not even "open-sided."

In OSHA standard 1910.23(c)(1) quoted earlier, three conditions are enumerated, any one of which calls for a "toeboard." Toeboards are vertical barriers along the exposed edges of the walking or working surface to prevent falls of *materials*. A standard toeboard is 4 inches high and leaves no more than a ¼-inch gap between the floor and the toeboard. If the toeboard is not of solid material, its opening must not be greater than 1 inch.

Although 4 feet is the maximum height for open platforms in general, special situations require protection *regardless* of the height of the walking or working surface. Examples are open pits, tanks, vats, and ditches. When the hazardous opening size is small, it may be more practical to place a removable cover over the opening than to place a guardrail around it. Some Safety and Health Managers have saved their firms a great deal of money by calling attention to this alternative. Another special situation is walking and working surfaces adjacent to dangerous equipment, where standard railings are appropriate.

Figure 5.2. New guardrails. Some employers tore down their guardrails and rebuilt them 3 or 4 inches higher just to meet OSHA's standard in the early years of OSHA enforcement. Note the paint marks on the wall where the old guardrail had been attached. OSHA has since relaxed its rule somewhat.

Floors and Aisles

The most important consideration for floors and aisles is not how they are built, but how they are maintained. Federal standards for housekeeping require areas to "be kept clean and orderly and in a sanitary condition." The obvious question is: How is the Safety and Health Manager to know what constitutes "clean and orderly and in a sanitary condition"? There is no clear-cut answer to this question, but some information can be gleaned from past cases of OSHA citations:

Example 5.1

In the railroad yard area of a steel manufacturing industry, piles of debris such as railroad ties and cables were lying about close to the track and presented tripping hazards to employees who must work in the area of the tracks. A complicating factor was that some of the debris was hidden by weeds.

Example 5.2

Dangerous accumulations of grain dust were found in many locations in a grain elevator. The dust was sufficiently concentrated so as to not only be a health hazard to cleanup personnel but also to present a serious explosion hazard. Since the condition had been cited before, the citation was established as Serious and Repeat, and the penalty was set at $10,000.

Example 5.3

> A cluttered workshop had obstructions that some of the employees had to step over or go around to do their jobs.

Example 5.4

> Leaking hydraulic cylinders dripped oil on the floor in a work area. No one was responsible for cleaning up the leaked oil.

Example 5.5

> Cutting oil was spilled on the floor of a machine shop and was not cleaned up.

Some conditions might *look* bad but not represent a hazard. In one example an OSHA Compliance Officer wrote a citation for piles (approximately 2 feet high) of sawdust around a planer and jointer in a woodworking shop. But the company resisted the citation because few employees would be in the area, and those few would be more inconvenienced than endangered. The company won its case.

It is also recognized that it is necessary in the course of many jobs for objects and materials to lie about in the work area. This is especially true of repair and disassembly work, and also of construction. What is generally considered excessive is material accumulation in the immediate work area in quantities in excess of what is actually required to do the given job, or scrap material lying hazardously about the work area in excess of a day's accumulation.

A prime measure of whether a housekeeping program is inadequate is the number of accidents such as trips and falls occurring in the area. Note that the preceding sentence was worded negatively in that too many accidents suggest an inadequate program, but "no accidents on record" does not prove that the work area is *free* of hazards. For every accident that is on file in the plant, the wise Safety and Health Manager will have documented what steps have been taken to remove or reasonably reduce the offending hazard, whatever that hazard has been determined by analysis to be.

Finally, Safety and Health Managers can look to their own industries for guidance. Any reasonable person knows that the definition of "clean and orderly and in a sanitary condition" is different in a foundry than in a pharmaceuticals plant. If the industry has any work practice guidelines, as put forth by trade associations, for instance, these guidelines would be very helpful in illustrating what is reasonable for a given industry.

Water on the floor is a problem in many industries, and constant surveillance becomes necessary to assure that the floor is kept clean and dry. In the design phase of buildings and facilities, attention to the problem of wet processes will suggest slopes and floor drainage systems to alleviate the problem. Another helpful device is to use a false floor or mat to provide a dry standing place for the worker who must work in a wet process.

Some buildings are not built to facilitate cleaning. The floors in others are not only poor from a cleaning standpoint but present trip hazards such as protruding nails, splinters, holes, or loose boards.

Trip hazards due to uneven floors can result in serious injury. In one case, a steel mill, a difference of 1 to 3 inches existed at a point where a grating and plate came together in the floor. The employee who worked in the area had the job of guiding red hot steel butts from the end of a roller line. A fall could have brought the employee into contact with the red hot butts of steel.

Aisles are important, and standards for aisles specify that permanent aisles be kept clear of hazardous obstructions and that they be appropriately marked. Ironically, the better aisles are marked, the more noticeable will be clutter or materials allowed to accumulate in the aisles. Conversely, the better materials or obstructions are kept clear for forklifts or other travel, the more conspicuous will be such a corridor if it is not appropriately marked as an aisle.

There is a lesson to be learned from the irony described above. There is a tendency among Safety and Health Managers to go out all over the plant, indiscriminately laying out aisles, and then to take pride in marking them according to appropriate standards. It is human nature for Safety and Health Managers to want to show everyone that they are taking positive action to enhance safety and comply with established standards. The unfortunate result is that aisles can become too regimented, and so much time is spent policing aisles that production efficiency is lost. In fact, the effect on safety and health can thus become negative, with employer and employee alike wanting to say, "Do we really need that Safety and Health Manager in the plant?" Aisle-marking procedures embody a prime illustration of the principle that overzealous safety and health action can do more harm than good. Every time a decision is made to mark an aisle, the Safety and Health Manager should stop and ask the question: "Can we keep this aisle free of material or other obstructions?"

Conspicuously absent from federal standards for aisles is a specific minimum width for aisles. State codes which reflect the National Fire Protection Association's Life Safety Code often specify a minimum exit access width of 28 inches. But the federal code is silent on this point except to use the language ". . .sufficient safe clearances shall be allowed. . . ." Aisles in forging machine areas receive special attention: ". . .sufficient width to permit the free movement of employees. . . ." But this standard also does not specify an actual minimum width dimension. These two standards are therefore good examples of "performance standards."

The OSHA aisle-marking standard has left a mark of its own upon industry. The term "appropriately marked" originally meant black or white or combinations of black and white for aisles. Untold thousands of dollars were spent by industries switching from the yellow stripes widely used for aisles to OSHA-acceptable white stripes. However, the white striping used was not as durable, and frustrated facilities superintendents would return in desperation to the more durable yellow stripe in order to maintain any stripe at all. The black/white rule became more and more unpopular until OSHA finally revoked the aisle color code as superfluous. In summary, the Safety and Health Manager should be sure that aisles are well marked, but the choice of color for the marking stripe is of little importance. Thus the aisle-marking standard was formerly a specification standard but is now a performance standard.

The discussion about floors thus far has concerned guarding of openings, surfaces, maintenance, and appropriate markings. Nothing has been said about the structural design of the floor itself or whether the floor can withstand the loads applied to it. One is reminded of the eight-story vehicle parking building in which the eighth floor collapsed upon the seventh. The seventh floor was well constructed but was unable to withstand the shock load of the failing eighth floor, and from then on, all the floors came down like dominoes. An even more grim reminder was the Kansas City hotel tragedy of 1981, mentioned in Chapter 3. Almost no one really pays attention to whether a floor is overloaded. Everyone has heard of some structural collapse somewhere, but the few incidents that do occur seem so remote and rare that few of us ever worry about the problem.

Federal standards require marking plates for floor loads approved by the "building official." One of the chief complaints from industry Safety and Health Managers is that the floor loading standard does not explain its use of the term "building official." Confusion over this term has led to phone calls to various agencies in attempts to identify some *public* official to come to the facility to make an engineering determination to use as a basis for the floor-load-marking plates. Since the term "building official" is not defined, a better course of action for the Safety and Health Manager is to secure the services of a competent Professional Engineer either inside or outside the company. This would show a good-faith effort to comply with the standard and would virtually eliminate the possibility of a hazard.

Floor-load-marking plates are of little value if they are ignored by employees. Adherence to floor load limits is an administrative or procedural matter and thus calls for vigilance in order to maintain compliance. A good system of records and inventories can make weights and locations part of the data base of an overall management information system. If this system is computerized, the computer can monitor loads at all times and print out a warning message in the unlikely event that a distribution of inventories exceeds a floor load limit.

Such a computerized system may seem to be an overreaction to the problem, and in fact it would be an overreaction for most companies medium to small in size. But in a large-scale computerized warehouse system, the floor load monitoring system would represent such a small sideline to the overall computerized operating monitor that it would require only a few seconds of computer time per month. The system could be programmed to print out status reports monthly or on an exception basis whenever loads exceed or perhaps approach limits. The monthly status reports would be preferable to the exception reports because the monthly status reports would provide evidence of positive adherence to approved standards.

A word of caution is in order here. If there actually exists a violation of floor load limits in the plant, no fancy computerized information system is going to hide it. Manual calculations can be used to determine whether a floor load limit has been exceeded, and symptoms such as crushed, bent, or cracked floor members can be as embarrassing as they are hazardous.

The whole problem of floor and aisle design and maintenance is magnified by the use of mechanical handling equipment such as forklift trucks. Forklifts both aggravate the problems and at the same time become more of a hazard themselves whenever problems with floors exist. This problem will be revisited in Chapter 10.

Stairways

Building codes and standards for stairs are well established, but many factories and businesses have stairways that do not meet code. One or two steps are usually not a problem, but if the stairways have four or more risers, they need standard railings or handrails and must be kept clear of obstructions. Note the use of the words "handrail" and "railing." *They are not the same.* A handrail, as used in this standard, is a single bar or pipe supported on brackets from a wall to furnish a handhold in case of tripping. A railing, however, is a vertical *barrier* erected along the exposed sides of stairways and platforms to prevent falls. A handrail and railing may of course be combined into the same unit, but the two terms are not interchangeable.

The OSHA standard is very flexible on the subject of the placement of stairway landings, using such language as "avoid" (long flights of stairs), and "consideration should be given to. . ."

The placement of stairway landings is a safety consideration. Most people think that the purpose of stairway landings or platforms is to give the climber a chance to rest, and it is true that this is a supplemental purpose. However, the main purpose of the stairway landing is to shorten the distance of falls, and thus landings play an important role in building and facilities safety. Extremely long flights of stairs are obviously more dangerous than stairs interrupted by landings. To be effective they must be no less than the width of the stairway and a minimum of 30 inches in length measured in the direction of travel.

Ladders

Ladders are not the simple devices most people think they are. Design is critical because the construction can be neither too strong nor too weak. A weak ladder is obviously dangerous, but a ladder that is overdesigned is difficult or impossible to handle safely. A long, heavy ladder can be as much hazard when it is carried as when it is climbed. It is easy to see why ladders must be manufactured to exacting standards.

Most firms buy ladders from regular ladder manufacturers, and the Safety and Health Manager can be quite sure that the ladders were constructed properly in the first place. What is more important to industrial safety is how the ladders are used and maintained. Defective ladders must be either repaired or destroyed, and while awaiting either fate they must be tagged or marked "Dangerous, Do Not Use." Personal interviews with Safety and Health Managers have indicated that they often saw defective ladders in half rather than risk the possibility of a defective ladder being returned to service. When a job needs to be done and there is no regular ladder in sight, the temptation is too great for maintenance or other personnel to remove the danger tag and make immediate use of a defective ladder. At best, a repair job on a portable ladder usually does not look very good and will likely arouse suspicion even if the ladder is safe. It takes a good engineer to convince anyone that the repaired ladder is just as good as new, even if the engineer has been able to convince himself or herself.

Portable metal ladders share common hazards with portable wood ladders; a major difference, however, is the fact that metal ladders conduct electricity. Most workers are aware of the increased hazards of electrocution present when using metal ladders. Rubber or otherwise nonconducting feet are a good precaution on metal ladders, but the hazard is still present, and the rubber feet should not be allowed to give rise to complacency.

Whether the portable ladder is made of wood or metal, it is the way the ladder is used that will chiefly determine its safety. Almost everyone knows the admonition that it is unsafe to ascend or descend a ladder with the climber facing away from the ladder. Less obvious is the fact that portable ladders are typically not designed to be used as platforms or scaffolds, and they become very weak if used at angles close to horizontal.

A common error in the use of ladders is to use ladders that are too short. For instance, when accessing a roof, the ladder needs to extend at least 3 feet above the upper point of support. For ordinary stepladders the top should not be used as a step. Another foolish practice is to place ladders on boxes, barrels, or other unstable bases to obtain additional height. Some people even try to splice short ladders together to make longer ones.

The first consideration in the use of a portable ladder is its condition, especially the rungs. Next to the hazard of broken rungs, the greatest hazard probably is a ladder that slips or tips because it is insecurely positioned. The proper slant is 4 feet vertical to 1 foot horizontal. A good rule is tie off the ladder at the top so that it cannot tip or slide down. However, this is not always practical, and alternative solutions can solve the slipping problem. Sometimes a ladder can be made stable by positioning it where the structure of the wall or building limits its movement and makes the ladder safe. Another solution is to use nonslip bases, but this method may not work on some surfaces, such as oily, metal, concrete, or slippery surfaces. Metal ladders in particular may be subject to slipping and need to be equipped with good safety shoes. Safety shoes are as much to prevent slipping as they are to prevent electrocution. Even spikes may be useful on some surfaces to make the footing secure. If the ladder is used without safety shoes on a hard slick surface, the ladder needs a foot-ladder board to prevent slipping.

Fixed Ladders

From a safety and health standpoint, fixed ladders require a somewhat different approach from portable ladders. With portable ladders the emphasis is on care and correct use as discussed above. With fixed ladders the emphasis is on design and construction. It is beyond the scope of this book to specify all of the details of fixed ladder design. If fixed ladders need to be constructed, the designer should follow the detailed specifications laid out in established standards, not this textbook.

This text will seek only to alert Safety and Health Managers to problems they may encounter with existing fixed ladders. Some fixed ladders may have been constructed long ago or perhaps without the benefit of standards. It is the duty of the Safety and Health Manager to alert the company if fixed ladders are found to be incorrectly constructed. The Safety and Health Manager can then recommend that design details be followed when the ladders are rebuilt.

Some knowledge of fixed ladder design principles is necessary, however, if the Safety and Health Manager is to know how to recognize common problems found. For brevity this text selects some easy-to-observe problems with fixed ladder design, illustrated in Figure 5.3. The rung offset is to prevent the foot from sliding off the end of the rung. Other means, such as siderails, would also be acceptable.

When the fixed ladder is more than 20 feet long but less than 30 feet long, ladder cages are needed, but many Safety and Health Managers do not realize that there is an alternative to installing ladder cages. On tower, water tank, and chimney ladders over 20 feet in unbroken length, ladder safety devices may be used in lieu of cage protection. Ladder safety devices usually take the form of a combination of fixed equipment on the ladder and personal equipment worn by the climber. One type utilizes a trolley which moves along a fixed rail. The trolley is attached to the climber's belt. The trolley moves easily up or down the rail when the climber ascends

1. Distance between rungs not more than 12 inches and uniform.
2. Minimum rung length is 16 inches.
3. Back clearance minimum is 7 inches. This dimension is the most frequently violated, according to frequency of OSHA citation.

Figure 5.3. Easy-to-recognize problems with fixed ladders; all three of the above are required by OSHA. Note the rung offset to prevent the foot from slipping off. 1, Distance between rungs should not be more than 12 inches and uniform. 2, Minimum rung length is 16 inches. 3, Back clearance minimum is 7 inches. This dimension is the most frequently violated, according to frequency of OSHA citation. (Based on OSHA standards.)

or descends, but a brake engages which stops the trolley in event of a fall. One thing to watch carefully with a rail-and-trolley device is to assure that the device is equipped to deal with ice on the rail if the ladder is located in a climate where ice is a possibility (most outdoor locations in the United States). The safety problem with ice on the rail is somewhat indirect; the trolley simply cannot move along the ice-covered rail, making the system useless. But this is when the safety device is most needed. The problem is not without solution because there are deicing devices for trolley systems. Safety and Health Managers, however, should choose their protection systems with care, being sure to consider all factors, including the problem of icing, in order to avoid later embarrassment and wasted investment on the part of the company. An example rail-and-trolley ladder safety device system is illustrated in Figure 5.4.

Dockboards

A "dockboard" or "bridge plate" provides a temporary surface over which loads can be transported, particularly during loading or unloading of a cargo vehicle. One of the main safety hazards with these boards is that they may shift while in use. It is

Figure 5.4. Example of ladder safety device system.

also possible that the surfaces connected by the dockboard can shift, such as that caused by movement of the cargo vehicle itself. Finally, the dockboard itself may not be strong enough to carry the load.

EXITS

Exits are usually considered doors to the outside and from a safety standpoint are considered a means of escape, especially from fire. Such thinking is correct but incomplete. The Safety and Health Manager should enlarge the concept by using the more general term "means of egress" to include:

1. The way of exit access
2. The exit itself
3. The way of exit discharge

By considering means of egress instead of simply exits, the Safety and Health Manager can analyze the entire building to determine whether every point in the building has a continuous and unobstructed way of travel to a public way. In this way, one must think of such building facilities as stairways, intervening rooms, locked interior doorways, and limited-access corridors. Outside the building one must think of yards, exterior storage of materials, fences, courtyards, and shrubbery. One thinks of shrubbery and landscaping as affecting neither safety nor health. However, employees might escape a burning building (or perhaps a building with a ruptured chlorine gas line) through an exit door, only to find outside that the exit empties into a courtyard tightly confined by a fence, dense shrubbery, or other obstruction.

Almost every Safety and Health Manager sometime during his or her career encounters the embarrassment of an exit that is *locked*. This can be a two-edged problem because many Safety and Health Managers are also responsible for plant security. In many cases the only practical solution is to provide panic bars or other mechanisms for locking doors from outside entry while maintaining free and unobstructed egress from the inside. Where unauthorized exit can be as much a security problem as unauthorized entry, the automatic-alarm-sound type of door may be the only alternative. Facilities designers are turning more and more to the use of unlocked, automatic-alarm-sounding emergency exit doors. Even more frequently encountered than locked exits are exits which are cluttered or blocked by obstructions or impediments. Stacks of material obstructing the door or way of travel defeats the purpose of the exit.

Exits must also be marked by a readily visible sign. Sometimes exits are marked, but the way to reach the exit may not be apparent. Intermediate signs along the access to the exit may be necessary to show occupants how to reach the exit.

A line of reasoning similar to the reasoning described earlier for aisles and aisle marking can be seen for exits. The Safety and Health Manager should not be too eager to go out and buy signs from the exit sign sales representative. It is important to weigh carefully a decision to name a door an "exit." Once a sufficient number of exits have been identified, the Safety and Health Manager should be careful about putting exit signs on any additional doorways. An exit sign is an invitation to start checking for obstructions, locked doors, inadequate signs along the exit access, incorrect sign dimensions, obstructions outside the exit, and even lighting. Provided that a sufficient number of marked exits have been provided, there is no law which says that an employee shall not escape the building by an unmarked passage or doorway to the outside. Such doorways should not even be called "exits."

ILLUMINATION

The subject of lighting appeared in the preceding paragraph. Lighting or the lack of it can be a safety hazard, but there is no code for minimum safe lighting except for specialized areas. For instance, if forklift trucks are operated in the plant area, the minimum general lighting level is 2 lumens per square foot unless the forklift trucks themselves have lights. Every exit sign should be suitably illuminated by a reliable light source giving a value of not less than 5 footcandles on the illuminated surface. This is not to say that the exit sign must be the kind that is *internally* lighted. An alternative to be considered is artificial lighting *external* to the sign. Also, there is nothing wrong with relying on *natural* illumination (sunlight) on the exit sign in an amount not less than 5 footcandles. Natural illumination can be a problem, however, if the area is accessed on second or third shifts. Incidentally, 5 footcandles is not very much illumination. Most plant areas are normally illuminated by much greater levels of illumination.

MISCELLANEOUS FACILITIES

Maintenance Platforms

It was pointed out earlier in this chapter how important it is to plan for maintenance activities when constructing a new building. Many modern buildings have built-in, safe suspension systems for exterior window cleaning and other exterior maintenance. Maintenance workers for buildings not so equipped are less fortunate and typically work from suspended scaffolds of the same type as construction scaffolds, which are discussed in Chapter 14. Not only are the maintenance workers less fortunate, but so are the employers of these workers and the Safety and Health Managers who must worry about the safety of the scaffolds, the proper securing of scaffolds on the roof of the building, and other items governed by applicable standards.

Safety and Health Managers who do find that their buildings are equipped with powered platforms for exterior maintenance should direct most of their attention to how these platforms are being used and maintained, not how they are made. The manufacturer of such equipment would normally be very careful to adhere strictly to standards when fabricating the powered maintenance platform. Typical problems with these platforms are missing guardrails, missing toeboards, missing side mesh, disabled safety devices, and inadequate inspections or records of inspections.

Regarding the equipment itself, some companies have been tripped up for not having load-rating plates on the platform. The load rating must be stated in letters at least ¼ inch in height. The wire rope suspending the platform must also be marked with a metal tag stating its maximum breaking strength and other data, including the month and year the ropes were installed.

For some types of powered platforms workers need to wear safety belts; for others they are safe without the belts. A platform supported by four or more wire ropes can be so designed that the working platform will maintain its normal position even if one rope fails. However, many powered platforms are suspended by only two wire ropes and will tip dangerously if one of the ropes fails. One of the more dangerous types of platform is known as "Type T," and workers on these platforms must wear safety belts attached by lifelines. If the platform qualifies as Type T, it will upset with a single wire rope failure, but it will not fall to the ground. Therefore, the lifeline may be attached either to the building structure or to the working platform. Compare this to construction industry standards (Chapter 14) which require that lifelines be secured to an anchorage or structural member instead of to the scaffold.

Public utilities and tree trimmers often use a type of platform that is vehicle mounted, such as aerial baskets, aerial ladders, boom platforms, and platform-elevating towers. Again the majority of accidents arise from improper use of the platform rather than from equipment failure or design. This is even more true of the vehicle-mounted platforms than of the building-mounted models discussed earlier.

The most serious hazard with vehicle-mounted platforms is contact with high-voltage power lines, and this kills workers every year. The hazard is so severe that a safety distance must be maintained at all times, except of course in the case of electric utility companies who by the nature of their work must approach closer. For safety the utility companies must insulate aerial devices that work closer than the standard safety distance. For nonutility companies the accepted standard is a 10-foot distance in the case of a 40-kilovolt line, for example. Different line voltages may need higher or lower safety distances, and these distances are covered in more detail in Chapter 14 in the discussion of mobile cranes.

Sometimes special sensors called "proximity warning devices" are installed on the boom to warn the operator when the basket is too close for safety. However, these warning devices do not provide positive protection and thus should not be considered an excuse for moving the boom closer to the line than authorized minimums.

Workers in aerial baskets often fail to wear a body belt and lanyard attached to the boom. Adding to the fall hazard is the possibility of unexpected contact with an object which might strike and perhaps sweep the worker out of the basket or off the platform. Echoing the hazards avoidance principles of Chapter 3, such unusual hazards point to the importance of training for personnel who work in aerial baskets. Other unsafe procedures are failure to secure the aerial ladder before travel-ing, climbing or sitting on the edge of the basket, or improvising a work position other than the floor of the basket.

Elevators

Elevators are everywhere, but when can you remember one falling? The catastrophic fall of an elevator is such a horrifying thought that the public long ago set up regula-tions for safe elevators. Jurisdiction was placed within the states, and most states administer elevator inspections through "Labor" or "Labor and Industry" com-missions. Next time you ride an elevator, look at the certificate of inspection posted inside the elevator car.

Elevators must be inspected both when new (or altered) and periodically. Many states even require construction permits from the authorized elevator inspec-tion agency before elevator construction is begun. Elevator operating permits and fees are also required by some states. Not every inspector must be an agency official, but state licensing procedures for elevator inspectors may be applicable.

Manlifts are used as elevators, but unlike elevators, there are federal standards for manlifts. Manlifts are much cheaper and more efficient than elevators for many plant operations and are thus sometimes used instead of elevators. As can be seen in Figure 5.5, however, a manlift is inherently more hazardous than an elevator. It is ironic that elevators—which are more safe than manlifts—are governed by strict state inspection, licensing, permits, and approvals. With manlifts, though, it is up to the Safety and Health Manager to interpret general standards and identify hazards. It is apparent from Figure 5.5 that the biggest hazard with manlifts is getting on and getting off. Exit is essential because the belt is continuous, and to stay on the belt past the top or bottom floor would be either impossible or extremely hazardous.

Boilers

Steam boilers and pressure vessels are so safe today that most people do not even think about them. It has not always been that way. Although steam boilers are not as popular today for building heating systems, many are still in use; in addition, in-dustrial processes use hundreds of thousands of boilers and pressure vessels. So the lack of familiarity with boiler accidents is not because the boilers themselves are rare. It is the accidents which are rare, not the boilers, and when an accident does occur, the energy released by the explosion is so devastating that it usually is a catas-trophe.

The extreme hazard of an unsafe boiler led to the early regulation and safe-guarding of these vessels. As with elevators, the historical development of boiler codes has placed their jurisdiction within the states. State control has been very ef-fective in keeping boiler accidents to an absolute minimum.

The Safety and Health Manager needs to assure that boilers and pressure ves-sels in the plant are being inspected and that state procedures are being followed.

Figure 5.5. Manlift: a continuously moving belt on which workers may ride both up and down.

One question that immediately comes to mind is: What pressure vessels are covered by the regulations? Most states exempt containers for liquefied petroleum gases (LPG), as these are covered by other regulations. The same can generally be said of vessels approved by the Department of Transportation for public highway transportation of liquids and gases under pressure. Some states also exempt vessels used in connection with the production, distribution, storage, or transmission of oil or natural gas. Note, however, that the foregoing does not exempt refineries or chemical plants that produce petroleum *products*. In the industry the term "oil production" refers to the drilling and extraction or "mining" of petroleum, not its refining. The only way to be sure about exemptions in a given state is to check with that state's agency of authority.

The time to stop and think about regulations for boiler and pressure vessel safety is whenever such a vessel is to be purchased, installed, modified, moved, or sold. Welding on such vessels may weaken them and is scrutinized carefully by the inspector, although welding is not absolutely prohibited. Even hot water storage containers should be installed or reinstalled by persons properly licensed to do the work.

SANITATION

The sanitation of lunchrooms seems straightforward and obvious, but sanitation decisions can be trickier than they appear. If a decision is made to allow employees to eat in the plant, principles of hygiene must be observed. The Safety and Health Manager should be sure that a sufficient number of waste receptacles are provided to avoid overfilling. But before going overboard, the Safety and Health Manager should realize that *too* many waste receptacles can be provided also. If too many receptacles are provided, maintenance personnel will become lax about emptying containers that receive little use, resulting in additional sanitation problems.

The presence of toxic materials complicates the whole problem of food service, consumption, and storage. Certainly, food and beverages must not be stored in areas where they will be exposed to toxic materials. This rule may seem obvious, but the Safety and Health Manager should not only consider the plant cafeteria or lunchroom but also the employee who brings snacks from home and stores them in areas in which they will be exposed to toxic materials.

Some toxic materials, such as lead, are particularly susceptible to exposure by ingestion during food consumption. Some toxic materials, such as vinyl chloride and arsenic, are of such concern that there are strict, specific standards for their control. Toxic substances will be discussed in more detail in Chapter 6.

SUMMARY

Safety and Health Managers who are willing to plan ahead can save their companies a great deal of money by heeding building and facility codes *before* commencing construction or expansion of plant space. Planning ahead is the key to compliance with standards for floors, aisles, exits, and stairways. Guardrails, ladders, and platforms may be added, but they, too, deserve some advance consideration to assure that installations meet requirements.

Safety and Health Managers need to be careful not to get too enthusiastic and provide for too many aisles or exits. An extra aisle or exit that is improperly maintained or marked can easily lead to problems and indeed is not even in the interest of safety.

The subject of buildings and facilities may not be the most exciting topic for the Safety and Health Manager's attention. However, even ordinary matters such as housekeeping and sanitation deserve careful consideration and judgment to promote safety and health at reasonable cost.

EXERCISES AND STUDY QUESTIONS

5.1. Explain why the title for a major OSHA subpart is "Walking and Working Surfaces" instead of simply "Floors." Name 10 different walking and working surfaces.

5.2. What is deceptive about the danger of an open-sided floor or platform only 4 feet high?

5.3. Suppose that in your plant welders must stand on top of a 10-foot-high tank to complete manufacturing operations. What should be done, if anything, to protect them from falling?

5.4. What is the purpose of a toeboard?

5.5. Explain under what conditions, and how, roofing workers should be protected from fall hazards.

5.6. When must an open-sided floor or platform be equipped with a railing?

5.7. When would OSHA require guarding for a service pit for use in vehicle maintenance? How would you guard such a pit?

5.8. Explain how poor housekeeping could result in an OSHA citation which classifies the violation as "serious."

5.9. As a Safety and Health Manager, how would you deal with the problem of housekeeping?

5.10 Explain the danger of marking too many permanent aisles.

5.11 Explain the difference between a "handrail" and "railing."

5.12. Describe the purposes and requirements for stairway landings.

5.13. Describe important points to consider in safety with portable ladders.

5.14. Under what conditions are ladder cages specified? When may alternate ladder safety devices be used? What are the advantages and disadvantages of these devices?

5.15. Define the term "means of egress."

5.16. What are the biggest problems with exits?

5.17. What are vehicle-mounted work platforms? What are their chief hazards?

5.18. Explain how either too many or too few waste receptacles can lead to sanitation problems.

6

Health and Environmental Control

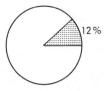 12%   Percent of OSHA
general industry citations
addressing this subject

Health hazards carry a great deal of impact because the potential harm to exposed employees is great, and the cost of correction of a single health hazard problem can run into millions of dollars.

Industrial hygienists have been saying for many years that health hazards deserve more attention, and in response to this pressure a shift from safety toward health activities has been in evidence almost since OSHA commenced operations. At first OSHA did not have a sufficient cadre of qualified health professionals to assess health hazards, and the natural focus was on safety. But the proportion of health specialists in OSHA has increased a great deal since the early 1970s.

Chapter 1 discussed the bases of competition and even friction between safety professionals and health professionals. The two groups have common objectives, but their backgrounds are usually vastly different, which results in conflict. The more successful professionals on both sides are busy emulating the characteristics typical of the opposite side, and thus the differences between the two professions are beginning to dissolve.

Health hazards will always tend to be more subtle to detect than safety hazards, by the very definitions of "health" and "safety." Remembering these definitions from Chapter 1, health deals with the long-term chronic exposure effects, whereas safety deals with the more obvious acute effects which do their damage immediately.

BASELINE EXAMINATIONS

Most everyone has taken a preemployment physical examination, but few understand the importance of the physical exam to the overall safety and health program. The baseline condition for each employee is established by this examination. This condition is important in placing the employee in the right job and in detecting any health deterioration due to job exposures. Occupational health exposures are perhaps the most important reason for preemployment physical examinations because of the chronic nature of health hazards. If an employee already has emphysema or other lung disorders, it is essential that this fact be established at the time of employment. The same can be said of hearing impairment, as will be discussed later in this chapter.

TOXIC SUBSTANCES

Exposure to toxic substances is the classic "health problem," and this topic will be used to model the entire subject of health and environmental control. Choice of words can be important with this topic. For instance, the term "hazardous materials" is sometimes used to refer to toxic substances, but the term "hazardous" is a much more general term that would include such safety hazards as presented by flammable and combustible liquids and explosives. This text follows a popular convention which tends to associate the term "materials" with safety hazards and the term "substances" with health hazards. In fact, the title of Chapter 7 is "Hazardous Materials," a chapter dealing almost exclusively with safety hazards.

The Safety and Health Manager needs to have a general knowledge of what various types of toxic substances can do to the body. Such general knowledge will be useful in convincing workers and management alike that toxic substances need to be controlled for the health of the workers as well as to avoid OSHA citation. This discussion will describe some of the various types of toxic substances by considering their effects on the body.

Irritants

Irritants inflame the surfaces of the parts of the body by their corrosive action. Some irritants affect the skin, but more of them affect the more moist surfaces, especially the lungs. Even a weak irritant to the upper respiratory tract will be easily detectable by the victim, but irritants to the lower respiratory tract may go unnoticed.

When the irritant is some type of dust, the lung disease that results is called pneumoconiosis. This is a general term which includes reactions to simple nuisance dusts as well as fibrosis, a more serious reaction which includes the development of fibrous scar tissue which impairs the efficiency of the lungs. Example pneumoconioses are siderosis (from iron oxide dust), stannosis (from tin dust), byssinosis (from cotton dust), and aluminosis (from aluminum dust). Examples of the more dangerous fibroses are asbestosis (from asbestos fibers) and silicosis (from silica).

Everyone is familiar with the strong odor of ammonia. Ammonia gas and moisture on the mucous membranes of the body combine to form ammonium hydroxide, a strong caustic agent. It is easy to understand why this would irritate and injure the delicate tissues of the nose, trachea, lungs, and other parts of the body that it might contact. By similar logic any of the gases that combine with water to produce acids would be irritants, as would airborne particles of the acids themselves.

Plating operations easily impart acid mists to the air since the plating tanks are often splashing, hot, and acidic. Particularly offensive is chromic acid mist, which causes an ominous-sounding malady, "chrome holes." Chromic acid also destroys the nose septum which separates the two nostrils.

Another well-known irritant is chlorine gas, a widely used industrial chemical. Chlorine's halogen relatives, fluorine and bromine, are also irritants, especially fluorine, the strongest of all halogens. Even the soluble *salts* of fluorine are poisonous. Less well known are those substances which are irritants deep within the lungs, such as the oxides of nitrogen and phosgene. Phosgene is best known as a chemical warfare gas, which is evidence of its toxicity. But phosgene can also be generated inadvertently in the workplace when chlorinated hydrocarbon solvents are exposed to welding radiation.

Chronic exposure to irritants over a long period can cause scar tissue to develop in the lungs. Some of these substances do not produce an appreciable immediate irritant effect but are dangerous in the long run. The most notorious of these scarring agents is asbestos fibers. Coal dust is also a scarring agent. Scarring agents are in the form of tiny solid particles, and their action on the lungs is *mechanical*, as contrasted with the systemic poisons discussed in the next section.

Systemic Poisons   *Lead*

More insidious than the irritants are the poisons that attack vital organs or systems of organs, sometimes with toxic mechanisms which are not understood. The chlorinated hydrocarbons common in solvents and degreasers, for example, are blamed for liver damage.

Perhaps the best known systemic poison found in occupational exposures is lead. Lead is disappearing from paint pigments because of its poisonous reputation, but it still exists in tetraethyl lead used in leaded gasoline. The author of this text formerly worked in a tetraethyl lead plant, and decades ago the workers in this plant had a high awareness of what lead can do to the body. Lead attacks the blood, the digestive system, and the central nervous system, which includes the brain. Autopsies have also shown damage to kidneys, liver, and reproductive systems, but these results are inconclusive. Other toxic metals are mercury, cadmium, and manganese. Magnesium, sometimes confused with manganese, is less toxic.

Another important systemic poison is carbon disulfide. Carbon disulfide is unusual in that its hazards are extreme from the standpoint of both safety (fire and explosions) and health. It is widely used in industry as a solvent, a disinfectant, and an insecticide. As a systemic poison carbon disulfide attacks the central nervous system. Methyl alcohol (methanol), a popular solvent, also is a systemic poison to the central nervous system but it a much milder poison than carbon disulfide. Indeed, methanol is even acceptable in small quantities as a food additive! Methyl alcohol is also a fire and explosion hazard.

Depressants

Certain substances act as depressants or narcotics on the central nervous system and as such can actually be useful as medical anesthetics. Unlike the systemic poisons discussed earlier, the effect of the action of depressants on the central nervous system is usually temporary. Some substances, such as methyl alcohol, however, are both systemic poisons and depressants. Besides affecting health, depressants can have an adverse effect on safety because they interfere with worker concentration while the worker operates machinery.

The most familiar depressant is ethyl alcohol (the "drinking" variety of alcohol), sometimes called ethanol in industry, although its harmful effects as an industrial hazard are minimal compared to its effects from drinking. In fact, the greatest hazard from ethanol on the job is without doubt from "voluntary ingestion" from bottles brought on the premises by employees. Ethanol is not as toxic as methanol.

Acetylene, the most widely used fuel gas for welding, is a narcotic, but its health dangers are minimal compared to its safety hazard from fires and explosions. Acetylene has been used in medical anesthesia.

Benzene is a very popular industrial chemical used principally as a solvent. Benzene is a depressant on the central nervous system, an irritant, a systemic poison, and has recently been recognized as causing leukemia. In addition, benzene is a dangerous fire and explosion hazard. OSHA has a movie which dramatically films the hospital bed testimony of a young benzene worker who is dying from leukemia.

Asphyxiants

Asphyxiants prevent oxygen from reaching the body cells, and in the general sense any gas can be an asphyxiant if there is enough of it to crowd out the essential proportion of oxygen in the air. Many people have committed suicide by breathing natural gas, which is essentially methane. But methane is merely a simple asphyxiant, in that it displaces the proportion of oxygen in inhaled air. Methane can be encountered in industrial environments, as it is a product of fermentation. Other simple asphyxiants commonly encountered are the inert gases such as argon, helium, and nitrogen used in welding gases.

It may seem incorrect to classify nitrogen as an air contaminant and an asphyxiant when it is the principal constituent (78%) of normal air. But too much nitrogen will reduce the normal proportion of oxygen (21%) in the air. Any proportion of oxygen less than 19.5% is considered oxygen deficient. Oxygen deficiency is very dangerous, a more serious condition than most people think. It is really more of a safety hazard than a health hazard because the effects are acute and often fatal. However, the condition is more likely to be monitored by an industrial hygienist than a safety specialist, even though it is an acute hazard.

Carbon dioxide is one of the most important simple asphyxiants, although it is a harmless constituent in air in normal quantities. Fire is the primary source of dangerous industrial concentrations of carbon dioxide. Carbon dioxide is heavier than air, which causes it to accumulate in low, confined spaces, increasing its hazard potential. Confined spaces have a great deal of hazard potential, not only for carbon dioxide but for all air contaminants.

The asphyxiants discussed so far are "simple asphyxiants," essentially nontoxic substances which replace the essential oxygen content in the air. The other type of asphyxiant is the "chemical asphyxiant," which interferes with the oxygenation of the blood in the lungs or later interferes with the oxygenation of the tissues. The most notorious chemical asphyxiant is carbon monoxide, a substance for which blood hemoglobin has a stronger affinity than it does for oxygen (over 200 times the affinity). The resulting compound, carboxyhemoglobin, is a very stable substance which prohibits the life-critical exchange of oxygen and carbon dioxide by their vehicle, blood hemoglobin.

Another well-known chemical asphyxiant is hydrogen cyanide, an industrial insecticide but better known as the gas used in prison execution chambers. The gas is formed by dropping sodium cyanide pellets into a small container of acid. Some workplaces have the potential for becoming execution chambers. In a workplace inspection in California a laboratory was found in which strong acids in glass bottles were stored on shelves located directly above sodium cyanide salts!

Carcinogens

Carcinogens are substances which are known to cause or are suspected to cause cancer. A great deal of attention has been given to carcinogenesis since the advent of OSHA, but the source of the emphasis is not OSHA alone. NIOSH, the Consumer Product Safety Commission (CPSC), and other agencies have concentrated on carcinogens. Public awareness is now high, and many workers and consumers alike are becoming very cautious about their exposures to carcinogens. One of the frightening things about carcinogens is that the cancer has such a long latency period. Sometimes a lapse of 20 or even 30 years occurs between exposure and appearance of the cancerous tumors.

New carcinogens are being labeled every year, and many of the substances indicted are very commonly used industrial materials, such as benzene and vinyl chloride. The familiar saying "Everything I do is either illegal, immoral, or fattening" might be modified to include "or causes cancer." So many substances have been indicted by laboratory tests on animals that it is possible that a sort of complacency could be developing in the minds of the public.

Vinyl chloride, mentioned earlier as an example carcinogen, is extremely dangerous from many aspects. It is a severe explosion hazard, and when it burns it is very difficult to extinguish. Added to this is the hazard of highly toxic phosgene being liberated during vinyl chloride fires. Acute skin exposures can result in injury from skin freezing due to rapid evaporation of vinyl chloride. Chronic inhalation is now recognized to cause a form of cancer of the liver, angiosarcoma. Vinyl chloride is also believed to be a teratogen, a type of substance to be explained in the next section.

Before going to teratogens, a distinction should be made between the extremely hazardous vinyl chloride and polyvinyl chloride (PVC). PVC is a type of plastic synthesized by polymerizing the unstable vinyl chloride monomer into the stable polyvinyl chloride polymer. Figure 6.1 illustrates the arrangement of the atoms in the dangerous vinyl chloride monomer and in the harmless and very stable polyvinyl chloride polymer.

Figure 6.1. Comparison of vinyl chloride molecule with polyvinyl chloride molecule: (a) vinyl chloride monomer; (b) polyvinyl chloride polymer.

Teratogens

Teratogens affect the fetus, so their toxic effect is indirect. Women should be careful about exposures to certain substances during pregnancy, especially in the first "trimester" (first three months). Teratogens should not be confused with mutagens, substances that attack the chromosomes and thus the species instead of the individual. Teratogens do their damage after conception but before birth, whereas mutagens do their damage before conception. Mutagens can affect the chromosomes of either potential fathers or mothers.

A tricky legal question is whether an industry can prohibit women of childbearing age from working in jobs in which they may be exposed to teratogens. The question is whether this constitutes sex discrimination. The affirmative contingent says that the teratogens must be controlled by the industry so that pregnant women can have a safe and equal opportunity to work in the same jobs as those offered to men. The negative contingent says that it makes more economic, safety, and social sense to employ women in other jobs during pregnancy.

Routes of Entry

The term "toxic substance" can be considered a synonym to the term "poison," a word familiar to all of us, who were taught as children not to eat or drink poison. Poison is even in our fairy tales, such as "Snow White and the Seven Dwarfs." It may be true that the greatest danger from poisons in the home is from ingestion (swallowing), but on the job the greatest danger is from breathing. In fact, it has been said that the order of importance of routes of entry to the body for toxic sub-

Routes of entry	Importance At home	At work
Ingestion	Most important	Least important
Skin contact	Of moderate importance	Of moderate importance
Inhalation	Least important	Most important

Figure 6.2. Toxic substances' routes of entry to the body. Note reverse sequence of importance at work versus at home.

stances is the exact opposite on the job from the order at home, as shown in Figure 6.2.

The different routes of entry to the body for toxic substances are more related than most workers realize. Inhalation of toxic substances results in accumulations on the mucous membranes; mucus is later coughed up, and some of it is inevitably swallowed. Skin contact with toxic materials can also result in ingestion as substances become embedded beneath fingernails and on hands which later come into contact with food. Toxic dusts in the air are also picked up in the hair and are later deposited on the pillow during sleep and thus gain entry to the body indirectly.

From knowledge of routes of entry for toxic substances it is easy to see the importance of sanitation. Some of the principles of sanitation discussed in Chapter 5 take on more importance when toxic substances are considered. Proper storage of food and shower and washing facilities may be essential in the control of the quantities of toxic substances entering the worker's body.

Air Contaminants

The greatest concern with toxic substances is with air contamination, and this is as it should be (as shown in Figure 6.2). Air contaminants take on many physical forms, and most people confuse these forms in everyday language. The Safety and Health Manager should know the difference, for instance, between vapors and fumes. Although air is essentially gases, the *contamination* of that air can consist of any of the three states of matter: solids, liquids, or gases.

"Gases" easily contaminate the air because air consists of gases, and gases readily mix. The most familiar toxic gas is carbon monoxide. Also dangerous in the industrial environment are hydrogen sulfide and chlorine. Even "harmless" gases, such as carbon dioxide and inert nitrogen, can become dangerous if allowed to accumulate in large quantities so that they become asphyxiants by crowding life-giving oxygen out of the breathing zone.

"Vapors" are also gases, but these substances are normally liquids or even perhaps solids which release small quantities of gases into the surrounding air. Some of our most useful industrial liquids, such as gasoline and solvents, have a strong tendency to release these vapors.

"Mists" are really tiny droplets of liquids, so small that they remain suspended in the air for long periods, as in clouds. Since liquids are heavier than air, they eventually settle out or coalesce into larger droplets which fall in the manner of rain. Long before that happens, however, mists can be inhaled by the worker. Fine mists are generated as vapors condense into clouds. Coarse mists are produced by splashing or atomizing operations, such as from machine tool cutting oils or from electroplating. Pesticide spray is also usually a mist.

"Dusts" are recognized as solid particles. Technically speaking, dust particles range in size from 0.1 to 25 micrometers (0.000004 to 0.001 inch) in diameter. Everyone is exposed to dust, and some dust is relatively harmless. Dangerous dusts include asbestos, lead, coal, cotton, and radioactive dusts. Silica dust from grinding operations is also becoming recognized as a hazard, although ordinary earth dust is principally silica. Asbestos dust particles are fiber shaped instead of round, and this contributes to their hazard.

"Fumes" are also solid particles but are generally too fine to be called dusts. Actually, the sizes for fume particles and dusts overlap, as can be seen in Figure 6.3. Whereas dust particles are typically finely divided by mechanical means, fumes are formed by the resolidification of vapors from very hot processes such as welding. Chemical reactions can also produce fumes, but gases and vapors also produced by chemical processes are not to be confused with fumes. Metal fumes are the most dangerous, especially those of the heavy metals.

"Particulates" are really a general classification which includes all forms of both solid and liquid air contaminants (ie., dusts, fumes, and mists). Particle sizes thus can vary greatly, some of which are visible to the naked eye, but most of which are not. Figure 6.4 displays some example particle sizes over the range from tiny visible particles down to submicroscopic large molecules.

It is probably already obvious that industry and technology do not eliminate exposures to toxic substances; they merely control these exposures to keep them within acceptable levels. It is both naive and unnecessary for the Safety and Health Manager to adopt a strategy of complete elimination of worker exposure to toxic substances. There is no known poison to which a human being cannot be exposed without significant harm, provided that the exposure is tiny enough and distributed over a long enough interval for the body to assimilate or counteract it. On the other hand, even the mildest of poisons can become lethal if the worker is constantly exposed to it in massive doses. And just such massive doses are the kind that can be found in industrial exposures, more than in any other environment.

Figure 6.3. Dust and fume particle sizes compared (not to scale).

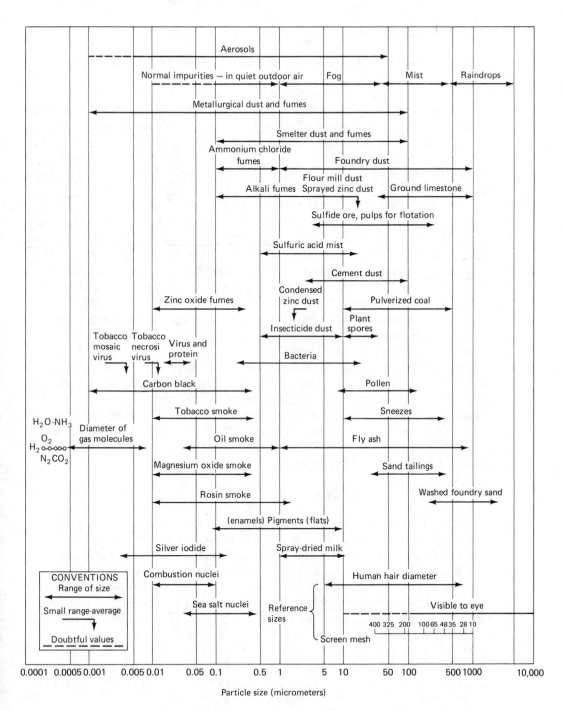

Figure 6.4. Size of airborne contaminants. (Note the logarithmic scale.) (Courtesy of the Mine Safety Appliances Company.)

Threshold Limit Values

Since no poison is lethal in small enough doses, and all poisons are lethal in large enough doses, no clear-cut line separates the harmful from the benign worker environment. But some line has to be drawn to serve as a basis for action to control toxic substances. Particularly with regard to airborne contaminants, it becomes necessary to identify some level of concentrations below which one need not worry about worker exposures. Thus the term "threshold limit value" (TLV) evolved to mean that level of concentration to which the worker could be exposed during the entire workday without significant harm. The TLV varies drastically with the toxicity of the contaminant, of course, and every toxic substance has its own TLV.

So far this discussion of TLVs is in a generic sense, and thus every toxic substance has a TLV even if no one knows what it is and even if no one knows that the particular substance is toxic. But for _known_ toxic substances there is a "listed TLV," which is a value agreed upon by a committee of the American Conference of Governmental Industrial Hygienists (ACGIH) and is listed in the "TLV booklet." When the term TLV is used instead of spelling out the words "threshold limit value," the ACGIH listed value is usually the meaning intended.

Just because a committee decides on the TLV for a given substance, it does not mean that employers must control worker environments to comply with that level. Some TLVs are based on hard scientific data and are very robust criteria for action. Others are based on much more sketchy data, and thus professional judgment is necessary to determine what actions should be taken to control the worker environment. The TLVs themselves even change from year to year as more information becomes available. Thus to be strictly correct one should refer to the date of the list when naming a given TLV: for example, "the 1983 TLV for carbon monoxide was. . . ." Incidentally, when TLVs change, their levels usually decrease as new information about hazards is uncovered.

Permissible Exposure Levels

The focus of this book is on what enforcement agencies require, not what ACGIH lists, but the two are definitely related. At OSHA's beginning, when it was permitted to adopt "national consensus standards" without formal promulgation, the agency adopted hundreds of TLVs, most of which represented the TLV levels published in 1968 by ACGIH. Since the OSHA-published list assumed a regulatory character, the term "Permissible Exposure Level" (PEL) was used to distinguish between the OSHA-prescribed level and the ACGIH term "TLV." For the most part, PELs have remained static, as OSHA has had difficulty getting the public to accept stricter and stricter levels of control. However, the TLVs remain dynamic, as ACGIH continues to revise the list whenever the collective professional judgment of the committee determines that a new TLV should be added to the list or an old one adjusted, usually downward.

Time-Weighted Averages

An even more familiar term than TLV and PEL is TWA, time-weighted average. The TWA is a weighted average concentration over an 8-hour shift, and recognizes that it is permissible to exceed the TLV for a short period if the concentration is lower than the TLV for other periods during the workshift. Because an 8-hour shift is a practical period of exposure, TLVs have typically been set up as 8-hour TWAs (or 7-hour TWAs, recognizing a somewhat shorter than full-shift measurement). So close is this relation that such TLVs are often called TWAs, but the distinction should be remembered.

Since concentrations of air contaminants change over time, it is necessary to compute a weighted average, by using the following formula:

$$E = \frac{\sum_{i=1}^{n} C_i T_i}{8} = \frac{C_1 T_1 + C_2 T_2 + \ldots + C_n T_n}{8} \tag{6.1}$$

where E = equivalent 8-hour time-weighted average concentration
  C_i = observed concentration of the contaminent in time period i
  T_i = length of time period i
  n = number of time periods studied

Example 6.1

Time period, t	Observed concentration, C_i	Length of period, T_i (hr)	$C_i \times T_i$
1	2	1 1/2	3
2	4	2 1/2	10
3	7	1	7
4	5	2	10
5	3	1	3
Total		8	$33 = \sum_{i=1}^{5} C_i T_i$

$$E = \frac{33}{8} = 4.125$$

The calculation shown above is quite adequate if there is only one toxic substance present in the industrial atmosphere. However, mixtures present a different problem. Suppose, for instance, that an industrial atmospheric TWA concentration of nitric acid was barely below the specified PEL of 5 milligrams per cubic meter. But suppose that the same atmosphere showed TWA concentrations of just under the prescribed limits of 1 milligram per cubic meter for sulfuric acid and 25 milligrams per cubic meter for acetic acid *at the same time*. Taken separately, none of these three acid concentrations violates the standard, but common sense says that the three concentrations present at the same time are dangerous.

The synergistic effect of combinations of toxic substances is a complicated subject. Most research has concentrated upon direct effects of substances acting alone. Some mixtures of contaminants would tend to neutralize each other and be beneficial. For instance caustics mixed with acids may produce benign salts. In the example described above, however, the three acids working together are bound to have a combined effect. For some mixtures the combined effect may be far worse than the sum of each of the individual effects.

OSHA takes a moderate approach by requiring simple combinations of toxic substances to be considered, but generally ignoring the complex synergistic effects. The method is to sum the ratios of concentrations of each substance to its own PEL. The resulting sum should not exceed unity. The following formula summarizes the computation:

$$E_m = \sum_{i=1}^{n} \frac{C_i}{L_i} = \frac{C_1}{L_1} + \frac{C_2}{L_2} + \cdots + \frac{C_n}{L_n} \tag{6.2}$$

where E_m = calculated equivalent ratio for the whole mixture
$\quad\ C_i$ = concentration of contaminant i
$\quad\ L_i$ = permissible exposure level (PEL) for contaminant i
$\quad\ n$ = number of contaminants present in the atmosphere

$$E_m \text{ may not exceed } 1.$$

Example 6.2

	Nitric acid	Sulfuric acid	Acetic acid
Contaminant, i	1	2	3
Concentration, C_i	4	0.9	22
Limit, L_i	5	1	25

$$E_m = \sum_{i=1}^{3} \frac{C_i}{L_i} = \frac{4}{5} + \frac{0.9}{1} + \frac{22}{25} = 2.58$$

Since 2.58 > 1, the concentration of the mixture exceeds the PEL, even though the individual PELs are not exceeded.

The Appendix displays the PELs published in the *Federal Register* in OSHA standard 1910.1000 as of June 1981. Besides the main table (A.1) there is a table of levels approved by ANSI for certain substances (A.2) and a table for mineral dusts (A.3). Mineral dusts are considered separately because solid particles are sampled and measured by different means than are toxic gases, mists, and vapors.

The main table has two columns, and these are really two different measures of the same level for each substance. Gases are usually more conveniently measured by volume, and thus the first column, labeled p/m (parts per million), is usually used for these substances. Liquids and some solids are more conveniently measured by weight, and thus the second column, labeled mg/m³ (milligrams of particulate per cubic meter), is preferred for these substances. If the molecular weight of the substance is known, conversion can be made by the following formula:

$$p/m = \frac{mg/m^3 \times 24.45}{M.W.}$$

where M.W. is the molecular weight of the substances. Parts per million is sometimes abbreviated ppm rather than p/m.

In addition to the substances listed in the OSHA tables, some toxic substances have whole standards devoted to their control. The standards for these substances have been issued as a series, and all the standards in the series have been issued several years after the passage of the OSHA act. This means that the standards were scrutinized by the public promulgation procedures and survived the controversy of opposing factions. Several of the substances experienced stormy promulgation, including the invoking of "emergency temporary" promulgation procedures and subsequent court challenges. Much of the background research to justify the issuance of these separate standards for individual substances was conducted by NIOSH. The effort has been called the "standards completion project" because it has been envisioned by some that every toxic substance should eventually have its own unique standard instead of simply being included in a list of TLVs. A list of standards completed appears in Table 6.1.

TABLE 6.1 OSHA STANDARDS FOR SPECIFIC TOXIC SUBSTANCES
("STANDARDS COMPLETION PROJECT")

1910.1001	Asbestos	1910.1014	2-Acetylaminofluorene
1910.1002	Coal tar pitch volatiles; interpretation	1910.1015	4-Dimethylaminoazobenzene
	of term	1910.1016	N-Nitrosodimethylamine
1910.1003	4-Nitrobiphenyl	1910.1017	Vinyl chloride
1910.1004	alpha-Naphthylamine	1910.1018	Inorganic arsenic
1910.1005	[Reserved]	1910.1025	Lead
1910.1006	Methyl chloromethyl ether	1910.1028	Benzene
1910.1007	3,3'-Dichlorobenzidine (and its salts)	1910.1029	Coke oven emissions
1910.1008	bis-Chloromethyl ether	1910.1043	Cotton dust
1910.1009	beta-Naphthylamine	1910.1044	1.2-Dibromo-3-chloropropane
1910.1010	Benzidine	1910.1045	Acrylonitrile
1910.1011	4-Aminodiphenyl	1910.1046	Exposure to cotton dust in cotton
1910.1012	Ethyleneimine		gins
1910.1013	beta-Propiolactone		

Source: Code of Federal Regulations 29 CFR 1910.

Ceiling Levels

Note carefully in the Appendix that some of the substances are preceded by the letter C. This notation refers to "ceiling value." Most of the PELs listed in the tables are to be considered TWAs, but for those substances preceded by a C, the tables list values that are not to be exceeded even for a short period. For these toxic substances the concern is for single acute exposures, not 8-hour exposures. Thus a "maximum acceptable ceiling" (MAC or simply C) is specified. Logically, ceiling values are higher than TWA values if both are listed for a given substance.

Another type of ceiling level is the "short-term exposure limit" (STEL). The STEL states a maximum concentration permitted for a specified duration. The specified duration is usually only a few minutes. For instance, Table A.2 lists the following PELs for benzene:

Benzene

TWA   10 ppm

MAC   25 ppm

STEL   50 ppm for 10 minutes

Note that the STEL for benzene is twice as high as the MAC. One would think that the STEL would lie somewhere between the TWA and the MAC, but the standards consistently list STELs higher than the MACs. This suggests that if the exposure duration is shorter than the duration specified by the STEL, there is no limit to the allowable concentration! This seems to be a contradiction of the definition of the MAC, but in reality it is not feasible to measure the concentration for so short a time period except for "grab samples," a very unreliable method of measurement. Indeed, even the STELs are impractical to check with current state-of-the-art instruments, so STELs are virtually ignored by industry and enforcement officials alike.

Action Levels

One other level, the Action Level (AL), deserves mention. If control measures are taken only after TLVs are exceeded, it may be too late to prevent serious harm and also perhaps too late to prevent citation by authorities. ALs are somewhat a lock-the-barn-before-the-horse-runs-away measure which anticipates the problem before TLVs or other measures are exceeded. ALs are set arbitrarily at ½ PEL. A great deal of statistical and instrument variation prevents precisely accurate surveys. The difference between the AL and the PEL provides a margin for error to assure that worker exposures do not exceed the PEL.

Detecting Contaminants

It is nice to have a list of toxic substances with permissible exposure levels for each, but more is needed to determine whether a problem exists. There are simply too many substances on the list to keep tabs on all possibilities. Safety and Health Managers need to have knowledge of the processes within their plants so that they know where to look or at least whom to ask. Air sampling and testing are the way to determine concentrations as accurately as possible, but before the test is made, some suspicion of possible contamination needs to be provided by other evidence.

One of the most common ways of first detection of a potential problem is by sense of smell. People seem to think they can smell an air contaminant, and they *usually* can smell either the toxic substance or some odorous agent which usually accompanies the toxic substances. But sense of smell is not enough to detect some of the most dangerous contaminants. The most notorious example is carbon monoxide, but carbon dioxide, nitrogen, and methane are also essentially odorless and can be dangerous simply by displacing the oxygen in the air. Some readers will question the statement that methane is odorless because they know that methane is the principal ingredient in natural gas. But the odor in "natural" gas is from a stenching agent deliberately introduced as a safety precaution so that people will detect leaks by sense of smell. Even hydrogen sulfide, a gas that is both dangerous and has a very strong rotten odor, cannot be detected reliably by sense of smell. The odor is so foul that it quickly overcomes the olfactory system, and the victims begin to block out the smell sensation so that they are not aware of the degree of their exposure.

Another approach is to examine technical literature to determine what industries might release what substances. Table 6.2 provides some information from NIOSH literature regarding possible contaminants in various industries.

TABLE 6.2 POTENTIALLY HAZARDOUS OPERATIONS AND AIR CONTAMINANTS

Process types	Contaminant type	Contaminant examples
Hot operations		
Welding	Gases (g)	Chromates (p)
Chemical reactions	Particulates (P)	Zinc and compounds (p)
Soldering	(dust, fumes, mists)	Manganese and compounds (p)
Melting		Metal oxides (p)
Molding		Carbon monoxide (g)
Burning		Ozone (g)
		Cadmium oxide (p)
		Fluorides (p)
		Lead (p)
		Vinyl chloride (g)
Liquid operations		
Painting	Vapors (v)	Benzene (v)

(continued)

TABLE 6.2 (CONTINUED)

Process types	Contaminant type	Contaminant examples
Degreasing	Gases (g)	Trichloroethylene (v)
Dipping	Mists (m)	Methylene chloride (v)
Spraying		1,1,1-Trichloroethylene (v)
Brushing		Hydrochloric acid (m)
Coating		Sulfuric acid (m)
Etching		Hydrogen chloride (g)
Cleaning		Cyanide salts (m)
Dry cleaning		Chromic acid (m)
Pickling		Hydrogen cyanide (g)
Plating		TDI, MDI (v)
Mixing		Hydrogen sulfide (g)
Galvanizing		Sulfur dioxide (g)
Chemical reactions		Carbon tetrachloride (v)
Solid operations		
Pouring	Dusts	Cement
Mixing		Quartz (free silica)
Separations		Fibrous glass
Extraction		
Crushing		
Conveying		
Loading		
Bagging		
Pressurized spraying		
Cleaning parts	Vapors (v)	Organic solvents (v)
Applying pesticides	Dusts (d)	Chlordane (m)
Degreasing	Mists (m)	Parathion (m)
Sandblasting		Trichloroethylene (v)
Painting		1,1,1-Trichloroethane (v)
		Methylene chloride (v)
		Quartz (free silica, d)
Shaping operations		
Cutting	Dusts	Asbestos
Grinding		Beryllium
Filing		Uranium
Milling		Zinc
Molding		Lead
Sawing		
Drilling		

Source: NIOSH, Ref. 54.

Yet another approach is to analyze the processes within the plant to determine potential leaks to the industrial atmosphere. Such an analysis can become quite technical and involves not only an understanding of the machines, pumps, valves, sumps, and tanks, but also a knowledge of the materials used, quantities, intermediate phases, volatilities, and other characteristics which affect the amount of the contaminants that become airborne. A chemical engineering qualitative flow process chart and perhaps a quantitative flow process chart may be necessary to determine whether air contaminant potentials are present. The Safety and Health Manager should not hesitate to involve a chemical engineer in the analysis of process potential air contaminations. A technical analysis at this phase may preclude a lot of expensive and laborious sampling experiments.

Measurement Strategy

Once an air contamination risk has been determined to exist, a procedure is needed to go about taking samples, measuring employee exposures, and instituting controls. NIOSH recommends a strategy for this purpose, and this strategy is displayed in the form of a decision chart in Figure 6.5.

Measurement Instruments

Federal regulation by both OSHA and EPA of permissible exposure levels for toxic substances in the atmosphere has stimulated the electronics and instrumentation industries to develop new and more precise instruments for determining concentrations. Interest in parts per million is shifting toward closer scrutiny to detect parts per *billion*. Such demands are pressing the physics of the instruments, and large-scale inaccuracies are the result.

With such demands by high technology on atmospheric measurement devices, one would think that air contamination monitoring would be a new field. But historically there have been other, more crude, means for monitoring breathing air. Animals formerly were used to test for toxic gases or oxygen deficiency. A canary or a mouse in a cage was often carried into mines. If the canary or mouse died, the workers were alerted to the hazard. A flame safety lamp was used to test for oxygen deficiency; the flame would die if the oxygen proportion in the atmosphere was too low. A brighter-burning lamp was supposed to be an indication of the presence of methane.

These methods were crude, but they did provide some essential indications— for *acute* exposures. With the recognition of threshold limit values and the rise in importance of chronic exposures, the canary test or flame test is totally inadequate. By the time the canary developed cancer or scarred lung tissue, the workers also would be victimized. Besides, animal life spans are too short to deal with chronic effects to which humans might be susceptible.

There are basically three approaches to measuring air contaminant exposures:

1. Direct-reading instruments
2. Sampling and subsequent laboratory analysis
3. Dosimeters

Direct-reading instruments are used principally for oxygen deficiency, flammable gases or vapors detection, and certain commonly encountered contaminants. Direct-reading instruments are of course more convenient than sampling and analysis, so instrument manufacturers are trying to meet demands for these devices for as many

Figure 6.5. NIOSH exposure measurement strategy. Each individual substance health standard and should be consulted for detailed requirements. AL, action level; PEL, permissible exposure limit. (From Ref. 54.)

toxic substances as possible. Really the only way an atmosphere can be tested for a dangerous acute exposure just prior to entry is by a direct-reading instrument.

For more obscure contaminants and for rarer concentrations, sampling devices must be used. These devices either pump a prescribed quantity of air past a filter or sorbent which collects the contaminant or they merely collect a precise volume of the air itself. The filter, the sorbent, or the air sample is then sent to a laboratory for analysis.

Dosimeters are the most convenient device of all, especially for gathering TWA data. A dosimeter is a small collector worn on the worker's body or clothing and which collects a time-weighted average exposure over a specified time period, such as a full shift. Unfortunately, accurate dosimeters are still beyond the state of the art for most toxic substances. Figure 6.6 shows one type of dosimeter for air monitoring.

VENTILATION

Ventilation is a good way to deal with toxic air contaminants, but it is not the best way. Before investing in an expensive ventilation system, the Safety and Health Manager should first ask some probing questions about the sources of the contamination.

The most desirable way to deal with an air contaminant is to change the process so that the contaminant is no longer produced. This is so obvious that it is sometimes overlooked. It may be that the process cannot be changed, but if it can, there may be tremendous gains in store, not only in health and safety, but in production cost and efficiency as well. For instance, it may be found that machined parts can be machined dry, avoiding cutting oil exposures to the machinist's skin as well as solvent contamination of the air as the parts are later cleaned. Forging, die casting, or powdered metal technology may eliminate several machining processes, changing the process in beneficial ways. Some of these ideas may generate more disadvantages than advantages in given situations, but each application should be checked for potential benefit. There is no way the Safety and Health Manager can win better recognition from top management than by generating a clever idea which cuts costs or increases production while it enhances safety or health.

For some more idea generators, consider the hazards of toxic air contaminants from welding operations. Sometimes the principal source of the contaminant is the surface coating on the metal to be welded. Perhaps as a process change, this surface coating can be removed prior to the commencement of welding. Better yet, perhaps the material does not require welding at all. Possibly a crimping operation could produce an effective joint, eliminating the need for welding or soldering.

Chemical processes are classified as batch or continuous. Choice between the two usually involves many considerations including investment cost, length of production run expected, and volumes to be produced. But an important factor to be considered among these is air contamination. Continuous processes generally reduce the exposure of materials to the air because open handling is reduced, and

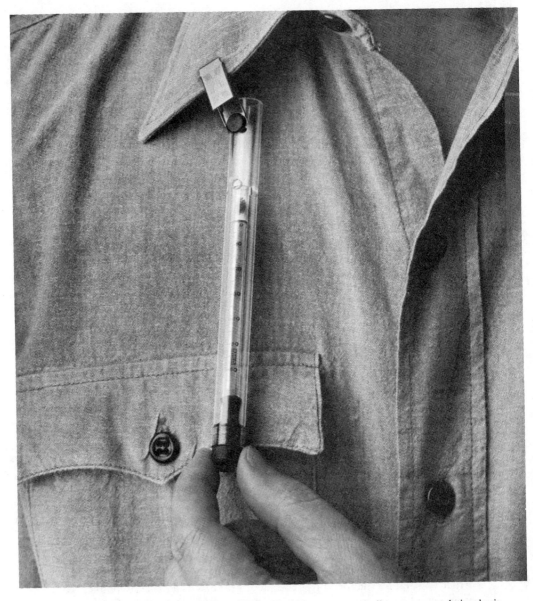

Figure 6.6. Dosimeter for gathering TWA data automatically on a cumulative basis. (Courtesy of the Mine Safety Appliances Company.)

with continuous processing, batches of materials are not sitting idle awaiting processing. However, the mechanical handling equipment used for continuous processes may increase the contamination levels. Each situation must be studied to determine the best solution, but the safety and health aspects need to be considered.

One way a process can be changed is to isolate it or enclose it. If a particularly contaminating process is in the plant, perhaps it should be located in a separate building so that it does not contribute to the overall ventilation problem.

A slight variation to changing the process is to change the materials used. Carbon tetrachloride has been found to be a health hazard, so other solvents have been substituted. The chlorinated hydrocarbon solvents substituted, such as trichloroethylene and perchloroethylene, are also being found to be hazardous, but fortunately not as hazardous as carbon tetrachloride. New solvents may be found to reduce hazards even further.

In another example, silica sand is often used for blasting to improve surface characteristics. But airborne silica causes the lung disease called silicosis. Perhaps the silica sand could be replaced by steel shot, removing the silica contamination.

A classic example of changing materials to reduce hazards is the switch from hazardous lead-based paints to substitute materials such as iron oxide pigments. Another classic was the switch from Freon to propane as a propellant for aerosol cans. In this case the materials switch was intended to protect the environment (the ozone layer), but the solution may be more hazardous to the individual because propane is a flammable gas.

Design Principles

If the process cannot be changed or materials substituted, a well-designed ventilation system may be the best solution to the problem. OSHA has a standard that deals with this subject, but it must be emphasized that ventilation is a very technical subject, and the Safety and Health Manager may want to turn to a professional engineer to design an adequate ventilation solution to an air contamination problem. Exhaust ventilation is not the same thing as ordinary heating and air conditioning, and design errors can be made if this difference is not considered. An example is shown in Figure 6.7. Most heating and air-conditioning ducts have right-angle bends which may be fine for gases but greatly impair the ability of the ducts to transport particulates.

Another questionable ventilation system is an ordinary household fan used to blow away smoke from a contaminant source. It is true that a fan can dilute the concentration of a contaminant in a given place, and dilution ventilation is a recognized method of reducing concentrations to levels lower than the PEL. But the question is: Where is the fan blowing the contamination *to*? It is adding to the overall back-

Not this . . .   . . . This

Ordinary heating
and air-conditioning
practice

No more than 45°

Figure 6.7.   Avoid sharp angles at duct entry.

ground level of air contamination in the plant and may later have to be dealt with if other processes are also producing contaminations.

A basic objective of exhaust ventilation is to isolate and remove harmful contaminants from the air. The more these contaminants are concentrated into limited plant areas, the easier they are to separate from the air. Dilution ventilation, although expedient in some cases, is counterproductive to the goal of removing the contaminant. Dilution ventilation can be likened to "sweeping dirt under the rug."

Focus is an important consideration for ventilation systems. Sheer volume or flow velocity is not enough. Ventilation technology is producing some very fine local exhaust systems which focus the intake right on the contaminant, much as a vacuum cleaner is designed to do. Even if sufficient flow could be achieved with a general ventilation system, the flow could be a nuisance because such a wind might blow away papers and other materials, making the job awkward and inefficient.

The mention of a vacuum cleaner in the paragraph above might give some Safety and Health Managers the wrong idea. Particle size for fume or other contaminants will usually be too small to be effectively trapped in the bag of an ordinary vacuum cleaner. If the process does not trap the contaminant, it is merely blowing it around, perhaps increasing exposure.

The best exhaust ventilation systems are the "pull" types, not the "push" types. Even within the exhaust duct the fan should be placed at the end of the duct if possible, as shown in Figure 6.8. Leaks in the duct then merely draw in more air rather than pumping contaminated air back into the plant environment.

Makeup Air

With an exhaust ventilation system or systems, some source of "makeup air" is essential. Open windows and doors usually serve this purpose. But with the advent of the energy crisis in the 1970s, adequate supply of makeup air has become a real problem. The increased cost of energy has greatly enhanced the attractiveness of recirculating the exhaust air after filtration and decontamination. Not only does such a solution save energy, it reduces external atmospheric pollution, a very important

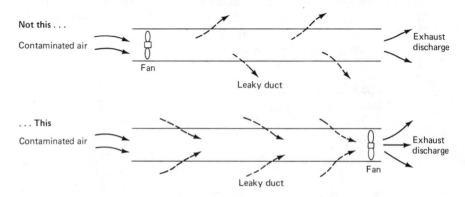

Figure 6.8.   Fan location: keep negative pressure within duct.

point considering regulations of the Environmental Protection Agency (EPA) and the general environmental concerns of the public.

Figure 6.9 diagrams a recirculating system in which the objective is to remove dust by means of a high-efficiency filter. It is essential to recognize the importance of the condition of the filter to the overall effectiveness of the system. The filter will clog up over time if it is doing its job, and thus it must be serviced, cleaned, or changed. Note the bypass damper, which permits selective proportioning of the system from full recirculating to full exhaust. This bypass can save energy when weather conditions are mild and recirculation is unnecessary. Also note the inclusion of a manometer to detect pressure differential across the filter as well as an alarm to sound if this differential becomes too great. Both are intended to provide indication or warning to the operator that the filter has become fouled and needs service.

Safety and Health Managers are warned to monitor recirculation systems closely. Frequently in industry a sophisticated ventilation system is installed for a

Figure 6.9. Example of recirculation from air cleaning devices (dust). (From the American Conference of Governmental Industrial Hygienists Committee on Industrial Ventilation.)

process, and then it is ignored. Filter alarms sometimes are merely red lights, which operators often ignore. Even the audible alarms such as buzzers and horns are sometimes disabled by disconnecting electrical leads. Both operations and maintenance personnel have been found guilty of disabling these devices, which are designed to protect the workers' health.

Besides using recirculating systems, there are some other ways around the problem of energy losses due to the introduction of makeup air into the building. One method is to introduce the makeup air right at the point at which the contamination is taking place. With this strategy the makeup air may need no air conditioning—no cooling in summer and no heating in winter. The exhaust system will merely suck up the unconditioned makeup air together with the contaminants, and workers will have little exposure to either.

Another solution to the problem is to utilize a heat exchanger to recapture the energy of the exhaust air and transfer it to the incoming makeup air. It is difficult to make this solution practical, however, because the heat differential between makeup air and exhaust air is usually too low to make the heat exchanger effective. Also, the heat exchanger approach necessitates passing the makeup air duct close to the exhaust air duct, which introduces the possibility of cross-contamination. Finally, heat exchanger systems can be expensive—both to install and to maintain effectively—too expensive in many cases to amortize by means of energy cost savings over the life of the system.

Besides the energy problem, another problem with supplying makeup air is the presence of contaminated air from the outside. This is an unusual problem, but it has on occasion presented itself. For one plant the makeup air inlet was adjacent to a major freeway, and carbon monoxide and other automobile exhaust emissions were being drawn into the building. In another poor design the makeup air inlet was so close to the exhaust system discharge that contaminants were being drawn back in and recirculated around the plant. That way if workers managed to escape breathing the contaminated air the first time through, they got another chance to become exposed!

A quick check to determine whether makeup air supply is sufficient is to check atmospheric pressure both inside and outside the plant. The pressure inside should only be slightly lower than the pressure outside. If the pressure inside is substantially lower, the makeup air supply is insufficient. The basic relationship between makeup and exhaust is illustrated in Figure 6.10. The cross-sectional area of makeup open-

Figure 6.10. Balancing makeup and exhaust air.

ings multiplied times the velocity of flow through those openings must equal the cross-sectional area of the exhaust openings multiplied times the velocity of flow through the exhaust.

Purification Devices

If the exhaust air is clean enough to meet external standards, no filtration or purification may be necessary once the air gets outside the plant. But often some type of purification device is necessary on the outside as is required indoors for recirculating systems, particularly for removal of particulates. The following paragraphs describe some of the basic types of particulate removal devices.

"Centrifugal" devices, often called "cyclones" (see Figure 6.11) take advan-

High-efficiency centrifugals

Low-pressure cyclone   Dry-type dynamic precipitator

Figure 6.11. Cyclone and other dry-type centrifugal collectors for removal of particulates from exhaust air. (ACGIH Committee on Industrial Ventilation.)

tage of the mass of the contaminant particles, causing them to collect on the sides of the cyclone in the swirling air and then slide to the bottom and settle in the neck of the funnel where they can be periodically emptied. Another type of centrifugal device causes the dirty air to strike louvers, causing particles to separate from the air. A typical application for cyclones is grain dust removal for grain elevators and mills. Cyclones are also used for woodworking sawdust, plastic dusts, and some chemical dry particulates.

"Electrostatic precipitators" place a very high (e.g., 50,000 volt) electrical charge on the particles, causing them to be attracted to an electrode of opposite electrical charge. The collecting electrode can consist of plates, rods, or wires, all of which can be agitated to shake off the collected dust and cause it to settle in the bottom of the chamber. Figure 6.12 illustrates one type of high-voltage electrostatic precipitator. Electrostatic precipitators are used in the steel, cement, mining, and chemical industries. Electrostatic precipitators are also used in smokestacks to reduce fly ash.

"Wet scrubbers" include a wide variety of devices which employ water or chemical solution to wash the air of particulate or other contaminants. Some types pass the dirty air through standing water or solution. Another type forces dirty air to rise in a tower packed with a filler media through which water falls, as shown in Figure 6.13. Note in the figure that the cleanest water is at the top of the tower, where the exiting air is at its cleanest. The bottom of the tower is the dirtiest, where the water vehicle is exiting the packing and is contacting the dirtiest, untreated exhaust air. Wet centrifugal types, like dry centrifugal types, take advantage of the mass of the particles by causing them to impinge on blades, plates, or baffles. Wet scrubbers are seen in the chemical industries, where they are able to remove gases and vapors in addition to particulates. Other industries using wet scrubbers are rubber, ceramics, foundries, and metal cutting.

"Fabric or bag-type" filters are essentially like a vacuum cleaner bag. Some are huge and are located in separate buildings called "bag houses." Figure 6.14 shows three types of fabric filters. Fabric filters are used in the refining of toxic metals such as lead and in woodworking, metal cutting, rubber, plastics, ceramics, and chemical industries.

INDUSTRIAL NOISE

Noise exposure is another classic health problem because the chronic exposures are the ones that typically do the damage. A single acute exposure can do permanent damage, and in this sense noise is a safety problem, but noise exposures of this type are extremely rare. As with other health hazards, noise has a threshold limit value, and exposures are measured in terms of time-weighted averages. To understand the

High-tension support frame

High-voltage insulators

Gas seal

Collecting plate

Rapper

Discharge electrodes

Discharge wire weights

Dust hopper

Ionizer   Collector plates

Low-voltage design

12,000 to 13,000 volts on ionizer
6,000 to 7,000 volts on collector
plates

High-voltage design

Discharge electrode (normally negative)
30,000 to 75,000 volts. Collecting plates
are usually grounded

AMERICAN CONFERENCE OF GOVERNMENTAL INDUSTRIAL HYGIENISTS
Electrostatic precipitator

Figure 6.12. Electrostatic precipitators for removing particulates. (ACGIH, Committee on Industrial Ventilation.)

Water in

Distributor

Outlet

Suitable packing media

Steel cylindrical jacket

Corrosion lining where required

Support plate

Inlet

Dust and water out

Packed tower

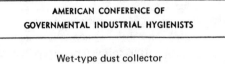

AMERICAN CONFERENCE OF
GOVERNMENTAL INDUSTRIAL HYGIENISTS

Wet-type dust collector

Figure 6.13. Packed tower type of wet scrubber (ACGIH Committee on Industrial Ventilation.)

Cloth tube or stocking type

Envelope- or bag-type

Compartments 1, 2, and 3
under air load
Compartment 4 closed off
for screen cleaning

Multiple section continuous automatic.

AMERICAN CONFERENCE OF GOVERNMENTAL INDUSTRIAL HYGIENISTS
Fabric collectors

Figure 6.14. Fabric filters for exhaust air. (ACGIH Committee on Industrial Ventilation.)

units of these measures, some background is needed in the physical characteristics of sound.

Characteristics of Sound Waves

Noise can be defined as unwanted sound. In the industrial sense noise usually means excessive sound or harmful sound. Sound is fairly well understood by everyone as a pressure wave in the atmosphere. In liquids, sound is also a pressure wave; in rigid solids the sound takes the form of a vibration.

Two basic characteristics of sound waves important to the subject of noise control are

1. The amplitude or pressure peak intensity of the wave
2. The frequency in which the pressure peaks occur

Our sense of hearing can detect both of these characteristics. Pressure intensity is sensed as "loudness," whereas pressure frequency is sensed as "pitch." Figure 6.15 illustrates the wave form of sound and also graphs the relationship between pressure and time. Note in Figure 6.15b that the "period" is the length of time required for the wave to complete its cycle. In the graph it is measured at the point at which the pressure differential becomes zero and is starting to become negative. However, it could have been measured from peak to peak, valley to valley, or at any other convenient reference point in the cycle. These "periods" are always short, too short to count their occurrences while listening to the sound. If we could count the occurrences of these wave cycles, the resulting count per unit time would be the "frequency," usually measured in numbers of cycles per second (called hertz). A typical sound is at a frequency of 1000 cycles per second, that is, 1000 hertz (abbre-

Figure 6.15. Characteristics of sound waves: (a) pressure wave is longitudinal (in the direction of travel of the sound); (b) relationship between pressure and time at a given point of sound exposure; (c) low pitch, soft sound; (d) high pitch, soft sound; (e) low pitch, loud sound; (f) high pitch, loud sound; (g) short-duration sound; (h) sustained sound.

viated 1000 Hz). Obviously, we could never count 1000 pulses of pressure in a single second, but our ears have a surprising sensitivity to variations in this frequency count. The sensation is known as "pitch," and skilled musicians have trained their ears to hear very slight variations in sound-wave frequency. Frequency is important in analyzing the sources of occupational noise exposure.

Even more important than pitch in industrial settings is the pressure intensity of the sound wave. High peaks of pressure in the waves can do permanent damage to the delicate mechanisms in the human ear and cause permanent hearing loss. The ear must be delicate to pick up the tiny pressures of the faintest audible sounds, but it is able to withstand an incredibly large range of pressures. The human ear can withstand without damage a sound pressure 10,000,000 times as great as the faintest sound it can hear! A necessary result of this incredible range of pressures is that the ear is not very sensitive to shades of differences in these pressures, especially as the pressures get into the upper part of the range. In other words, as sounds get fairly loud, the human ear cannot readily detect a large increase in the intensity, even a doubling or tripling of the intensity.

Decibels

It is difficult to talk sensibly about such a large range of audible pressures, and it is especially difficult to set standards. Imagine a noise meter reading in the millions. The situation is further complicated by the lessening of the ability of the human ear to detect pressure differences as sounds get louder. To deal with these problems a unit of measure called the decibel (abbreviated dB) has been devised to measure sound pressure intensity. The decibel has a logarithmic relation to the actual pressure intensity, and thus the scale becomes compressed as the sound becomes louder, until in the upper ranges the decibel is only a gross measure of actual pressure intensity. But this is appropriate because as was mentioned before, the human ear only hears gross differences anyway when the sound gets very loud. Figure 6.16 relates the decibel to familiar sound levels.

The logarithmic decibel scale is convenient, but it does give rise to some problems. If a machine in the plant is very loud, putting a second machine just like it right beside it will make the sound twice as loud—right? Wrong! Remember the tremendous range of sound pressures and that the human ear hears only a slight increase in loudness, when the actual sound pressure may have doubled due to the addition of the extra machine. The decibel scale recognizes the addition of the new machine as an increase in noise level of only 3 decibels. Conversely, if the noise level in the plant exceeds allowable standards by very much, shutting off half the machines in the plant—an obviously drastic measure—may have very little effect in bringing down the total noise level on the decibel scale. Table 6.3 provides a scale for combining decibels to arrive at a total noise level from two sources. If there are three or more sources, two sources are combined and then treated as one source to be combined with a third, and so on, until all sources have been combined into a single total. An example is helpful in illustrating the method.

Noise sources	Sound level (dB)	Speaking effort required
Pneumatic chipper (at 5 ft)	115	<u>Nearly impossible</u> to communicate by voice
Chain saw (at ear) Teenage rock n roll band Riveting machine Nail machine Casting shakeout area Wood planer Punch press Forging hammer	110	<u>Very difficult</u> to communicate by voice
	105	<u>Shout</u> with hands cupped between mouth and other person's ear
Pneumatic air hoist: 4000 lb	100	<u>Shout</u> at 1/2 foot
Tumbler 6 in. x 3 in., small castings	95	<u>Shout</u> at 1 foot
Automatic screw machine, Nut blanking	90	Normal voice at 1/2 foot, shout at 2 feet
Boiler room Arc welder	85	Normal voice at 1 foot, shout at 4 feet
Milling machine (at 4 ft)	80	Normal voice at 1-1/2 feet, shout at 6 feet
Pneumatic drill	75	Normal voice at 2 feet, shout at 8 feet
Inside a car (50 mph)		

(Telephone use impossible marked along the dB scale between 75 and 110)

Figure 6.16.   Decibel noise levels of familiar sounds. (From NIOSH.)

TABLE 6.3 SCALE FOR COMBINING DECIBELS

Difference between two decibel levels to be added (dB)	Amount to be added to larger level to obtain decibel sum (dB)
0	3.0
1	2.6
2	2.1
3	1.8
4	1.4
5	1.2
6	1.0
7	0.8
8	0.6
9	0.5
10	0.4
11	0.3
12	0.2

Source: NIOSH, Ref. 40.

Example 6.3

Suppose that noise exposure at a workstation is essentially due to four sources, as follows:

Machine A	86 dB
Machine B (identical to machine A)	86 dB
Machine C	82 dB
Machine D	78 dB

First the two identical noise sources, machines A and B, are combined to produce a noise level of 89 dB. Then machine C is added as follows:

$$\text{dB difference} = 89\ \text{dB} - 82\ \text{dB} = 7\ \text{dB}$$

From Table 6.3 a difference of 7 dB between two sources results in the addition of 0.8 dB to the larger source. Therefore, the combined sound of machines A, B, and C is

$$\text{combined sound (A, B, C)} = 89\ \text{dB} + 0.8\ \text{db} = 89.8\ \text{dB}$$

Adding Machine D, we have

$$\text{dB difference} = 89.8\ \text{dB} - 78\ \text{dB} = 11.8\ \text{dB} \approx 12\ \text{dB}$$

Returning to Table 6.3 a difference of 12 dB between two sources results in the addition of 0.2 dB to the larger source. Therefore, the combined sound of all machines is

$$\text{combined sound (A, B, C, D)} = 89.8\ \text{dB} + 0.2\ \text{dB} = 90.0\ \text{dB}$$

Figure 6.17 diagrams the computation of Example 6.3. It should be noted that the measurement of both the total combined sound and the individual machine contributions toward that total are made from the position of the operator. Otherwise, distance factors would affect the results. The computation of combined noise levels is useful in considering the potential benefits of removing machines, enclosing machines, or changing the process.

We have been emphasizing sound intensity measured in decibels, but frequency or pitch also plays a role in noise control. Industrial noise is typically a combination of sound frequencies from each of several sources. The total range of sound frequencies audible to the human ear is from about 20 Hz to about 20,000 Hz. The ear is more sensitive to some of these frequencies than others, particularly the upper middle range from about 1000 Hz to about 6000 Hz. Thus sound-level meters have been devised to bias the decibel reading slightly to emphasize the frequencies from 1000 to 6000 Hz. This biased reading is called the A-weighted scale,[1] and resulting readings are abbreviated dBA instead of simply dB. OSHA recognizes the A-scale, and OSHA PELs are expressed in dBAs.

---

[1]There are also a B-scale and C-scale, but these scales are seldom used.

Machine "B"
86 dB

Machine "C"
82 dB

89 dB

89·8 dB

90 dB

Machine "D"
78 dB

Machine "A"
86 dB

Figure 6.17.   Combining noise from several sources.

OSHA Noise Standards

Noise is unusual in that it is a hazard for which OSHA has set both a PEL and an AL. The best known is the PEL, which is set at 90 dBA for an 8-hour TWA. The Action Level (AL) was established in the early 1980s at 85 dBA for an 8-hour TWA, about 10 years after the 90-dBA PEL was established. It is widely recognized that workers can tolerate short periods of noise higher than the 8-hour TWA without damage, so OSHA specifies a range of decibel exposure levels for various exposure periods. The range of OSHA PELs for noise exposure is given in Table 6.4.

The range of permissible exposures in Table 6.4 makes possible a computation of a time-weighted average exposure, relating each exposure time to the limit permitted for that sound level. The procedure is very similar to the calculation used earlier where multiple contaminants are present in the atmosphere. The formula used is as follows:

$$D = 100 \sum_{i=1}^{n} \frac{C_i}{T_i} = 100\left(\frac{C_1}{T_1} + \frac{C_2}{T_2} + \cdots + \frac{C_n}{T_n}\right) \tag{6.3}$$

where D = total shift noise exposure ("dose") as a percent of PEL
$\quad C_i$ = time of exposure at noise level i
$\quad T_i$ = maximum permissible exposure time at noise level i (from Table 6.4)
$\quad n$ = number of different noise levels observed

An interesting computation is for total shift exposure exactly at the Action Level of 85 dBA. Using equation (6.3), the computation is as follows:

$$D = 100 \sum_{i=1}^{n} \frac{C_i}{T_i} = 100\left(\frac{8}{16}\right) = 50\%$$

Thus the Action Level is computed to be 50% of the maximum permissible PEL. However, the reader should note from the earlier discussion of sound intensity that 85 dBA represents less than one-half of the absolute sound intensity of noise at 90 dB.

As a second illustration of the computation of noise dose from equation (6.3), a more comprehensive example is undertaken next.

Example 6.4

Noise-level readings show that a worker exposure to noise in a given plant is as follows:

8:00 A.M.–10:00 A.M.	90 dBA
10:00 A.M.–11:00 A.M.	95 dBA
11:00 A.M.–12:30 P.M.	75 dBA
12:30 P.M.–1:30 P.M.	85 dBA
1:30 P.M.–2:00 P.M.	95 dBA
2:00 P.M.–4:00 P.M.	90 dBA

Adding up the noise durations for each level, we obtain

At noise level 90 dBA:	2 + 2	= 4 hours	
At noise level 95 dBA:	1 + ½	= 1½ hours	
At noise level 75 dBA:		1½ hours	(ignore) *9+ below*
At noise level 85 dBA:		1 hour	
Total		8 hours	

TABLE 6.4 OSHA's TABLE OF PELs FOR NOISE

A-weighted sound level	Reference duration time (hr)
80	32
81	27.9
82	24.3
83	21.1
84	18.4
85	16
86	13.9
87	12.1
88	10.6
89	9.2
90	8
91	7.0
92	6.2
93	5.3
94	4.6
95	4
96	3.5
97	3.0
98	2.6
99	2.3
100	2
101	1.7
102	1.5
103	1.4
104	1.3
105	1
106	0.87
107	0.76
108	0.66
109	0.57
110	0.50
111	0.44
112	0.38
113	0.33
114	0.29
115	0.25
116	*0.22*
117	*0.19*
118	*0.16*
119	*0.14*
120	*0.125*
121	*0.110*
122	*0.095*
123	*0.082*
124	*0.072*

TABLE 6.4 (continued)

A-weighted sound level	Reference duration time (hr)
125	0.063
126	0.054
127	0.047
128	0.041
129	0.036
130	0.031

Source: Code of Federal Regulations 29 CFR 1910, 95.

The reason that the 1½-hour exposure at 75 dBA was ignored is that 75 dBA is below the range of Table 6.4. In other words, workers may be exposed to noise levels of 75 dBA for as long as desired with no adverse effects, at least as far as safety standards are concerned.

Computing the ratios at each level and summing, in accordance with equation (6.3), yields

$$D = 100 \sum_{i=1}^{i} \frac{C_n}{T_n} = 100 \left[\frac{4}{8} + \frac{1\,1/2}{4} + \frac{1}{16} \right]$$

$$= 100\,(0.5 + 0.375 + 0.0625)$$

$$= 93.75\%$$

Since 93.75% is less than 100%, the PEL is not exceeded. However, since 93.75% is greater than 50%, the Action Level of 85 dBA (8-hour TWA) is exceeded.

Sometimes noise is percussive or intermittent so that technically speaking there are tiny intervals of silence between sharp reports. Some employers have followed the scheme that these tiny intervals of silence can be counted toward the quiet time, reducing the observed duration of noise in excess of 90 dBA, but this interpretation is incorrect. Any variations in noise levels where the maxima are less than 1 second apart are to be considered continuous. The "slow response" scale of modern noise-level meters tends to ignore such tiny interval variations, and thus "slow response" is specified in noise-level metering.

The standards do have a specification for peak impulse or impact noise at 140 dBA, but this of course is much higher than the PELs for continuous noise. Thus the 140 dB OSHA specification can be considered a ceiling or C value. The 140-dB ceiling should be considered a limit for acute exposure and is accordingly a safety hazard. However, such exposures are so rare and difficult to measure after the fact that ceiling violations are virtually never cited. Ordinary sound-level meters are not very effective in measuring impact noise. Even for continuous exposures, measurement can be a problem which will not be addressed.

Noise Measurement

As a first check for potential noise problems, the Safety and Health Manager should take a walk through the plant and listen. As a rule of *Thumb,* if you can reach out and touch someone with your *Thumb* but still cannot hear and understand that person's conversation (without his or her shouting), either your hearing is already damaged or defective, or there is excessive noise in the area. If the noise is continuous throughout the working shift but is no louder than a continuously running vacuum cleaner, there is probably no violation of standards. But if the noise is as loud as a subway passing through a station continuously throughout the full shift, a violation probably exists. If subways are unfamiliar to you, imagine a fast-moving freight train passing within 20 feet; such a noise level would easily constitute a violation if the exposure were continuous for a full 8-hour shift. Some exposures louder than the train might be permissible if they are of short duration, as was obvious in the calculation of average exposures in the preceding section. Anything between the vacuum cleaner and the train in sound level would be a gray area which ought to be measured with accurate meters.

Accurate measurement of sound levels requires instruments such as the sound level meter (SLM) illustrated in Figure 6.18. The meter registers the sound intensity in decibels. Sound-level meters are delicate instruments and must be handled carefully. Accuracy is a problem, and the Safety and Health Manager should not expect performance better than ± 1 dB. Calibration is extremely important, and no sound-level meter is complete without a calibration device (known sound source) nearby. Variations in battery level must be compensated, and humidity and temperature conditions can cause distortions.

Figure 6.18. Example of sound-level meter.

Some skill is required in using the sound-level meter to obtain reliable readings. Naturally, the microphone receiver of the instrument must be held in the vicinity of the subject's ear in order to be representative of the exposure. However, the instrument should not be held too close, as the subject's body may affect the reading. The instrument's microphone receiver should be shielded from wind currents, and the instrument itself should not be subjected to direct vibrations. The Safety and Health Manager may find the sound-level meter most useful in comparing various locations in the plant and in comparing different machines and operating modes.

Determining the full-shift, time-weighted average exposure with a sound-level meter is tedious and requires a great deal of effort. A convenient substitute is a cumulative device called a "dosimeter," which is worn on the person of the subject under study. Dosimeters may seem to be a panacea, but they have their drawbacks. The wearer can easily bias the dosimeter by holding it close to a loud mechanism, or by simply rubbing, tapping, or blowing on the microphone. They are useful survey devices, however, and are used to monitor exposures if the firm's noise levels have exceeded the Action Level (AL) of 85 dBA for an 8-hour TWA.

Once it is determined that a problem exists, a more sophisticated measurement of noise level may be necessary to isolate the sources of the undesirable noise. An octave-band analyzer permits decibel readings to be taken at various frequencies over the audible range as shown in the example in Figure 6.19. Various noise reduction media have characteristic frequencies at which they are most effective. The octave-band analysis will help to delineate the problem frequencies as well as provide evidence to identify the problem sources. Finally, octave-band analysis is useful in determining the frequency characteristics of a particular industry's noise sources so that damage from these sources can be distinguished from damage from noise exposures that have occurred off the job.

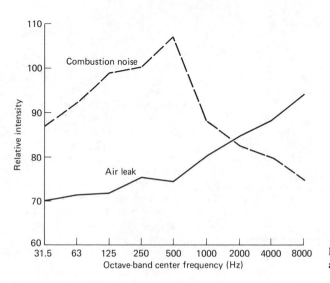

Figure 6.19. Example of octave-band analysis. (From NIOSH, Ref. 40).

Engineering Controls

Once the instruments have proven that a problem exists, the Safety and Health Manager needs physical solutions to the problems. If noise levels exceed the PEL, federal standards require that feasible engineering or administrative controls be utilized. If these measures fail to reduce noise exposures to within the PEL, personal protective equipment must be provided and used to reduce sound levels to within the PEL. Engineering controls should be considered a more thorough and permanent solution to the problem.

As with the control of toxic substances, the simplest solutions may be so obvious that they are overlooked. Process modification or elimination should always receive consideration. Another very simple solution is merely to move the operator away from the primary source of the noise if feasible. This idea has more merit than it would intuitively appear because the noise intensity from a given source goes down as the square of the distance, in the absence of reflective walls and other distorting factors. The reason for this relationship can be seen in Figure 6.20.

It must be remembered that it is absolute sound intensity, not decibels, that varies inversely with the square of the distance from the source. The logarithmic decibel scale results in a 3-dB change whenever the sound intensity is changed by a factor of 2 (doubled or halved). This leads to a rule of thumb for distance. Since sound intensity varies as the square of the distance from the source, a doubling of distance results in a *fourfold* reduction in sound intensity, which in turn reduces the decibel level by 6 dB.

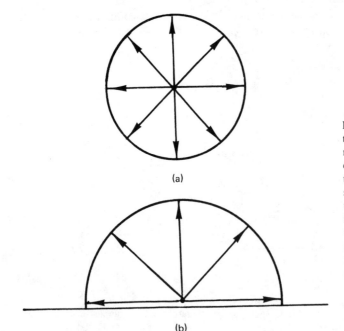

(a)

(b)

Figure 6.20. Distribution of sound intensity over the surface of a sphere as it radiates from a single source. (a) Sound emanates from a point source in all directions distributing over the surface of a sphere, the area of which is calculated by the formula $4\pi r^2$. Thus the sound intensity is reduced as the square of the radial distance from the source; (b) Sound emanates from a point source located on the floor or other surface. The floor either absorbs or reflects the sound, but the resultant sound is still distributed over a hemisphere, the area of which is $2\pi r^2$. The squared-distance relationship still holds approximately.

Example 6.5

A worker's machine is located at a distance of 2 feet from the operator and produces a noise exposure of 95 dB to the operator. How much is to be gained by moving the operator to a position 4 feet from the machine? How much reduction could be achieved by a move to 8 feet?

Solution   A move from 2 feet to 4 feet is a doubling of distance and results in a 6-dB reduction in sound level. The resultant level would be

$$95 \text{ dB} - 6 \text{ dB} = 89 \text{ dB}$$

which would probably be within the 8-hour PEL of 90 dB even after considering reflections and other sources, if these sources are not very significant.

A move to 8 feet would be a second doubling, resulting in a reduction of another 6 dB to a resultant 83 dB, ignoring reflections and other sound sources. This would reduce noise exposures to less than the Action Level of 85 dBA for an 8-hour TWA.

It is not likely to be feasible to move operators away from their own machines, and even if it is, to move away from one machine may place the operator close to a neighboring machine. Distance factors work best in separating operators from noise arising from adjacent machines or other processes in the area. A general spreading out of the plant layout can be beneficial in this regard.

If spreading out the plant layout is infeasible or too costly, the installation of sound-absorbing barriers between stations can increase their virtual separation as far as noise is concerned. The gain to be achieved by such barriers is variable and complicated to estimate in advance. An acoustics expert is recommended for advice in this area, and even the expert is likely to experiment with various temporary barriers, measuring the "before" and "after" sound levels with each. Heavy materials absorb sound vibrations, a fact that makes curtains or shields containing lead a popular choice.

Sheet metal surfaces on machines are susceptible to mechanical vibrations and may act as sound-amplifying surfaces on machines. Metal gear contact in the drive mechanism can sometimes be eliminated by substitution of nylon gear wheels for metal gears or by the use of belt drives instead of gears. Simpler still would be a stepped-up preventive maintenance schedule to lubricate the gears more often, perhaps reducing noise levels. Useful principles for engineering controls to reduce noise are illustrated in Figures 6.21 to 6.26.

More expensive than any of the engineering control approaches discussed so far would be the isolation of the offending machine by means of an enclosure. The effectiveness depends on the type of material used to construct the enclosure and also depends to a surprising degree on the number and extent of openings or leaks in the enclosure. Figure 6.27 on p. 137 shows the relationship between opening size and loss of effectiveness for an example noise enclosure which has a capability of a 50-dB reduction if there are no leaks. Note that most of the effectiveness of the enclosure is lost if there is a hole in the enclosure wall of less than one-half of 1% of the total area of the enclosure.

Figure 6.21. Application of welded stiffeners to reduce vibrations in a sheet metal component. (From Arkansas Department of Labor, Ref. 43.)

Figure 6.22. Enlarge or reduce size of a part to eliminate vibration resonance. (From Arkansas Department of Labor, Ref. 43.)

Administrative Controls

It was stated earlier that either engineering or administrative controls are specified where feasible for excessive noise levels and that engineering controls are preferable. However, the administrative controls alternative was left unexplained. Administratively, management can schedule production runs so that noise levels are split between shifts and individual workers are not subjected to full-shift exposures.

Other tricks are to interrupt production runs with preventive maintenance to give workers quiet time. During normal shift breaks workers can be removed to a quiet rest area. Sometimes workers can share a loud job and a more quiet one by trading jobs at midshift. All of these practices can be used to bring down noise exposure levels to within the PEL for the given time of exposure as determined from Table 6.4. Since the term "administrative control" is somewhat vague, the term "work practices control" has become preferable to refer to the various methods of shifting employee exposures to comply with Table 6.4.

Hearing Protection and Conservation

Personal protective equipment is required when engineering and administrative controls both fail to reduce noise to legal levels. Specifically, hearing protection is to be provided to all employees exposed to the 85-dBA TWA Action Level. In addition to providing protection, employers must permit employees to select protectors from a variety of suitable types and must train employees in the proper use and care of the protectors. One aspect of proper use is fit, and employers must assure proper fit.

The actions described above are mandatory when Action Levels are exceeded, but the reader may have noted that the actual wearing of the protectors by the workers is not among the mandatory steps to be taken if Action Levels are exceeded.

Figure 6.23. Rubber cushions on both sides of vibrating sheet metal surfaces where they are joined. (From Arkansas Department of Labor, Ref. 43.)

Figure 6.24. Flexible section in rigid pipe isolates vibrations. (From Arkansas Department of Labor, Ref. 43.)

Figure 6.25.   Resilient floor. (From Arkansas Department of Labor, Ref. 43.)

(a)   (b)

Figure 6.26.   Power press improvements to reduce noise levels: (a) design die to result in a dull, crunching sound instead of a sharp bang; (b) replace the sharp impact action of a mechanical press with the relatively quiet pressure action of a hydraulic press.

However, there are conditions under which workers must actually wear the protectors. These are as follows:

1. Whenever worker exposures are greater than the PELs (see Table 6.4)
2. Whenever worker exposures are greater than the Action Level of 85 dBA (TWA) *and* the worker has experienced a "permanent significant threshold shift"

When testing shows a "threshold shift" in a worker's hearing, the implication is that the worker's hearing has been damaged and needs special protection. In these cases

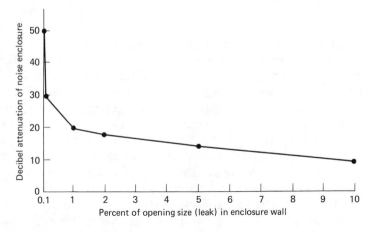

Figure 6.27. Loss in effectiveness of a noise enclosure due to leaks. For an example enclosure which has an ideal (airtight) noise reduction (attenuation) capability of 50 dB, leaks in the walls reduce the decibel attenuation by remarkable amounts as shown.

the noise levels are required to be reduced by the protectors to 85 dBA (TWA), not 90 dBA (TWA).

Personal protective equipment should not be considered a final solution because elimination of the source of the noise provides a more satisfactory work environment. Sometimes workers are lax in the wearing of personal protective equipment, and injurious exposures result. The careful selection and fitting of various types of personal hearing protective equipment is discussed in Chapter 8.

Whenever the Action Level of 85 dBA (8-hour TWA) is exceeded, a "continuing, effective hearing conservation program" should be administered, including audiometric testing, noise monitoring, calibration of equipment, training, warning signs for noisy areas, and recordkeeping of audiometric tests and equipment calibration.

Audiometric testing involves a small, perhaps portable, listening station in which a subject listens to recorded sounds and an audiologist measures the subject's hearing acuity at various frequencies. Audiometric testing can be very useful in determining sources of hearing loss, or more particularly in providing data to determine whether hearing loss is due to occupational or off-the-job exposure. If plant noise levels are high, it is foolhardy to hire new employees without first testing their "baseline" hearing acuity. Without any evidence of hearing deficiency at the time of employment, any hearing deficiencies that show up after employment appear strongly to be work related.

When observing the frequency profile of a worker's hearing acuity, the audiologist often looks for a "4000-Hz shift" for evidence of occupational exposure. Experience has shown that much industrial noise occurs in the 4000-Hz frequency range, which causes audiologists to suspect occupational exposure when hearing acuity becomes deficient in the 4000-Hz range.

RADIATION

A natural progression from the subject of noise is to the subject of radiation. Noise is, in fact, a form of radiant (wave) energy, but the term "radiation" is generally considered to mean electromagnetic radiation such as X-rays and gamma rays, or high-speed particles such as alpha particles, protons, and electrons. Federal standards separate radiation into the categories "ionizing" and "nonionizing."

Ionizing radiation is the more dangerous of the two types and is the type most associated with atomic energy. By far the most important category of ionizing radiation, from the occupational exposure standpoint, is the X-ray. X-rays are no longer the exclusive domain of the medical and dental professions. X-rays are being widely used in manufacturing operations, especially in inspection systems.

Nonionizing radiation is somewhat of a misnomer but applies to a more benign type of radiation in the electromagnetic spectrum. Included are radio and microwave frequencies. These phenomena are also finding increasing importance in industrial applications.

SUMMARY

Health hazards, although subtle, can be very serious. Worker health has historically taken a backseat to safety, but recent attention to long-term effects of worker exposures to toxic substances, noise, and radiation has led to rapid development of the field of industrial hygiene.

Threshold limit values for worker exposure to chronic health hazards are continually being studied and updated by the American Conference of Governmental Industrial Hygienists. TLV values do not necessarily become law, but federal standards have been strongly influenced by ACGIH's list of published TLVs, especially in the initial publication of permissible exposure levels. Many toxic substances and even noise have acute effects and should be considered safety hazards as well as health hazards. There are ceiling limits for many of these exposures, but some of the most common and dangerous substances have no ceiling values. Carbon monoxide is a notorious example, having a TLV for health considerations but no ceiling value for safety.

Fundamental to the assessment and control of health hazards and the industrial environment are instruments capable of measuring conditions. These instruments can be very sophisticated, and the underlying physics limit the accuracy that can be obtained. Calibration and care of the instruments are extremely important.

The Safety and Health Manager will probably find it beneficial to employ experts to take time-weighted averages with appropriate meters and sampling instruments. Subsequent control measures such as ventilation systems for toxic substances and acoustic panels for noise control may need the design capabilities of experts in their respective fields. Dealing with these experts demands a degree of understanding of their methods and terminology but does not require the Safety and Health Manager to duplicate the experts' capability in each field. Such an understanding of methods, terminology, and basic principles of occupational health and environmental control are what this chapter has attempted to provide.

EXERCISES AND STUDY QUESTIONS

6.1. Name at least five pneumoconioses. Which are most dangerous?

6.2. How do fibroses differ from other pneumoconioses?

6.3. Name the two basic classes of asphyxiants with examples of each.

6.4. What is the difference between depressants and systemic poisons? Name examples of each.

6.5. Explain the following terms.
  (a) mutagen
  (b) carcinogen
  (c) teratogen

6.6. In what ways are the threats of poisons different at work than at home?

6.7. What is the difference between "fumes" and "vapors"?

6.8. Compare the particle sizes of the following.
  (a) zinc oxide fumes
  (b) tobacco smoke
  (c) diameter of human hair
  (d) bacteria

6.9. Explain the following terms.
  (a) TLV
  (b) PEL
  (c) TWA
  (d) MAC
  (e) STEL
  (f) AL

6.10. Name some traditional methods of detecting the presence of dangerous air contaminants and explain advantages and disadvantages.

6.11. Name three basic approaches to measuring air contaminant exposures.

6.12. What are some desirable alternatives to industrial ventilation to remove air contaminants?

6.13. What is "makeup air"?

6.14. Ten machines all contribute equally to the noise exposure of one worker, whose exposure level is 99 dB for a full 8-hour shift. When all the machines are turned off, the noise level is 65 dB. How many of the 10 machines must be turned off to achieve a full-shift noise exposure level which would meet standards if the worker wears no personal protective equipment?

6.15. A worker stands on a factory floor, and a sound-level meter shows a reading of 55 dB at that point. A machine 3 feet away is turned on, and the meter jumps to 90 dB. What will the SLM read if the machine is moved to a point 12 feet away?

6.16. Suppose that an industrial process produces the following air contaminant concentration for the periods shown:

Period	Methanol (ppm)	Nitric oxide (ppm)	Sulfur dioxide (ppm)
8:00 A.M.–10:00 A.M.	50	5	0
10:00 A.M.–11:00 A.M.	150	10	1
11:00 A.M.–1:00 P.M.	100	5	1
1:00 P.M.–4:00 P.M.	200	10	1

Taken collectively, would these concentrations exceed permissible exposure levels?

6.17. In Question 6.16, suppose that the solvent ethanol could serve to replace the methanol in this process, but at the expense of double the concentrations in the atmosphere. Would this help or hinder matters? Explain.

6.18. An industrial process liberates 2 cubic feet of chlorobenzene per hour into a room which measures 20 feet × 40 feet and has a ceiling height of 12 feet. What minimum general exhaust ventilation in cubic feet per minute is necessary to prevent a general health hazard in this room?

6.19. In Question 6.18, what would be the effect of increasing the ceiling height to 16 feet?

6.20. A woman spills a bottle of 150 proof rum on the kitchen floor in her small one-bedroom apartment. The total apartment area is 600 square feet and has standard 8-foot ceilings. By the time she cleans up the spill, approximately 5 cubic feet of alcohol vapor has entered the air through evaporation. Noting the strong smell of alcohol in the air, she opens the window and, feeling drowsy, goes to bed and sleeps all night (8 hours). The open window permits a gradual dilution of the alcohol in the air, and by morning the concentration of alcohol in the air is down to 500 parts per million. Assuming a constant rate of decline in alcohol content of the air all night, was the alcohol concentration a hazard? If the exposure had been occupational, would PELs have been exceeded?

6.21. In Question 6.20, the woman's original intent was to bake rum cake. Had she gone ahead and made the cake, suppose that the hot oven had caused an additional 25 cubic feet of alcohol vapor to be liberated into the apartment. How would this concentration compare with the PEL?

6.22. A worker exposure to noise in a given plant is measured, resulting in the following readings for various time periods during the 8-hour shift:

8:00 A.M.– 9:00 A.M.	86 dBA
9:00 A.M.–11:00 A.M.	84 dBA
11:00 A.M.–noon	81 dBA
noon–1:00 P.M.	101 dBA
1:00 P.M.– 4:00 P.M.	75 dBA

(a) Perform computations to determine whether maximum PELs have been exceeded.
(b) Have Action Levels been exceeded?   85 or above
(c) Given the noise exposure described above, would the employer be required to furnish hearing protectors?
(d) Would employees be required to use the hearing protectors?
(e) Suppose that an engineering control could be devised that would cut the noise level (sound pressure level) in half either in the morning or in the afternoon, but not both. Which would you select? Why?

a) $D = 100 \left[\dfrac{1}{13.9} + \dfrac{2}{18.4} + \dfrac{1}{27.9} + \dfrac{1}{1.7} \right]$

$= 100 \, (\,.072 + .109 + .036 + .588)$

$\dot{=} \, 80.5 \%$

Maximum PEL has not been reached ✓

b) Action Level is 50%

80.5% exceeds the Action Level

c) Yes the employer should definitely furnish hearing protection.

d) Employees could or could not be required to wear hearing protection depending on the hearing protection program set up by the employer considering only the 101 db (A) really the employer should definitely required H.P.

e) Morning — it would make a lot more sense to cut in half the higher levels

7

Hazardous Materials

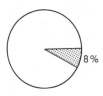 Percent of OSHA
general industry citations
addressing this subject

A simple listing of all known hazardous materials together with a concise summary of the characteristics of each would require a book larger than this textbook. Furthermore, additional hazardous materials are being developed or recognized daily. The focus of this text is on only a few materials, those for which individual national consensus standards already exist. Also, the concept of the term "hazardous" in this usage tends more to refer to hazards of fire and explosion, rather than to health hazards. Materials for which the hazard is primarily one of health were discussed in Chapter 6.

FLAMMABLE LIQUIDS

The most familiar hazardous substances are flammable liquids, and the new Safety and Health Manager might expect applicable standards for flammable liquids to be the easiest to learn and apply. But unfortunately, the standards are quite complicated as a natural result of the fact that flammable liquids occur so frequently in industry and in such widely varying quantities and applications. To illustrate this point, procedures for handling gasoline in a petroleum refinery where gasoline is *manufactured* are vastly different from procedures for storing and handling flammable liquids in an ordinary factory or office. Therefore, no simple set of rules for flammable liquids is appropriate.

Familiar as flammable liquids are, most people do not really understand many of the commonly used terms, such as "flashpoint," "Class I liquid," "flammable," "combustible," and "volatile." Much confusion also surrounds the sources of ignition for flammable liquids and the circumstances under which gasoline, for example, will burn, explode, or not burn at all. This chapter will therefore focus first upon definitions and principles of flammable liquids ignition and then will turn to discussion of some of the problems in complying with appropriate standards.

Perhaps the most basic term should be defined first, and this is the term "liquid." Almost everyone knows what a liquid is, but on the other hand, virtually every flammable substance exists both as a liquid or gas, depending on temperature or pressure. A good practical rule to follow is that if the substance is normally a liquid, it is defined as a liquid. Where one can get into trouble is in classifying propane and butane, which are gases and were not intended to be considered flammable liquids, although they can be liquefied. The NFPA standard definition of flammable liquid excludes propane and butane by excluding all "liquids" having a vapor pressure in excess of 40 pounds.

The term "flashpoint" is very important to the Safety and Health Manager because it is the principal basis for classification of flammable and combustible liquids. Therefore, it is primarily flashpoint that determines the amounts of the liquid that are permitted to be stored in various types of containers. Flashpoint is the point to which a flammable liquid must be heated so that it will give off sufficient vapor to create a flash at the surface of the liquid when there has been a spark or flame applied to it. Flashpoint is not the same as "firepoint"; firepoint is a higher temperature and is the temperature at which a fire on top of the liquid is sustained.

Three principal testing methods are in use for determining flashpoint. The Cleveland Open Cup Test is simple but not often used because it is intended for heavy oils. The most frequently used method is the Tag Closed Tester method. The term "Tag" is simply an abbreviation for a French name "Tagliabue." The third method is the Pensky-Martens Closed Tester method. This method utilizes a small stirring rod and is used for viscous liquids and liquids which form a film on the surface. This method is also more rare than the Tag test. The open cup method best simulates the in-plant situation where open vats are being utilized. The closed cup situation best simulates the situation for flammable liquids in storage.

Classification of flammable liquids also depends on "boiling point," but even this can be confusing because liquid does not normally boil at only a single point of temperature. This range is acknowledged in the standards by designating the "10% point" as the key. The "10% point" is the temperature at which 10 percent of the liquid has become gas. "IBP" refers to initial boiling point and is the temperature at which the first drop of liquid falls from the end of the distillation tube in a standard ASTM[1] Test Distillation.

"Volatility" refers to how readily a liquid will evaporate; it is closely related to boiling point. "Light" and "heavy" refer to high volatility and low volatility, respectively.

"Flammable liquid" is a term that means the same thing as "Class I liquid." However, such liquids are further divided into IA, IB, and IC. "Combustible

[1]American Society for Testing Materials.

liquid'' is the general term for both Class II and Class III liquids. The entire classi-fication scheme, which is based on flashpoint and boiling point, is explained in Figure 7.1.

Although it has become increasingly scarce, gasoline is the most widely used and plentiful flammable liquid. Because of its widespread use and because of horrible explosions and fires in its history, gasoline is blamed for a great deal of the fire hazard in the workplace and elsewhere. Some of the fears of gasoline hazards arise from ignorance rather than knowledge-based prudence, and this is detrimental to the cause of safety. In Chapter 3 it was explained that overzealous safety rules contribute to worker apathy for the rules and in turn work against the cause of safety rather than helping it. Operational and safety rules for working around gaso-line and other flammable liquids are sometimes in this unfortunate category. Ad-mittedly, gasoline can be extremely dangerous, but there is no substitute for know-ledge of the mechanism of its hazards so that the worker can intelligently take sensible precautions.

Knowledge begins with exposing the falsehood of the many myths surrounding the subject. Gasoline and other flammable liquids have their share of such myths, and this text will attempt here to dispel some of them.

Perhaps the wildest myth about gasoline is as follows:

FLAMMABLE LIQUID MYTH 1

A lighted cigarette when brought into contact with the surface of a container of gasoline is sure to ignite it.

To the contrary, it is almost impossible to ignite a tank of gasoline *at the surface* with a lighted cigarette. As with any ordinary fire, three ingredients are essential to support combustion:

1. Fuel
2. Oxygen (usually from the air)
3. Sufficient heat

There is plenty of fuel at the surface of a container of gasoline, but both of the other two ingredients are generally insufficient to support combustion. Concentration of gasoline vapors in excess of 7.6% are too rich and will not burn, and at the surface of gasoline standing in still air the concentration is much higher than 7.6%. Also, a glowing cigarette is not hot enough in most cases to permit ignition.[2] In fact, dramatic demonstrations have been performed in which a lighted cigarette is ex-tinguished by drowning it in a cup of gasoline! Incidentally, there are hazards in-volved in such demonstrations, and experimentation is not recommended. Things can go wrong such as tiny flame on the cigarette paper being hot enough for ignition. Also, there is the problem of getting the cigarette *through* the region in

---

[2]The ignition temperature for gasoline (536 to 853°F) is higher than for wood (approximately 400°F)!

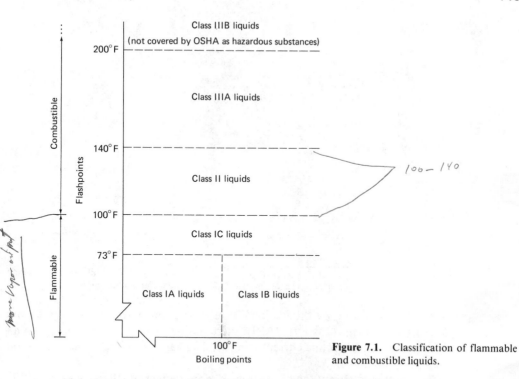

Figure 7.1. Classification of flammable and combustible liquids.

which the vapors are not too rich and causing ignition before the rich area near the surface can be reached. Also, small quantities of gasoline in the surrounding area can cause vapor/air mixtures just right for combustion. These are the reasons for "no smoking" rules around gasoline.

Gasoline has a burnability range of 1.4 to 7.6% gasoline vapors in dry air. Some other flammable liquids have wider burnability ranges and are more easily ignited than gasoline. Burnabilities for some commonly used highly flammable liquids are shown in Figure 7.2. Note that though gasoline is easier to ignite at "lean" concentrations than alcohol, alcohol will ignite at much richer concentrations. Also note the extremely broad and dangerous burnability range of carbon disulfide.

Another myth regarding gasoline has to do with service station fires and other fires around underground tanks.

FLAMMABLE LIQUID MYTH 2

Fires in underground gasoline tanks burn or explode with such intensity as to destroy much life and property around service stations.

Actually, fires do not burn in underground tanks of gasoline, even when a serious fire occurs above the ground. John A. Ainlay[3] states that in a study of 45 years of

[3]John A. Ainlay, Evanston, Illinois, is a nationally recognized authority on the chemistry of petroleum fires.

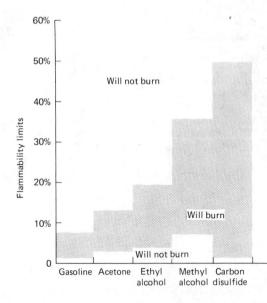

Figure 7.2. Burnability ranges of some popular flammable liquids.

petroleum fire reporting by the National Fire Protection Association (NFPA) and the American Petroleum Institute (API), there has never been a fire reported in an underground tank of gasoline currently in use. The vapor mixture in the tank is too rich to support combustion.

An abandoned empty tank is curiously more dangerous than a full or nearly full tank in use. The abandoned empty tank has had a chance to dry out, and vapors have gradually dispersed until the mixture may have become lean enough to result in an explosive mixture. The same is true of any *empty* gasoline drum. But a drum that is full or partially filled with gasoline will normally have a vapor/air ratio that is too rich to support combustion. Figure 7.2. shows that fire has much greater hazard potential inside drums of alcohol or carbon disulfide.

With aboveground tanks the hazard takes on a different dimension. The aboveground tank is exposed to intense heat and possible rupture during a service station fire. When a tank ruptures or explodes, tremendous quantities of fuel are suddenly added to the fire in the presence of abundant supplies of oxygen and heat.

The preceding paragraph describes ample reason for the standards that prohibit aboveground tanks for service stations, except when special conditions are met. There is, however, still another hazard for aboveground tanks. The vapor density[4] of gasoline is higher than 3 to 1. This means that unlike natural gas or other lighter-than-air materials, gasoline vapors will settle in low places. An aboveground tank is an invitation for gasoline vapors to settle in low areas of the service station, such as in service pits. The vapor density of gasoline is the reason basements in service stations are now illegal.

Despite the hazards just described, a large number of violations of the stan-

[4]"Vapor density" is the ratio of the weight of the vapor to the weight of the same volume of air.

dard prohibiting aboveground tanks have been found. Small independent service stations and private, in-plant service stations are the most frequent violators. Many of these installations were made prior to the writing of the standard, and the owners are asking: "Why didn't you tell me the rule before I built my station?" Some feel that such a standard was intended to act as a "building code" and apply to all future installations but not require a remodeling of all existing noncomplying stations.

Misconceptions about "octane rating" are at the root of the third myth.

FLAMMABLE LIQUID MYTH 3

High-octane "aviation gasoline" or "premium gasoline" is much more hazardous than regular gasoline.

Octane rating has to do with the preignition characteristics of gasoline within internal combustion engines and has nothing to do with fire safety. High-octane gasoline should be given the same fire precautions as regular gasoline—no more, no less.

SOURCES OF IGNITION

Dispelling myths about flammable liquids should make personnel more cautious. Understanding the chemistry of a petroleum fire gives explanation to some of the seemingly peculiar instances of nonoccurrence of petroleum fires. But at the same time, this understanding can highlight the extreme hazard involved when conditions are right for a fire. The seeming innocence of a broken light bulb can result in a disastrous fire. In the instant before a hot light bulb filament burns out after the bulb glass breaks, that filament is hot enough to ignite gasoline vapors. This is why it is important to protect light bulbs in the presence of flammable vapors as required in both the *National Electrical Code®* and further references in federal standards.

Another important ignition hazard for flammable vapors is welding sparks. The usual temptation is to speed up a repair operation, and welding is often commenced before sources of flammable vapors are removed from the area and existing vapors purged. Welding around flammable vapors has cost the lives of many inexperienced personnel who were not aware of the hazards involved.

Associated with welding is the grinding of the finished weld. Sparks generated are not to be trusted. It is true that many grinding sparks do not reach the required ignition temperature for gasoline vapors, but some sparks do reach ignition temperature, and spark-producing grinding should not be performed in the presence of flammable vapors.

The hazards of static electric discharge around flammable vapors are well publicized. After all, it is electrical discharge that ignites gasoline vapors with precise reliability in most internal combustion engines. Electrical arcing or static electric discharge is easily a source of ignition.

An understanding of the principles underlying the hazards with flammable liquids is useful when applying appropriate standards. Having covered some of the basic definitions and principles, we now turn to a direct analysis of the standards and helpful procedures to be followed by Safety and Health Managers to bring their facilities into compliance.

Federal codes for tank storage are quite complicated and are mostly the concern of layout designers of petroleum tank farms, bulk plants, dykes and drainage schemes, refineries, and service stations. The codes also cover such elements of design as tank construction and proper venting. Most Safety Managers do not need to concern themselves with mastering the details of tank construction. To the contrary, it is sufficient to know where to find the requirements and to alert designers and other planners that strict codes must be followed in specifying tanks for flammable liquids.

One criterion for distance requirements between tanks is whether the tank roof is fixed or floating. The general public does not usually realize that the roofs of many petroleum tanks rise and fall with the level of the liquid inside (see Figure 7.3). A tank with a fixed roof will not fill unless it is vented, and such venting causes a costly loss of vapors. But the Safety and Health Manager should understand that the floating roof also protects against the fire hazards of releasing vapors to the atmosphere. The vapor/air space inside an empty or nearly empty tank with a fixed roof is also more hazardous than for the floating roof tank, which has little or no vapor/air space. The floating roof tank is a dramatic example of an industrial improvement that saves production cost while promoting a safer workplace.[5]

To prevent ignition hazards, electrical interconnection is required between nozzle and container when dispensing Class I liquids. But this raises questions regarding dispensing Class I liquids into containers made of plastic or other non-

Figure 7.3. Two types of storage tanks: (a) floating roof type; (b) conventional vented type. The floating roof rises and falls with the level of the liquid inside the tank, eliminating the necessity of costly and dangerous venting.

[5]The disastrous effects of improper design of flammable liquid tanks are illustrated in Exercise 7.1 at the end of this chapter.

conductive material. It makes no sense to bond a plastic container to the nozzle because such a bond would be ineffective in equalizing the static charge (see Figure 7.4). The National Fire Protection Association recognized this fact when they exempted nonconducting containers from electrical bonding requirements in NFPA standards.

Some readers may wonder at all the fuss about static electricity during dispensing operations. But filling tanks or containers *generates* static electricity due to the flow of the liquid—a little known phenomenon regarding fluid flow. The rapid flow of carbon dioxide from a fire extinguisher in use can cause static electrical shocks which make the extinguisher very uncomfortable to hold. During the loading of fuel oil into tank trucks at night, an astonishing phenomenon has been observed in which "lightning displays" are seen flashing around inside the tank!

To prevent buildup of static electricity during loading operations, the flow should be kept as quiet and smooth as possible. Filters are big static generators and are best placed as far back in the line as possible, remote from the fill spout. Another measure that can be taken to reduce static electricity is to retard the flow. Also, "splash loading" should be avoided; that is, the fill spout should extend down to a point close to the bottom of the compartment to avoid excessive splashing, which generates static electricity. When loading multicompartment trucks, the front and rear compartments are the most likely to present problems from splash loading because of the arrangement of some types of loading apparatus. The reason for this can be seen in Figure 7.5. Another static eliminator is to place a "rest area" in the delivery line, as shown in Figure 7.6. The rest area is an expansion area in the pipe which allows static charge to bleed off the liquid before rapid flow continues.

"Switch loading" is a danger from the standpoint of ignition by static electricity. With gasoline, static electricity is not as serious a problem in loading because the vapor concentration is generally far too rich to permit ignition. Even when loading gasoline into a tank that previously contained fuel oil, the vapor concentration becomes too rich as soon as loading commences. The real danger is when fuel oil is loaded into a compartment that previously contained gasoline. This is called "switch loading" and is very hazardous. The vapor concentration in such an

Nonmetallic safety can
needs no bonding wire
during filling operation

Figure 7.4. Bonding not necessary for nonconducting container.

Figure 7.5. "Splash loading" occurs more often while loading front and rear compartments than while loading middle compartment.

operation is just right for ignition and a static discharge or any other source of ignition can result in an explosion that will rip the truck apart. Remedies for the problem when switch loading is necessary are to (1) fill the tank with carbon dioxide (the cardox method), (2) use vacuum cleaners to purge the tank of gasoline, or (3) cut the loading speed down to about 30% until the tank is about one-third full.

One curious provision of appropriate safety standards requires accurate *inventory records* for Class I liquid storage tanks. Inventory records are usually for accounting and cost control, so what business has a *safety* standard in dealing with accurate inventory records? The answer is that accurate inventory records can be used to detect dangerous leaks. Unfortunately, inventory discrepancies are often attributed to "clerical error" or "unexplained loss" and are simply ignored. After a serious fire has occurred, an investigating team sometimes probes the past inventory records only to discover that the evidence of a dangerous leak had been in the records for many days.

In summary, the Safety and Health Manager should approach the flammable liquids problem with a knowledge of the principles of ignition. Any new facilities should be designed and constructed with recognition of the hazards of flammable liquids. After construction and installation, commonsense rules for eliminating sources of ignition and preventing dangerous leaks should be established and enforced by the Safety and Health Manager. Training of personnel to understand the underlying principles of the hazards will help a great deal with the problem of safety around flammable liquids.

Figure 7.6. Static eliminator "rest area."

SPRAY FINISHING

A direct concern of Safety and Health Managers, especially in manufacturing plants, is the installation of facilities and proper procedures for spray painting areas or booths. This subject is a logical extension of the subject of flammable liquids covered previously because most spray painting materials are flammable. The subject is important not only from a compliance standpoint but also greatly affects insurance rates and basic insurability.

The construction and operation of a spray painting area that meets applicable code is a quite costly undertaking, and a weak Safety and Health Manager sometimes attempts to circumvent the rules to appease top management. The most common such circumvention of the rules is to call the paint facility a "small portable spraying apparatus not used repeatedly in the same location" and thus achieve exemption from spray painting standards. But if the small "temporary" setup becomes a more or less permanent arrangement, a continuing and serious fire hazard will result. Furthermore, neither the insurance company inspector nor an experienced government inspector will be fooled by a so-called temporary arrangement which has eventually become permanent, because spray paint residues will accumulate throughout the area in large quantities.

There are both health and safety considerations in spray finishing operations, but standards for spray finishing principally concern safety aspects, particularly fire. The most frequent violations of spray finishing standards are in the following categories:

- Improper wiring type for hazardous location
- Exhaust air filter deficiencies
- Cleaning and residue disposal
- Quantities of materials in storage
- Grounding of containers
- "No Smoking" signs

Also frequently violated but of somewhat less emphasis are the physical construction requirements for spray booths and the mechanical ventilation requirements.

Applying the rule of attacking the easiest problems first, the Safety and Health Manager should take steps to immmediately install No Smoking signs in spraying areas and in paint storage rooms. This advice may sound superficial, but thousands of firms have received OSHA citations simply for failure to post No Smoking signs. The cost of complying with this rule is almost negligible.

After assuring that No Smoking signs are installed, the Safety and Health Manager should investigate the wiring in the spray area to see if it conforms with *National Electrical Code®* specifications for hazardous areas. A competent electrician, knowledgeable in the provisions of the *National Electrical Code®*, is useful for this phase of the problem.

Appropriate wiring and electrical equipment classification in and around spray areas may be simplified somewhat in the decision diagram of Figure 7.7. A great deal of controversy has arisen regarding enforcement of electrical requirements around spray areas, particularly regarding interpretation of the legendary "20-foot distance" from the spray area. It is interesting to note that the *National Electrical Code®*, 1975 edition, reduced the required distance from 20 feet to 5 feet from the spray area.

Besides distance of travel, another issue is the *direction* of travel of the flammable vapors after they leave the paint spray booth. Since flammable vapors can travel in any direction, it is recommended that the Safety and Health Manager

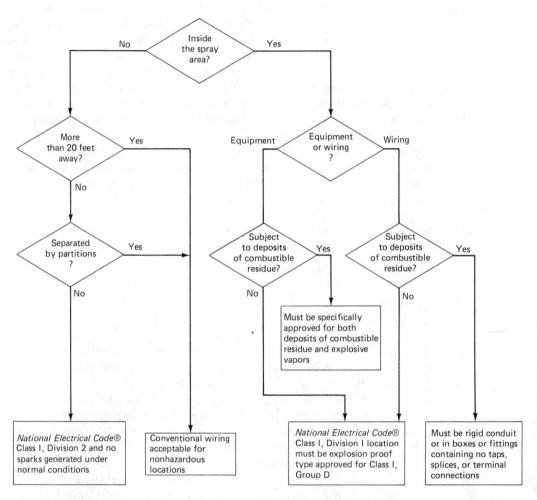

Figure 7.7. Decision diagram: How to follow OSHA standard 1910.107(c)(5) and (c)(6) specifications for spray area wiring and electrical equipment.

take the cautious rule of using Class I, Division 2 wiring in all directions from the open face of the spray area or booth, including vertically and around the corners alongside the booth. Additional guidance on the various classifications for electrical wiring is contained in Chapter 13.

Some people construe the standards to require automatic sprinkler systems for all paint spray areas, but the standards do not really say this. If an automatic sprinkler system is used, however, it must meet National Fire Protection Association (NFPA) requirements. The vendor who installs the sprinkler system should be required to assure that the system conforms to all applicable codes for the installation to which it will be applied. If the system is installed inside the ducts, sprinklers are needed on both sides of the filter system.

Combustible residues contribute to the largest proportion of spray booth fires. Reactions between different materials can add to this hazard, especially when peroxides are used. The control of overspray residues involves both engineering and administrative controls and is an item that deserves the attention of the Safety and Health Manager. Residue accumulation is easy to recognize and is an embarrassing testament to poor maintenance and lack of hazard control. Once removed, the residue and debris must be properly disposed to prevent spontaneous combustion and other fire hazards from oily rags and residue debris.

It was stated earlier that automatic sprinkler systems are not required for *all* areas. However, for fixed electrostatic systems automatic sprinklers are required "where this protection is available." Automatic sprinkler systems are preferred, and if such a system is already available close by (within approximately 50 feet), it should be extended to the electrostatic spraying area. In the absence of "available" automatic sprinkler systems, the standard requires "other approved automatic extinguishing equipment" for electrostatic spray areas. Such alternate systems would include fixed CO_2 or dry chemical systems, and these systems will be discussed further in Chapter 9.

The use of heating devices for drying in the spray area increases the hazard by raising the temperature of overspray residues and also by increasing the vapor level in the air. The standards are not very clear concerning the use of the spray area for a drying area also. The use of the spray area as a drying area is prohibited unless the arrangement does not "cause a material increase in the surface temperature of the spray booth, room, or enclosure."

DIP TANKS

Dip tanks often contain hazardous materials and are treated separately in federal standards. Care must be exercised when consulting the standard, however, because it applies only to those dip tanks containing flammable and combustible liquids. Plating dip tanks containing hazardous acids are *not* covered by the dip tank standard unless the acid is flammable or combustible.

Principal problems with dip tanks are:

- Automatic extinguishing facilities
- No Smoking signs
- Dip tank covers

The lack of dip tank covers is the most frequent violation of the three. One problem with covers is that the covers must be "kept closed when tanks are not in use." It is unreasonable to expect dip tank covers to be closed during short intervals of nonuse, such as coffee breaks and other short interruptions. However, an idle period of as long as one-half shift would be considered a "not in use" period. Also, if the dip tank is discovered to be idle for any period in which the work crew and supervisor have left the area, the dip tank should be considered "not in use." Automatic closure devices for actuation in event of fire are desirable but are not specifically required. Such closure devices "shall be actuated by approved automatic devices and shall also be arranged for manual operation." Automatically-closing dip tank covers are considered among the most appropriate means of automatic extinguishing facilities specified under conditions described in Figure 7.8.

Figure 7.8. Decision chart: Automatic extinguishing facilities for dip tanks.

EXPLOSIVES

Everyone knows that explosives are dangerous, and the general public completely avoids contact with them. Only the well-trained professional knows which procedures are safe and what to do in each situation. The body of existing codes governing explosives pertains almost entirely to storage or to the construction of the magazines in which the explosives are stored.

As in flammable liquids, explosives are classified according to degree of hazard. Class A is the most hazardous, and most of the materials that the general public considers "explosives," such as nitroglycerin, black powder, and dynamite, fall into the Class A category. Class B explosives include propellants, photographic flash powders, and some special fireworks. Class C explosives are manufactured articles which contain explosives in restricted quantities. The Safety and Health Manager should take advantage of manufacturers' labeling to determine explosive classes where a questionable item is concerned.

The storage magazines for explosives are divided into two groups, also called "Classes," but in the case of magazines the class designation is a Roman numeral instead of an alphabetic letter. The magazine class depends chiefly on quantity (weight) of explosives stored, not the explosives class. Class I magazines are for quantities more than 50 pounds, and Class II magazines are for 50 pounds or less.

Most facilities do not utilize explosives, and most Safety and Health Managers can ignore the standard. But those who do have explosives within their facility should take the precaution of assuring that handling and especially storage procedures are in compliance with applicable code.

LIQUEFIED PETROLEUM GAS

Liquefied petroleum gas (LPG) is a commonly used fuel gas, especially in areas remote from piped natural gas utilities. All petroleum gases can be liquefied if the temperature is lowered sufficiently, but natural gas, which is chiefly methane, is very difficult to liquefy, although it is otherwise cheaper than LPG. LPG is principally a mixture of propane and butane, both of which can be liquefied more easily than methane and therefore can be transported more compactly. The expansion ratio is approximately 1:270; that is, 1 gallon of liquid converts to 270 gallons of gas at normal temperature and pressure.

The exact mix between propane and butane is a matter of climate and economics. Butane has a higher boiling point and is unsuitable for cold climates because it will not convert to a gas at low temperatures. Butane has historically been cheaper, however, so it has been used in climates such as those in the southern United States. Recently, propane has been used almost exclusively because butane prices have increased due to butane's application in the manufacture of artificial fabrics.

Propane is a product of the refinery cracking[6] process and is odorless in the natural state. For safety's sake, an odorant (stenching agent) is applied before delivery to the customer. This stenching agent facilitates leak detection.

One of the safety hazards of propane is that it is heavier than air (approximately 1½ times the density of air). This contrasts with natural gas (methane), which is lighter than air. This difference in properties can lead to problems when one attempts to convert a piece of equipment from natural-gas powered to propane powered.

Another hazard with propane is that the extreme cold of its liquid state can burn the flesh. In fact, the treatment is the same as for third-degree burns. One way such a burn can occur is by opening the valve too quickly and by placing the hand over the valve to feel the flow. Injury from liquid propane while opening the valve does not indicate a defective valve.

Unlike gasoline and other atmospheric-vented tanks, LPG tanks are closed, and there is no opportunity for water vapor to accumulate inside the tank. Therefore, no "bleed-off" valve for moisture is used. If water should happen to be introduced into the system, it would probably freeze the valve during expansion of the gas. The closed tank of LPG then contains no air and is a mixture of liquid and gaseous propane. The vapor pressure inside the tank depends on temperature. At 0°F the pressure is 28 pounds per square inch (psi), and at 100°F the pressure is almost 200 psi. The pressure relief valve for tank trucks is set at 250 psi and for cylinders at 375 psi.

Although flesh burn from the extreme cold is a hazard to be considered, the principal hazard with LPG is fire, and when an LPG fire occurs, it is usually a disaster. The fire expands quickly, and portable fire extinguishers are generally useless, although they may be useful in extinguishing other burning materials that may be threatening LPG facilities. Once a tank itself ignites, however, it is strictly a matter for professional firefighters and special techniques using large volumes of high-pressure water spray to protect firefighters who must approach the tank to close valves or otherwise control the fire. Large tanks, such as railroad cars, have resulted in spectacular fires, including the phenomenon called "BLEVE" (rhymes with "heavy"), which means "boiling liquid expanding vapor explosion."

The federal standard requiring the use of laboratory-approved equipment ("listed" by an approved testing laboratory) seems like so much red tape. But without this requirement people will try all kinds of makeshift arrangements. One individual decided to use an old hot water tank for storage of LPG. The tank exploded, killing one person and injuring another. Another temptation is to use ordinary water hose instead of approved piping. The insidious nature of this hazard is that the water hose will often withstand the LPG pressures and will seemingly work, but the LPG will attack the rubber in the hose and will eventually result in a rupture. Similar problems result from using ordinary plumbing fittings and valves which have rubber seals.

---

[6]Large molecules are "cracked" into simpler molecules of more economically useful products.

Another frequent misuse of equipment is to interchange tanks for storage of anhydrous ammonia and LPG. Anhydrous ammonia attacks brass and copper fittings in LPG tanks, making them unsafe. Particularly dangerous is damage to the tank relief valve.

Fire, welding operations, and other sources of intense heat can weaken LPG cylinders and make them no longer capable of meeting laboratory tests. The Safety and Health Manager should be on the alert for this danger and have LPG equipment recertified after a plant fire or other exposure to heat. As for welding, no welding should be permitted directly to the tank shell; it is permissible, however, to weld to existing brackets, plates, and lugs, which in turn were already welded to the tank in its original manufacture in the configuration which has laboratory approval.

Fire control for LPG tanks is quite different from fire control for flammable liquid tanks. Dykes are built around flammable liquid tanks to contain the burning liquid in event of tank rupture. But such dykes are hazardous to LPG tanks because they can cause burning near or beneath the tank which might result in explosive rupture.

In any high-pressure cylinder with a valve on the end, there is danger should the valve be accidentally ruptured or even broken off. Like a torpedo in size and shape, the cylinder becomes a dangerous missile. The most dangerous are the very high pressure oxygen cylinders used in welding (see Chapter 12). However, LPG cylinders at 200 psi can also be quite dangerous. This danger is in addition to the fact that the gas liberated in the rupture can ignite explosively. Therefore, valves must be protected, and there are two acceptable methods for this: by recessing the valve into the tank or by a ventilated cap or collar.

Safety and Health Managers often misinterpret the provision of the federal standard for LPG, which reads as follows:

> Engines on vehicles shall be shut down while fueling if the fueling operation involves venting to the atmosphere.

In-plant refueling of forklift trucks that operate on LPG normally does not involve venting to the atmosphere. It is not required that the engines be shut down in such refueling operations.

CONCLUSION

A final suggestion for Safety and Health Managers is to take advantage of community resources, public and private, for advice and assistance in dealing with hazardous materials. Local fire departments and state fire marshals can be of assistance, particularly with regard to flammable liquids, spray finishing, and codes. Some fire or police departments have explosives experts on hand. Compressed gases, LPG, and anhydrous ammonia problems can sometimes be alleviated by consulting the local distributor for these materials. Some of these distributors are backstopped by extensive training resource centers at their company headquarters and can supply booklets, guides, warning labels, films, and slide-tape programs for in-house training to combat hazards.

EXERCISES AND STUDY QUESTIONS

7.1. Figure 7.9 shows a storage tank arrangement for storing tetrahydrofuran on the second-story level inside a building located in Chicago. The first day the tanks were loaded, the following events took place. A tanker delivery truck hooked up at the receiving valve at ground level and proceeded to deliver 500 gallons. About halfway through the filling operation a company employee inside the building yelled out the window saying that the tank was overflowing. Soon thereafter a tremendous explosion and fire killed both the company employee and the tank truck driver. Study the circumstances and identify potential causes of this terrible accident. What is the flash-point of tetrahydrofuran? $6°F$ $(51°F$

7.2. Compare the flammability hazards of gasoline with ethyl alcohol.

7.3. A manufacturing process utilizes the powerful solvent acetone. One phase of the process is to dry out the acetone in a "drying area." The drying area evaporates 2 gallons of acetone per hour. Every gallon of liquid acetone produces 41 cubic feet of vapor. How much ventilation (ft³/hr) would be needed to maintain vapor concentration below the burnable level? How many times per hour would the ventilation system exhaust a room 9 feet x 12 feet with a ceiling height of 10 feet?

7.4. In a one-time inspection of a facility, how will an insurance representative be able to determine that a temporary spraying area or a so-called "touch-up area" is really a more permanent arrangement?

7.5. Under what circumstances are automatic sprinkler systems required for paint spraying areas?

Figure 7.9. Tank configuration.

7.6.   When are dip tank covers required to be closed? Must they close automatically in event of fire?

7.7.   Would it be proper to use a Class II magazine for storage of Class A explosives? Explain.

7.8.   Compare liquefied petroleum gas (LPG) with natural gas in terms of safety hazards.

7.9.   Are fire extinguishers appropriate for LPG facilities? Why or why not?

8

Personal Protection and First Aid

Percent of OSHA general industry citations addressing this subject

In a way it is unfortunate that this chapter is even necessary. The need for personal protection implies that the hazard has not been eliminated or controlled. And the need for first aid implies something even worse! When feasible, engineering control of the hazard is preferred over the use of personal protective equipment. As discussed in Chapter 1, we realize that there will always be some risks remaining, but our goal is to eliminate *unreasonable* risks, not all risks. The job of improving safety and health in the workplace will never be completely finished, so we must concern ourselves with the need to provide personal protection against hazards which have not been completely eliminated and to provide first aid when an accident does occur.

The problem of providing personal protective equipment seems straightforward and easy enough to understand. But the simplicity of the problem is an illusion, and many industrial Safety and Health Managers fall into its trap. For instance, it would seem that if the noise level in the production area is too high, the solution to the problem would be to provide ear plugs for the workers. But anyone who has actually confronted this problem knows that the solution is not that simple. For a variety of reasons, most people do not want to wear ear protection. They may be shy about the appearance of ear protection equipment, they may feel discomfort or perhaps even pain, they may feel it interferes with their necessary hearing or efficiency, or they may feel that use of personal protective equipment is their own business, not their employer's. The bases for these complaints and what to do about them will be taken up later in this chapter, but first to be covered are the hazards for which personal protection may be needed and the types of equipment available to meet these needs.

The matter of personal protective equipment becomes very delicate when employees bring their own equipment to work. If the equipment is not properly maintained, who is responsible—employer or employee? As Safety and Health Manager, consider the following logic. If employees bring their own personal protective equipment to work, is it not possible that the equipment itself could represent a hazard? Personal protective equipment must be properly selected to match the hazard, and employees bringing their own equipment might falsely think they are safe when their equipment could actually be malfunctioning or even be of the wrong type.

Even if the employee works in a job that does not need personal protective equipment, the use of faulty or inadequate equipment might tempt the worker to take chances. For instance, suppose that a maintenance worker who has no need to even approach the edge of a rooftop while maintaining air-conditioning equipment wears his or her own personal protective rope attached to his or her leather belt as added protection, not required by company policy. Then suppose that some unusual situation tempts the employee to approach the edge of the building and the false sense of security from the improvised "safety belt and lanyard" causes the worker to proceed without caution. An accidental fall could cost the worker's life when the leather belt breaks under a 2000-pound shock load due to the fall. Alternatively, even if the leather belt withstands the shock load, the worker could be killed by the shock itself if the lanyard is too long and is nonelastic. These are facts workers might not know, and when they improvise or bring their own personal protective equipment to work, the employer should assure that the equipment is adequate for the situation and properly maintained.

HEARING PROTECTION

As might be expected, the greatest emphasis in providing personal protective equipment coincides with the principal problem in environmental control, covered in Chapter 6—the problem of noise. If engineering or administrative control measures are unsuccessful in eliminating the hazard of noise in the workplace, management must turn to personal protective equipment to shield the worker from exposure.

The most important factor in selection of the type of noise protection is probably effectiveness in reducing decibel level of noise exposure. However, this is by no means the only important factor, and selection can be somewhat complicated. Economics is always a factor, and if limited effectiveness is all that is necessary in a given situation, cheaper devices can be selected. Employee comfort is probably at least as important a factor as economics. The worker comfort factor goes beyond the simple goal of promoting worker satisfaction; it affects the amount of protection the worker will receive. If workers find a type of ear protection uncomfortable or awkward to wear, they will use every excuse not to wear it, with a resulting loss in protection. The following paragraphs will consider the merits of various types of ear protection.

Cotton Balls

Ordinary cotton balls, without addition of a sealing material, are virtually worthless as a means of personal protection from noise.

Swedish Wool

Similar in feel to cotton, "Swedish wool" is a mineral fiber which has much better attenuation values than cotton. Swedish wool is somewhat effective as is, but is much more effective when impregnated with wax to make a better seal. One problem with Swedish wool is that it can tear when it is pulled out. To alleviate this problem, Swedish wool sometimes comes in a small plastic wrapper which is inserted with the wool. Swedish wool can be considered only fairly reusable; reuse will depend on personal hygiene, quantity of earwax, and worker preference.

Ear Plugs

The most popular type of personal protection for hearing is the inexpensive rubber, plastic, or foam ear plug. Ear plugs are practical from the standpoint of being easily cleaned and reusable. Workers often prefer ear plugs because they are not as visible as muffs or other devices worn external to the ear. But within this advantage lies a pitfall: workers may be more complacent about using the ear plugs when it is not immediately obvious to the supervisor whether the ear plugs are being worn. The noise attenuation for properly fitted ear plugs is fairly good, falling somewhere between that of Swedish wool and the more effective earmuffs.

Molded Ear Caps

Some ear protectors form the seal on the external portion of the ear by means of a mold to conform to the external ear and a small plug. Since human ear shapes vary so widely, fit is a problem. Molded ear caps are more visible than ear plugs, which has both advantages and disadvantages, as discussed earlier. Molded ear caps may be more comfortable to the wearer, but are more expensive than ear plugs.

Earmuffs

Earmuffs are larger, generally more expensive, and more conspicuous than Swedish wool, plugs, or caps, but they can have considerably better attenuation properties. The attenuation capability depends on design, and more variety in design is possible with earmuffs. Although some workers object to wearing conspicuous earmuffs, it is interesting to note that some workers prefer them, stating that they are more comfortable than ear plugs.

Helmets

The most severe noise exposure problems may force the Safety and Health Manager to consider helmets for personal protection against noise. Helmets are capable not only of sealing the ear from noise, but also of shielding the skull bone structure from sound vibrations which can be transmitted to the sense of hearing. Helmets are the most expensive form of hearing protection but have the potential of offering protection from a combination of hazards. Properly designed, the helmet can act as a hardhat and a hearing protector at the same time.

It must be remembered that fit is very important for all types of hearing protectors. As in noise enclosures or sound barriers, the material itself might have excellent sound attenuation properties, but if there is a leak or crack, most of the effectiveness of the device is lost.

EYE AND FACE PROTECTION

The use of safety glasses has become so widespread and so many different styles are now available that many Safety and Health Managers establish a rule that they must be worn throughout the plant. A general custom in industry is to require visitors to wear safety glasses during plant tours.

There is a difference between "street safety" glasses and "industrial safety" glasses. Visitors or employees who claim their prescription glasses are "safety glasses" probably mean that they have "street safety" lenses. Industrial safety lenses must pass much more severe tests to meet ANSI standards. This is not to say that street safety lenses are not adequate for some industrial environments. Eye protection standards are not specific about which jobs do and which do not require safety lenses—street or industrial. It is good to have exacting standards to be sure that safety glasses will meet uniform performance standards. But responsibility for deciding when eye protection equipment is necessary usually falls on the Safety and Health Manager—not the optics industry, not the standards, and not enforcement agencies.

A word of caution is in order for the Safety and Health Manager who is setting forth a policy on the wearing of safety glasses. It can be as bad a mistake to require safety glasses in those areas of the plant where there is no eye hazard as it is to not require safety glasses in areas where they are needed. The peril is that workers will not respect the safety glasses policy, and use of the glasses will not be uniform. Eye injuries can be the result—in addition to violations of code. It is easy for inspectors to establish violations when they observe workers not wearing eye protection in the presence of an established company rule requiring eye protection. But some Safety and Health Managers say that a simple plantwide rule is easiest to enforce. The costs on both sides of the question are great, and the decision to require eye protection must be made with care.

There are some jobs for which industry and enforcement agencies alike have seemed to come to a consensus on the need for eye protection. Machining operations that produce chips or sparks are almost universally conceded to necessitate the use of eye protection. Notable among these operations are those of grinding machines, drill presses, and lathes. Both metal and wood materials can produce dangerous eye hazards when machined. Corrosive liquids or other dangerous chemicals also represent eye hazards when poured, brushed, or otherwise handled in the open. When working with such materials, face protection may be needed in addition to eye protection.

More important than when to *require* eye protection is how to *educate* workers to be alert to eye hazards and the long-term consequences of eye injury. The National Safety Council has some films that bring this point home. One effective film shows emergency eye surgery in progress. Another is a poignant personal testimony of a man who was blinded on the job and the hardships suffered by him and his family ever since.

RESPIRATORY PROTECTION

Of even more vital importance (in the literal sense of the word "vital") than eye and hearing protection is the need for respiratory protection from airborne contaminants. Chapter 6 discussed the different types of problems with industrial atmospheres, and determination of the type of atmospheric problem is essential in selecting the correct respiratory equipment. A well-designed and expensive gas mask is useless and might be more properly designated a "death mask" if the atmospheric problem to be tackled is oxygen deficiency, for example.

Particularly hazardous atmospheres may be referred to as IDL or IDLH, which stand for "Immediately Dangerous to Life" and "Immediately Dangerous to Life and Health," respectively. Recently, the term IDLH has become more widely used. If a single acute exposure is expected to result in death, the atmosphere is said to be "IDL." If a single acute exposure is expected to result in *irreversible* damage to health, the atmosphere is said to be "IDLH."

It should be apparent to the reader at this point that there is more to respiratory protection than simply handing out respirators to workers who might be exposed to hazards. Effective protection demands that a well-planned program be implemented, including proper selection of the respirators, fit testing, regular maintenance, and employee training.

Some firms supply respirators to employees without bothering with a comprehensive respirator program, relying on the excuse that the respirators are not really required anyway because contaminants in the plant atmosphere do not exceed maximum permissible exposure levels (PELs). But the Safety and Health Manager is really asking for trouble, though, by toying with a partial program. The atmospheres are no doubt marginal, or the question of a partial program would never have arisen. The marginal atmospheres might later deteriorate without warning. Employees would be lulled into a false sense of security by the superficial respirator program. Bad habits, such as negligent maintenance, inadequate fit testing, or improper equipment usage, could be developed. A whole feeling of complacency toward the use of respirators can be engendered by the use of such equipment when it is not really needed.

As with general personal protective equipment discussed earlier, the Safety and Health Manager is often placed on the spot when an employee brings his own respiratory protection equipment onto the job site. In this situation the employer should take a responsible position and assure that the employee uses his respiratory equipment properly, even if it is employee-owned. If the employee becomes belligerent due to the fact that he owns the equipment and feels it is his own business how he uses it, he should be reminded of the employer's continuing responsibility to eliminate hazards in the workplace, including misuse of personal protective equipment. If the equipment or its improper use can be hazardous, the employer can forbid the employee from bringing his own equipment onto the premises. The reader may feel some hesitation in exercising such authority regarding the employee's own personal property, but instances have already occurred where the employer has exercised such authority when necessary to prevent hazards.

Before going any further into the subject of respiratory protection, a classification of the various devices is in order. The two major classifications are "air-purifying" devices versus "atmosphere-supplying" devices. The air-purifying devices are generally cheaper, less cumbersome to operate, and the best alternative if they are capable of handling the particular agent to which the user will be exposed. But some materials simply cannot be reduced to safe levels by air-purifying devices, and a supplied air system is required. Another important consideration is oxygen deficiency. No amount of filtering or purifying is going to make an oxygen-deficient atmosphere safe. The only way to go in this situation is with atmosphere-supplying respirators. A summary classification of respiratory protection devices is as follows:

1. Air-purifying devices
   a. Dust mask
   b. Quarter mask
   c. Half mask
   d. Full-face mask
   e. Gas mask
   f. Mouthpiece respirator
2. Atmosphere-supplying respirators
   a. Air line respirator
   b. Hose mask
   c. Self-contained breathing apparatus

Further description of each type of device follows.

Dust Mask

The most popular respirator of all is also the most misused. Approved only for particulates (suspended solids), the dust mask (Figure 8.1) is not approved for most painting and welding hazards, although it is frequently seen used improperly in this way. Some dust masks are approved for mild[1] systemic poisons, but generally these masks are limited to "irritant dusts," those that produce pneumoconiosis or fibrosis (see Chapter 6). One of the principal limitations of the dust mask is fit. Even the best-fitting models have approximately 20% leakage. A rule of thumb is that approval is valid for particulates no more toxic than lead.

Despite its disadvantages, the dust mask is popular because it is inexpensive, sanitary, and can be thrown away after each use. Its low cost and general availability make the dust mask attractive for purchase at the local drug store and for personal use. Therefore this is the type of respirator the Safety and Health Manager is most likely to find employees bringing from home for use on the job. Care should be taken to train employees in the limitations of the dust mask.

Quarter Mask

The quarter mask, sometimes called the Type B half mask, is shown in Figure 8.2. The quarter mask looks very much like a half mask except that the chin does not go inside the mask. The quarter mask is better than the dust mask, but it, too, is generally approved for toxic dusts no more toxic than lead.

Half Mask

The half mask, shown in Figure 8.3, fits underneath the chin and to the bridge of the nose. This mask must have four suspension points, two on each side of the mask, connected to rubber or elastic about the head.

Figure 8.1.   Disposable "dust" mask.

[1]Dusts that have a TLV of not less than 0.05 mg/m^3.

Figure 8.2.   Quarter mask.

Figure 8.3.   Half mask.

Full-Face Mask

Actually, the gas mask is also "full face," but the name "full-face mask" generally refers to a mask in which the filtering chamber attaches directly to the chin area of the mask. The filters may be either dual "cartridges" or single "canisters." Both types are shown in Figure 8.4. Canisters contain granular "sorbents" which filter the air by adsorption, absorption, or chemical reaction.

(a)   (b)

Figure 8.4. Full-face masks: (a) dual-cartridge full-face mask; (b) canister-type full-face mask.

Gas Mask

The gas mask is designed for filtering canisters that are too large or too heavy to hang directly from the chin. In the gas mask, the canister is suspended by its own harness and is typically connected to the face mask by a corrugated flexible breathing tube. The gas mask is shown in Figure 8.5.

Mouthpiece Respirator

Perhaps the mouthpiece respirator should be omitted from this discussion because this type of device is not intended for normal use. But emergencies will occur from time to time, and being prepared to escape when emergency does occur is the purpose of the mouthpiece respirator. Breathing is accomplished through the mouth by means of a stem held inside the teeth. A nose clip must be used to prevent inhaling through the nose. It is possible to form a good seal with the mouth and lips, but the effectiveness of the mouthpiece respirator is greatly dependent on the knowledge and skill of the user.

Figure 8.5. Gas mask.

Air Line Respirator

The air line respirator is an atmosphere-supplying respirator and derives its name from the way in which air is supplied to the respirator mask. The air is supplied to the mask by a small-diameter hose (not over 300 feet long) which is approved together with the mask. Ordinary garden hose is not acceptable. The air is supplied by either cylinders or compressors. The method of delivery of air to the user results in three different modes of air line respirator: "continuous flow," "demand flow," and "pressure demand" mode.

In the continuous flow mode, the air line respirator receives fresh air without any action on the part of the user, that is, the flow is forced by the apparatus. The airflow must be at least 6 cubic feet per minute to qualify a hood for use with this mode, but the flow should not be much greater because it can be responsible for a high noise level inside the hood. One of the advantages of the continuous flow mode, however, is that it permits use of a somewhat leaky, loose-fitting hood. The positive pressure differential between the inside and outside of the hood keeps the flow outward, preventing toxic agent entry. The continuous flow mode needs an unlimited supply of air, so a compressor is used instead of tanks.

In the demand flow mode, air does not flow until a valve opens, caused by a negative pressure created when the user inhales. Exhalation, in turn, closes the valve. This mode has the advantage of using less air, so it is feasible with cylinders. But the disadvantage is the need for a tight-fitting facepiece. Since inhalation causes a negative pressure differential to develop, a leaky facepiece will readily draw in the toxic ambient atmosphere. In fact, if the facepiece is too leaky, the inhalation valve will fail to open, making use of the facepiece even more hazardous than using no personal protective equipment at all. For this reason the demand flow mode is becoming obsolete and is being replaced by the third mode: pressure demand.

The pressure demand mode has features of both the continuous flow and demand flow modes. As in the continuous flow, a positive pressure differential is maintained. The positive pressure is maintained by a preset exhalation valve. Despite its advantages, the demand flow mode still requires a good-fitting mask; use by a person with a beard is not acceptable.

Hose Mask

A hose mask is a somewhat crude form of an air line respirator. The diameter of the hose is larger than in the air line respirator, permitting air to be inhaled by ordinary lung power. A blower is sometimes used as an assist. The hose mask is declining in popularity.

Self-Contained Breathing Apparatus

In this type of respiratory protection, the user carries all of the apparatus with him, usually on his back. This has the advantage of increasing the distance the user can roam because there is no umbilical cord to drag along and perhaps sever or crush. But a disadvantage is that the large pack on the back may restrict the passage of the user through a manhole or other close passage. Engineers should consider this problem when designing manholes for vessel entry when toxic atmospheres may be a problem. Many multiple fatalities have occurred when rescue breathing packs were rendered useless because the rescuer was unable to enter the vessel with a breathing pack on his back.

Most self-contained breathing apparatus units at the present time are "open circuit," that is, the exhaled breath is discharged to the atmosphere (see Figure 8.6). "Closed-circuit" units recycle the exhaled breath, restoring oxygen levels. The advantage of the closed-circuit types is that the pack can be much smaller and lighter per minute of maximum permissible use. Some closed-circuit types (see Figure 8.7)

Figure 8.6. Open-circuit self-contained breathing apparatus.

Figure 8.7. Closed-circuit self-contained breathing apparatus; oxygen bottle type. (Courtesy of Rexnord Electronic Products.)

have a small high-pressure oxygen tank for restoring oxygen levels after carbon dioxide is removed. Other closed-circuit types utilize a chemical reaction to restore oxygen levels, making possible a very compact unit. The source of the oxygen is a superoxide of potassium in which oxygen is liberated by mere contact with water. The water is supplied by the moisture in the user's exhaled breath. It takes a little while for the chemical process to function and become balanced, so the user should prime the system while still in clean air before entering the hazardous zone. A hazard with the chemical oxygen-generating unit is that the potassium superoxide must be kept sealed from moisture, except for the small amount of moisture in exhaled breath. A water flooding of the potassium superoxide unit's interior is almost sure to cause an explosion. Another hazard with the chemical oxygen-generating unit is that it supplies an oxygen-rich atmosphere, and this can be a fire hazard.

Respirator Plan

The long list of available types of respiratory protection may bewilder the Safety and Health Manager who must choose which type of protection is optimum for each application. Expert consultants can help in this situation, but one simplifying factor is that usually the hazard dictates the choice of device or at least greatly narrows the field of choice. Devices are approved for particular concentrations of particular substances, and unapproved devices must not be used.[2] Consultants can help to select between approved respirators, considering efficiency, cost, convenience, and other factors.

Sometimes the excuse is offered that no approved device exists for a particular toxic substance. But this excuse is a poor one because the worker must be protected from dangerous atmospheres, some way or somehow. There are many different kinds of respiratory equipment just as there are many different types of respiratory hazards. It may be true that there is no approved device of a particular type for a given hazard. The employer may want to use some type of respirator, for instance, for a hazard in which there is no approved respirator as such, but another type of device *could* be used. An example would be the hazard of mercury vapor. No air-purifying device is acceptable for protection against the hazards of mercury vapor. Mercury vapor offers no warning odor or other means to detect when the canister is no longer effective. So with mercury vapor the employer must turn to another type of protection, such as air line breathing equipment. Thus for every hazardous atmosphere in which an employee must work, there must be an approved device of *some type* to protect the worker. If there is no approved device for the given hazardous contaminant, workers must be prohibited from the area.

Determining the approval status of a device for a particular respiratory hazard is no easy task. The approval authority is complicated and has frequently changed from agency to agency. Furthermore, the equipment approval lists are continually being updated. A practical solution to the problem is to seek the advice of the manufacturer of the equipment, for the obvious reason that the equipment manufacturers are likely to be most aware of the approval status of their own equipment. Not as worthy of trust is the local equipment distributor. With no reflection on the character of the local equipment distributor, such persons often carry so many lines of safety equipment that it is impossible for them to be as intimately aware of the various toxic substance approval limitations of each respiratory protection device as will be the manufacturer of the given device.

Complicating the situation is the misleading nomenclature. For instance, "organic vapor respirators" are not necessarily acceptable for commonly encountered organic vapors. Methanol, for instance, is not readily absorbed by any cartridge, and only air line equipment is effective for protection against hazardous concentrations of methanol. It is unfortunate that federal regulations require a respirator to be identified as "organic vapor respirator" because it has passed a certain prescribed test, even though the respirator may be useless for certain organic vapors. Workers are not even protected by warning labels that their "organic vapor respirators" may not be as effective for the organic vapors with which they are working. The same is true for gas masks.

[2]For approval of respirator devices, see *NIOSH Certified Equipment*, Cumulative Supplement, (NIOSH 77-195), June 1977, and subsequent issues available periodically from NIOSH Laboratories in Morgantown, West Virginia.

Why does such a confusing and misleading system exist? Part of the problem is the complexity of organic chemistry and the myriad hydrocarbon compounds that exist. If the manufacturer attempted to label each cartridge with all the organic compounds for which it was effective, there would soon be no space on the cartridge. Any kind of shorthand coding system might confuse users in the field. Furthermore, an attempt to itemize would imply that the list was exhaustive, and the user might check no further to see whether a particular cartridge might be effective against a given substance. So the best course of action again is to check with the manufacturer for more complete and detailed lists of substances for which the organic vapor respirator cartridge is effective.

A basic principle of respirator selection is never to select a gas-absorbing respirator to be used for a gas that has no distinguishing warning properties. A moment's reflection will reveal the logic behind this principle. All respirator cartridges eventually become saturated or loaded to the extent that they no longer are effective. The users will know automatically when that point is reached through their own senses of smell, taste, or perhaps irritation or whatever the sensory warning property of the dangerous gas may be. But if the toxic gas has no warning property, the users will never know and may be subjected to dangerous exposures while wearing an approved respirator.

A good example of such misuse is the typical roofing company use of half-mask respirators when applying expandable foam. But unfortunately, the organic substances present in expandable foam do not have adequate warning properties, and there is no air-purifying device approved for such use. Therefore, air line respirators should be used.

There is one exception to the rule that air-purifying devices cannot be used with gases that do not have warning properties capable of being sensed by the user. That exception is where canisters can be equipped with an effective end-of-service-life indicator. Currently, such canisters are available only for carbon monoxide (Type N canisters).

"Paint, lacquer, and enamels" cartridges offer protection from the particulates and also from the organic solvents in the paint. Actually, the organic vapors represent the real hazard with paint, lacquer, and enamels respirators. Except for lead-based paint, the particulates in paint are not really a problem.

Respirator cartridges are not like vacuum cleaner bags; they cannot be changed from manufacturer to manufacturer without approval. The cartridge and the respirator must be approved together as a unit, although it would be possible for a respirator of one make and a cartridge of another make to be approved together if the specific combination was subjected to approval testing. Public controversy over cartridge refill approvals has charged that manufacturers have monopolistically promoted safety regulations to protect sales of their own products. On the other side of the issue is the argument that cartridge fit and proper seal are critical to the effective operation of the respirator.

For jobs requiring the use of respirators a determination should be made to be sure that persons are "physically able to perform the work and use the equipment.

The local physician shall determine what health and physical conditions are pertinent.'' This sounds costly and time consuming, but there is a shortcut that can expedite the personnel screening process, eliminating a large majority of the problem cases. The shortcut is to use a questionnaire which can identify obvious problems before employment.

A variety of physical conditions makes a person unfit for respirator use, whereas that person might be quite well suited to a job that does not require the use of respirators. The U.S. Atomic Energy Commission lists some of these physical conditions which should be included in a questionnaire for screening workers from jobs requiring the use of respirators. Asthma or emphysema are pulmonary problems which may result in problems with respirators. If the environment requires self-contained breathing apparatus, the potential employee may not be fit for carrying the heavy equipment due to a back injury history or heart disease. Another physical problem is ruptured eardrums. A person with a ruptured eardrum actually breathes through his or her ears. Thus a face mask respirator that does not cover the ears may offer little in the way of protection. Epilepsy is another potential problem. Epileptics are often on medications, and these medications can present some hazards. Also, the possibility of a seizure presents the threat of the respirator being removed at the wrong time. Diabetics may also be on medications, and the use of respirators may aggravate such medications. These individuals, for instance, would probably be a poor choice for a rescue team. Even psychological factors such as claustrophobia may be important. The employee questionnaire can be a source of protection from future liability in case physical problems are present. The questionnaire should be developed with the assistance of a physician. If physical problems or variations are revealed by the preliminary questionnaire, the employee should be examined by the physician for a final determination.

An important element of an effective respirator program is fit testing or leak testing. Facial characteristics vary significantly, and fit tests are necessary to determine which particular model fits each worker best without appreciable leaking. Even the best respirators on the market will fit only a percentage of workers. Other respirators must be selected for other facial shapes.

One facial shape that is impossible to fit is the bearded one. A beard should be prohibited for any worker who must wear a fitted respirator for his safety and health. Two firemen in Los Angeles filed a discrimination suit because they were laid off because they wore beards. Their job required a possible use of self-contained breathing apparatus. The firemen lost their case. The argument against facial hair on jobs requiring respirators is well documented.

Respirator fit testing seems to imply expensive environmental chambers, but such chambers are not required. Some consultants recommend a simple plastic clothes bag suspended over the wearer's head. The subject's respirator should be equipped with an organic vapor cartridge, and the air inside the bag is contaminated with isoamyl acetate (available from the local pharmacy). If the subject smells the familiar ''banana oil'' odor, his or her respirator is leaking. Once fit-tested for a particular make and model of respirator, every respirator of the same make and model is acceptable for that user.

One ironic development in the field of respiratory protection is NIOSH's discovery of evidence suggesting that the substance di-2-ethylhexyl phthalate (DEHP) is a potential carcinogen. The irony is that DEHP had been used as a test agent for determining the fit of respirators!

For air supplied by compressors, care should be taken to select a "breathing air" type of compressor. What to avoid is the use of the ordinary mechanical tool air compressor for this purpose. Pneumatic tools are sometimes lubricated through a system of injection lubricants into the supplied air. This of course would not be satisfactory for breathing purposes.

Some confusion exists over "alarms" required for compressor failure or overheating. The reason alarms for overheating are needed is that a hot compressor can deliver deadly carbon monoxide to the air line respirator user. Some breathing air compressors have automatic shutdowns instead of alarms, and this is entirely satisfactory for some types of industrial environments. For instance, abrasive blasting atmospheres are not IDLH (immediately dangerous to life or health), and if the compressor overheats and shuts down, workers can merely take off their respirators *acute* and immediately leave the area. But in an IDLH atmosphere a compressor shutdown could result in a fatality, and an alarm is needed.

Respirators are subject to deterioration from improper maintenance, and besides, they can simply become unsanitary. A "routine" inspection before and after each use can be performed by the users themselves. These routine inspections could include a check for cleanliness, deterioration, and obvious function.

Self-contained breathing apparatus should have an effective monthly inspection. The regulator should be pressurized to check to see whether the low-pressure warning device operates. Cost for loss of pressurized air should not be an important factor because it should take less than 50 pounds of air to perform a complete checkout. Tanks should be pressurized to at least 1800 psi. Both the cylinder and the regulator have a gauge. Instead of trusting the cylinder gauge, the regulator should be pressurized and the gauge verified.

Self-contained breathing apparatus is usually reserved for emergency use, not ordinary on-the-job use. The whole idea of emergency respirators is serious business because in an emergency there is a high probability of IDLH atmospheres. Emergency rescue equipment in IDLH atmospheres demands high-quality equipment, carefully inspected and maintained, and effectively utilized by trained rescue teams.

Welding operations present special problems because toxic gases and/or fumes are often encountered, but at the same time the welder must be protected from harmful rays. Snap-on lenses are available for full-face respirators to protect against harmful rays. Another alternative is to wear the respirator beneath the hood. This requires a special welding hood, and only a half-mask respirator can be worn under this hood.

HEAD PROTECTION

A primary symbol for OSHA, corporate safety departments, and just about anything related to occupational safety and health is the familiar profile of the "hardhat." So important is the symbol that many zealous safety managers have established sweeping hardhat rules throughout large general work areas. There is nothing wrong with such rules if a genuine hazard exists. But when workers sense that no hazard exists and that the hardhat rule exists as a promotional device or for window dressing, they often show their defiance by refusing to wear the hardhat.

Hardhat rules should be carefully formulated with ample consideration for the consequences both ways. Once it has been decided that a hardhat rule is necessary, the Safety and Health Manager should take steps to assure its implementation. The evidence that was used to prove the need for the hardhat rule should be compiled into organized training packages to convince workers. After training and launching of the implementation phase, follow-up checks should be used to assure that the rule is being followed. Corrective steps should be taken to overcome individual violations of the rule, including disciplinary actions if necessary.

Hardhats seem to have won wider acceptance than hearing protection. Besides being a symbol for occupational safety and health, the hardhat has become a symbol for rugged, physical jobs. This image has appealed to males for centuries and is becoming an increasingly appealing image for the female worker also. Management personnel have also coveted the image conveyed by the hardhat. A hardhat worn by a manager seems to imply that the manager is well founded in the operations of the enterprise and is action oriented, doing more than talking on the telephone, attending meetings, and sitting behind a desk.

MISCELLANEOUS PERSONAL PROTECTIVE EQUIPMENT

Safety Shoes

Safety shoes are a more expensive undertaking than hardhats because safety shoes are worn out faster and are more expensive per item. Employees may buy their own shoes at attractive discounts in some arrangements, and this encourages actual use. Safety shoes come in a wide variety of appealing styles, and employee resistance to wearing safety shoes is largely a thing of the past.

The Safety and Health Manager is usually saddled with the decision as to which jobs require safety shoes and which do not. Although applicable national standards are explicit about the design and construction of safety shoes, the decision of where such shoes must be worn is left up to the user or to management.

One place safety shoes are clearly needed is on and around shipping and receiving docks. This should be obvious, but there has been some legal controversy over this issue. The courts have settled the question, and Safety and Health

Managers should assure that shipping and receiving dock personnel wear safety shoes.

Protective Clothing and Skin Hazards

Occupational skin disease represents approximately 70% of all occupational diseases, and over 80% of the skin disease reported is classified as contact dermatitis from irritants to the skin. The Safety and Health Manager should be alert to several skin hazard sources, such as welding, special chemicals, open surface tanks, cutting oils, and solvents.

Most welders know the value of heavy-duty protective aprons and flameproof gauntlet gloves. Leather or woolen clothing is more protective than cotton from a burnability standpoint. Nomex™ is a treated fabric which is a flame retardant.

Another concern for protective clothing is chemical exposures from open-surface tanks. Gloves must be impervious to the liquid being handled and must be long enough to prevent the liquid from getting inside. If the gloves are not long enough, they can be more hazard than benefit. Many workers' hands have become more irritated than their unprotected arms simply because the gloves they were wearing permitted liquids to get inside, and then the gloves themselves became "dip tanks for the hands!"

Discussion of personal protective equipment for open-surface tanks is not complete without mention of "chrome holes," an ominous-sounding affliction which comes by its name honestly. Vapors and mists from chromium plating tanks can cause open ulceration, particularly in moist tender parts of the skin. The inner septum of the nose, that part which divides the nostrils, is particularly susceptible to ulceration from chromic acid, sodium chromate, and potassium dichromate. In plating plants it is not unusual to find workers whose nose septums have been completely destroyed by these chromium compounds. Proper ventilation and engineering controls are the most appropriate means of prevention. Worker education and regular examination can also be helpful, including periodic examination of the nostrils and other parts of the body for workers exposed to chromic acids. As with other hazards, the last line of defense would be personal protective equipment in the form of respirators with cartridges to remove the dangerous chromic acid mists.

Easily the most common hazard among skin irritants are the cutting oils used in metal machining operations. Cutting oils are useful and sometimes essential to lubricate the tool, reduce cutting temperature, remove chips, permit a higher-quality cut, and lengthen tool life. But cutting oil is not always required and is not even desirable in some situations. The manufacturing engineer usually makes the cutting oil decision for the various manufacturing processes, but there is no reason why the Safety and Health Manager should not have some input in the cutting oil decision, especially since the decision affects safety.

Cutting oils are basically of two types: natural and synthetic. The natural oils are petroleum based and are the chief culprits in a very common industrial skin disease: oil folliculitis. Oil folliculitis is basically a clogging of the hair follicles in the skin which results in acne-like lesions. The synthetic oils are easily recognized by their familiar milky white appearance, and these oils have gained widespread use. Although the synthetic oils are not likely to cause folliculitis, they do have a nasty reputation for becoming contaminated with bacteria and presenting the hazard of skin infections. If antibacterial agents are added, these, too, can be skin irritants. Technology is working to improve cutting oils, but the job is not complete, and it appears that personal protective equipment is still in order.

Protective barrier creams for the skin are an alternative to gloves or protective clothing, but these creams are by no means a panacea. The creams must be removed and reapplied at least every break, every lunch hour, and every shift. Applying and reapplying creams cost labor time as well as the cost of the creams themselves. Also, creams are not considered as effective as gloves, even when properly applied. Creams can be used, however, when the requirements of the job make gloves infeasible.

A chief concern of the Safety and Health Manager with regard to protection of the skin is the use of various solvents within the plant. Solvents are essential for removal of grease and cutting oils, and herein lies another reason for doing without cutting oils if practical. A familiar solvent is trichloroethylene, and it is bad practice for workers to wash piece parts in trichloroethylene with their bare hands. Alternatives would include the use of wire baskets for handling the parts in the solvent, or perhaps simply substituting soap and water for trichloroethylene in some situations. For the most part, soap and water will not be effective against the oils and greases encountered, although in some situations soap and water washing is effective. The Safety and Health Manager is not doing the job unless he or she seeks out these situations and calls them to the attention of management and engineering. When alternative solvents, wire baskets, and other engineering controls are infeasible, personal protective equipment such as gloves is in order.

It seems appropriate to conclude a discussion of skin hazards, gloves, and protective clothing by mentioning one of the simplest personal protective measures of all—personal cleanliness and hygiene. Workers who are fastidious about washing their hands and bodies frequently have been found to enjoy a lower incidence of skin disease. It is easy to understand why. With skin irritants the amount of injury is generally directly related to the duration of exposure, other factors being equal. It is easy to forget how effective soap and water are in removing injurious elements of all kinds.

What can the Safety and Health Manager do to motivate workers to adopt good habits of washing and personal hygiene? The obvious answer is training and motivational reminders in the form of signs and posters around the plant. But this obvious answer is not the only answer and perhaps is not even the best answer. The Safety and Health Manager can attempt to influence the selection and layout of convenient, well-maintained, pleasant restrooms which will boost employee morale while encouraging workers to wash regularly. There is nothing more discouraging to employees than an encounter with a washroom with no warm water, no soap, and no towels.

FIRST AID

The Safety and Health Manager will frequently be responsible for the first aid station and may supervise a plant nurse, a major portion of the Safety and Health Manager's job. However, the first aid station may satisfy several additional functions besides providing immediate care for the injured. The first aid station is often used for medical tests, screening examinations, and monitoring of acute and chronic effects of health hazards. Also, the plant nurse or other first aid personnel may be responsible for performing some of the recordkeeping and reporting functions discussed in Chapter 2.

One adequately trained first-aid person is required in the absence of an infirmary, clinic, or hospital "in near proximity" to the workplace. No one seems to be able to determine authoritatively what constitutes "near proximity" in this context. Various interpretations around the country have been compared on this point, and the opinion has for the most part varied from 5 minutes' to 15 minutes' driving time. The interpretation has sometimes depended on whether the route to the hospital crosses a railroad track. If the workplace is not itself a hospital or clinic or is not directly adjacent to one, the Safety and Health Manager is advised to be sure that at least one, preferably more than one, employee is adequately trained in first aid.

A first aid kit or first aid supplies should be on hand, and the Safety and Health Manager should seek a physician's advice regarding selection of these materials. Unfortunately, medical doctors are hesitant to give such advice, probably because they fear subsequent involvement in litigation should an accident occur for which adequate materials are not available. Safety and Health Managers should do their best to obtain this advice and then document what was done.

Another first aid consideration is the provision of emergency showers and/or emergency eyewash stations where injurious corrosive material exposure is a possibility. Almost everyone has seen the deluge-type shower, which is activated by grabbing and pulling a large ring attached to a chain which activates the valve. Eyewash facilities are similar to a drinking water fountain in which two jets are provided, one for each eye.

Noting the cost and permanence of a conventional eyewash facility, some enterprising and innnovative individuals have marketed a simple plastic water bottle and plastic tube hanging on a frame which identifies its purpose as an emergency eyewash. Such a solution to the problem is probably not what the drafters of the standard had in mind, but the little water bottle is not without merit. For one thing water bottles can more easily be deployed to the best spots to permit instant use in case of an accident. Most workplace layouts change rather frequently, making the permanent eyewash installation cumbersome, expensive, and often in the wrong place. Furthermore, the little water bottle approach permits convenient introduction of antidotes or neutralizing agents for specific corrosive materials. However, woe be to the injured who uses an antidote for acid when the exposure was caustic, or vice versa. Also, the volume of water in the bottle approach is usually limited to a quart or perhaps 2 quarts. This may not be enough water if a first aid manual specifies flushing the eyes "with copious amounts of water for 15 minutes."

CONCLUSION

Reflection on this chapter leaves the impression that the proper provision and maintenance of personal protective equipment and first aid is not an easy task. Even after the proper equipment has been selected and procedures set up for maintenance, employees must be trained and disciplined to use the equipment properly. Inspectors can easily recognize simple violations such as failure of individual employees to wear prescribed personal protective equipment. One pitfall for the Safety and Health Manager is the blanket specification of personal protective equipment when its necessity is marginal. Such blanket specification is a trap that will lead to employee apathy and subsequent violation of the rule.

Returning to the principle that personal protective equipment is a last resort method of protecting the workplace, engineering the hazard out of the workplace is definitely preferred. Personal protective equipment seems to be the easy and less costly way out, but an examination of the principles and pitfalls of personal protective equipment is a strong motivation toward engineering controls.

EXERCISES AND STUDY QUESTIONS

8.1. In a not-so-unusual accident, two workers were killed when an employee was cleaning out a tank accessed by a manhole. A second employee watching saw the first employee collapse. While trying to rescue the first employee, the second employee was also overcome, and both died. What preventive measures can you suggest to prevent accidents of this type?

8.2. What are the two basic types of "safety lenses?" Which is more durable?

8.3. What is the argument against requiring safety glasses and hardhats throughout all areas of the plant, including areas where they are not needed?

8.4. Name some jobs for which eye protection should be provided.

8.5. Why are "organic vapor respirators" not to be trusted to protect against all organic vapors?

8.6. In an industrial plant a prominent sign reads:

> PREVENT ACCIDENTS: WEAR YOUR HARDHAT

What principles of personal protective equipment does this sign call to mind?

8.7. Why is the provision of personal protective equipment not a very satisfactory solution to the problem of protecting workers?

8.8. What are the hazards of employee-owned personal protective equipment being brought onto the premises at work?

8.9. In what way have undersized manholes for vessel entry resulted in a large number of multiple fatalities in the United States?

8.10. Why is there no air-purifying device approved for roofing workers who apply expandable foam?

8.11. Name several alternatives to washing piece parts in trichloroethylene by workers placing their bare hands in the solvent.

9

Fire Protection

Percent of OSHA
general industry citations
addressing this subject

2%

This chapter deals with perhaps the oldest topic in occupational safety and health. But modern developments in the field of fire safety place it in a very dynamic phase. Unlike other categories of safety and health, fire safety presents the Safety and Health Manager with a wide variety of alternatives for dealing with hazards.

The field of industrial fire protection formerly consisted of little more than dealing with fire extinguishers, their selection, placement, marking, inspection, and maintenance. True, there were a few obscure standards about such topics as "standpipe and hose systems," but almost all of the activity centered around the fire extinguishers. Today the field of industrial fire protection is much more sophisticated, recognizing such alternatives as emergency action plans, fire prevention, fire brigades, fire alarm signaling systems, fixed extinguishing systems, and automatic sprinkler systems. Rather than blindly following old specific standards for fire extinguishers, Safety and Health Managers now have the opportunity to explore alternative strategies or combinations of strategies to accomplish the most cost-effective method of fire protection for their own circumstances. Even the federal standards have been changed to delegate such decision-making authority to industrial managers.

It is easy to oversimplify fire protection as meaning only fire extinguishment, but it really encompasses three fields: fire prevention, fire suppression, and personal protection (escape). Much of the criticism of old fire safety standards has been that they emphasized fire suppression and extinguishers, which may not be the safest alternative in some fire situations. Some firms wanted no extinguishers at all and desired to instruct employees to escape without attempting to extinguish fires. Their argument was that fire extinguishers are mainly for property protection, whereas the safest thing for workers to do is escape. The current standards recognize this rationale together with several shades of combination strategies. Before discussing these strategies in detail, some facts about fires should put the fire protection problem in perspective.

INDUSTRIAL FIRES

Fire is the third-ranking cause of accidental death in America, according to statistics reported by the National Fire Protection Association (NFPA) (Ref. 50). One might think that the high-technology fire detection, protection, and suppression systems utilized in the United States would make possible one of the lowest fire death rates in the world. Exactly the opposite is true! The United States has one of the worst fire death rates in the industrial world, about 40 deaths per million people.

The U.S. fire death rate is a bad one, but anyone who attempts to pin the responsibility on industry, using NFPA statistics, is in for a disappointment. The vast majority of fire deaths in the United States are in residential occupancies. The relative insignificance of industrial fires is made obvious by Table 9.1. Table 9.1 does not tell the entire story because it reports only "multiple-death fires," but other data corroborate the low percentages of injury, death, and damage which arise from the "industrial" category of fires. It should be acknowledged that some industrial fires are not reported to the NFPA because they are extinguished by automatic systems or by internal fire brigades. But this fact only strengthens the conclusion that industry, more than most other elements of our society, has done a great deal to control fire hazards. Considering the incredible exposure to flammable liquids in refineries and chemical plants and the billions of labor hours spent in industrial plants every year, it is amazing that the number of fire deaths in all industrial plants

TABLE 9.1   DEATHS IN MULTIPLE-DEATH FIRES IN 1979

Occupancy category	Deaths	Percent of total
Residential	939	86.6
Stores, service, and repair outlets	15	1.4
Taverns	7	0.6
Industry, mines	18	1.7
Prisons	11	1.0
Vehicles	81	7.5
Miscellaneous	13	1.2
Total	1084	100.0

Source: Compiled from statistics published by NFPA, Ref. 50.

is no more than the number who die in fires in taverns and prisons, as shown in Table 9.1. In fact, more people died in a single fire in a supper club in Kentucky in 1977[1] than in all the multiple-death industrial fires in the nation in 1977, 1978, and 1979.

The most dangerous industries from a fire hazard standpoint are mines, grain elevators, grain mills, refineries, and chemical plants. The fire fatalities from these four industries dwarf the total for all remaining industries combined. For general manufacturing industries the number of fire fatalities is extremely low. The last major tragedy in an ordinary manufacturing plant was the Triangle Shirtwaist Company fire, which claimed 145 lives in New York in 1911. Perhaps it is the magnitude of this tragedy that has led to superior control of industrial fires and has virtually eliminated the hazard today.

FIRE PREVENTION

The best way to deal with fires is to prevent their occurrence, just as engineering controls were found in Chapter 8 to be preferable to personal protective equipment. Effective fire prevention requires anticipation of fire sources. Each facility is different and requires an individual analysis of potential fire sources. Once the hazards are identified, decisions must be made as to who has responsibility for controlling the hazards. These decisions should be documented in a fire prevention plan.

A principal cause of industrial fires is overheated bearings or hot machinery and processes. Some of these causes can be averted by adopting an effective preventive maintenance program. Such a program, while decreasing the likelihood of fire, may also extend the life of equipment. The Safety and Health Manager may see an opportunity in this strategy to save production costs while furthering the cause of fire safety.

Another ingredient in a fire prevention plan is a strategy for housekeeping. Accumulation of combustible dusts in grain elevators and paint residues in spray painting operations are good examples of how poor housekeeping can contribute to fire hazards. Even ordinary combustible paper and material waste can be a fire hazard.

EMERGENCY EVACUATION

Using the "escape strategy" for dealing with fires or other emergencies, the employer must prepare a written "emergency action plan." The emergency action plan concept has been around for many years for hospitals, schools, and institutions, and in the 1980s this concept was extended to industries in general.

---

[1]Beverly Hills Supper Club fire, Southgate, Kentucky, 165 lives lost.

Alarm Systems

Crucial to an emergency action plan is an employee alarm system. But alarm systems are not as simple as they may seem. Searching questions that must be asked are: "Will persons recognize the signal as a fire alarm?" and "What about deaf or blind employees?" Audible, visual, and tactile systems must be considered, or perhaps combinations of these systems. In small workplaces even direct voice communication may be the best fire alarm medium. Public address systems may be used in larger facilities, but the system should provide that emergency messages take priority.

System reliability is important to fire alarms because a failure within the system may not be immediately obvious. Some sophisticated systems have built-in monitor circuits to supervise reliability. Such systems do not need testing as often as do simple alarm systems which have no such monitoring circuits. When repairs are being made, some type of backup system is needed to provide continuous protection. The backup system might even employ "runners" or telephones or other informal systems, but the Safety and Health Manager should document what the backup system is.

Fire Detection Systems

Smoke alarms and other detection devices may be used to trigger the alarm system. However, it should be noted that automatic smoke alarms are not mandatory in American industries in general. Even manual or visual systems can be considered alarm systems.

If automatic detection systems are employed, care must be taken to maintain and protect the equipment. Most detection systems are delicate instruments and will not withstand the rigors of the industrial environment. Conditions to be considered are dust, corrosive atmospheres, weather exposure, heat from processes, and mechanical damage.

People are sometimes reluctant to sound fire alarms, with tragic consequences. Hotel managers are particularly unwilling to alarm occupants. The typical response from a hotel front desk when an occupant calls in a fire alarm is to send a bellman to investigate. It is rational to consider the hazards of panic when a fire alarm is sounded. However, this rationale is sometimes permitted to be an excuse for failing to take action.

FIRE BRIGADES

Some firms may elect a strategy in which employees are organized into brigades to fight fires themselves. Such strategies should be carefully scrutinized because in the scramble to protect property, these fire brigades can be a danger to employees.

Employee Fitness

Volunteering to join the fire brigade is not sufficient to qualify the worker to fight fires. Conditions that may be hazardous include heart disease, epilepsy, or emphysema. Other conditions, such as ruptured eardrums or the wearing of a beard, may make the use of respiratory equipment ill advised. The Safety and Health Manager should be sure that the fire brigade volunteers are screened, and a physician's certificate may be necessary for questionable cases. Volunteers unfit for interior structural firefighting may be used in other tasks.

Firefighter Training

Many states have fire training academies, and Safety and Health Managers should find out what schools and academies are available for their fire brigade members. Interior structural firefighting is more demanding, and fire brigade members assigned to such tasks should be trained at least quarterly. Other fire brigade members should be trained at least annually. Also, firefighting equipment to be used by the fire brigade should be inspected annually, and fire extinguishers should be inspected monthly.

Protective Clothing and Apparatus

If the firm elects to have fire brigade members fight interior structural fires, protective clothing and respirators must be provided. This includes protective boots or shoes, fire-resistant coats, gloves, and head, eye, and face protection.

One concept stressed for self-contained breathing apparatus is the mode for airflow into the mask. At this point the reader may want to review the three modes of respirator airflow discussed in Chapter 8. The preferred mode for firefighting is one of the positive pressure types: pressure demand or continuous flow. The only valid argument for using simple demand flow is that this mode permits longer duration exposure for a given charge. If the employer believes that the demand flow mode is essential, quantitative fit testing is necessary for each firefighter.

FIRE EXTINGUISHERS

Fire extinguishers are still the most effective method of immediately controlling a very local fire before disastrous consequences ensue. The Safety and Health Manager needs to understand the various fire classes and the type of extinguishers appropriate for each.

Fire Classes

The fire protection field classifies fires into four categories. Application of the wrong extinguishment media to a fire can do more harm than good.

Table 9.2 describes the four classes of fires, example appropriate extinguisher

TABLE 9.2   FOUR CLASSES OF FIRES AND APPROPRIATE EXTINGUISHING MEDIA

Fire class	Description	Example extinguishing media	Maximum OSHA-authorized travel distance to nearest extinguisher
A	Paper, wood, cloth, and some rubber and plastic materials	Foam, loaded stream, dry chemical, water	75 feet
B	Flammable or combustible liquids, flammable gases, greases, and similar materials, and some rubber and plastic materials	Bromotrifluoro-methane, carbon dioxide, dry chemical, foam, loaded stream	50 feet
C	Energized electrical equipment	Bromotrifluoro-methane, carbon dioxide, dry chemical	No specific maximum; distribute "on the basis of the appropriate pattern for existing Class A or B hazards"
D	Combustible metals such as magnesium, titanium, zirconium, sodium, lithium, and potassium	Special powders, sand	75 feet

media, and the maximum travel distance specified for extinguishers for each type. Liquefied petroleum gas (LPG) fires, although technically Class B, are really not adequately addressed by any of the four classifications. Such fires are extremely dangerous, and fire extinguishers are not appropriate for their control. LPG fires should be extinguished by professional firefighters using powerful water spray systems.

The key to determining whether an extinguisher is appropriate for a given class of fire hazard is to check the approval marking on the extinguisher itself. Some types of extinguishers have been found to be hazardous and are forbidden regardless of prior approval markings. These types are listed in Table 9.3. Some extinguishers may be approved for more than one classification of fire. These multipurpose fire extinguishers typically employ a dry chemical medium. Although dry chemical extinguishers are growing in popularity, they are not a panacea. Expensive equipment such as computers can be fouled or even ruined by application of a dry chemical extinguisher when a CO_2 extinguisher would have been satisfactory. Foam or water extinguishers may also be cheaper for the more ubiquitous Class A fires.

TABLE 9.3 FORBIDDEN FIRE EXTINGUISHERS

1. Carbon tetrachloride
2. Chlorobromomethane
3. Soldered or riveted shell self-generating soda acid or self-generating foam or gas cartridge water-type portable fire extinguishers which are operated by inverting the extinguisher to rupture the cartridge or to initiate an uncontrollable pressure-generating chemical reaction to expel the agent

Source: Code of Federal Regulations 29CFR 1910.157

Inspection, Testing, and Mounting

In the 1970s OSHA found thousands of violations in which fire extinguishers did not have appropriate tags indicating their inspection status. OSHA has since abandoned the rule requiring such tags but still requires the employer to maintain records of annual maintenance for each fire extinguisher for up to 1 year after the last entry or the life of the shell, whichever is less. In addition, a visual inspection is required monthly. Many employers have found it expedient to retain the tag system even though it is no longer required. Thus when anyone requests to see the inspection record for a given fire extinguisher, it is immediately available on the tag attached to the extinguisher.

In addition to the monthly and annual inspections, fire extinguishers must receive a hydrostatic test according to a prescribed test schedule for various types of fire extinguishers. Fire extinguisher shells deteriorate from mechanical damage or corrosion and may be unsafe for containing pressures inside. The hydrostatic test places the extinguisher under a test pressure to determine whether it can safely contain the pressures to which it will be subjected in use. The test has technical specifications and must be done by a trained person using suitable equipment and facilities. The Safety and Health Manager invariably leaves the hydrostatic tests for the professional fire extinguisher service to perform.

Another big OSHA enforcement issue of the 1970s was the mounting height and identification of mounting of fire extinguishers. OSHA citations for mounting, identification, and inspection tagging of fire extinguishers amounted to more than half of all OSHA violations named in fire protection citations throughout the 1970s. These requirements have all been eliminated or drastically changed by rewording them in performance language. The current standard permits the employer the latitude of mounting extinguishers high on the wall, out of the way of forklift truck traffic, and accessed by rope and pulley. The employer may select any convenient mounting scheme provided that the extinguishers are readily accessible without subjecting employees to possible injury.

Training and Education

A facility looks well equipped and protected when fire extinguishers are placed about the workplace, readily accessible for use in an emergency. But the appalling reality is that few employees know how to use a fire extinguisher effectively, and some would even be afraid to use the extinguishers if they knew how. This is especially true of fire extinguishers or hose systems located behind glass doors. For most people there is a great reluctance to break glass even in an emergency. A study conducted among nursing staff in an Ohio hospital revealed that most knew nothing about and were afraid to use fire extinguishers. Accordingly, the Safety Manager in charge set out to train the nurses to use fire extinguishers in an emergency. A hospital bed was taken outdoors and set afire to provide a simulated fire emergency for the nurses to extinguish. The field of industrial safety has now recognized the need for fire extinguisher training, and such training has been accepted as standard for general industry. Training is required upon initial employment and at least annually thereafter.

STANDPIPE AND HOSE SYSTEMS

Some employers choose to install standpipe and hose systems for firefighting, and these systems can be employed in lieu of general distribution of fire extinguishers in most cases. Standpipe systems come in various ratings or classes and are summarized in Table 9.4.

Equipment

Some standpipe and hose systems can be quite ancient. These old systems can usually be retained as long as they are serviceable and meet annual test requirements. But when replacements are made, new equipment should conform to all current standards. Examples of modern system changes are as follows:

1. *Shut-off*-type nozzles
2. *Lining* for hose
3. Dynamic pressure *minimums at the nozzle*
4. Hydrostatic *testing* upon installation

The water supply for standpipe and hose systems can be provided either by elevated

TABLE 9.4   STANDPIPE AND HOSE SYSTEM CLASSES

Class	Hose size	Description
I	2½ inch	For professional fire departments and persons trained to handle heavy systems
II	1½ inch	For incipient-stage fires
III	1½ and 2½ inch	For trained employee use on all levels of fires, including the more advanced stages
	"Small ⅜-inch to 1½-inch hose"	For incipient-stage fires

water supply tanks or by pressure tanks. The supply must be sufficient to provide 100 gallons per minute flow for at least 30 minutes. This calculates to 3000 gallons per use period, but remember that it takes considerably more than 3000 gallons to provide the head pressure to maintain the 100-gallon-per-minute flow throughout the 30-minute period. One seemingly clever idea is to forget about standpipes or pressure tanks and connect hose systems to city water supplies. But the flaw in the theory is the adequacy (pressure and flow) of the supply, which usually cannot meet the 100-gallon-per-minute flow requirement.

Maintenance

The ancient systems described earlier can create some unpleasant surprises when the hose is deployed for use. Hemp and linen systems are especially subject to deterioration. After hanging on the rack for many years without use, the hose may break apart or disintegrate when taken off the rack for use in an emergency. Hose systems should be checked annually and after each use. In the case of hemp or linen hose systems, the hose must be reracked using a different fold pattern.

AUTOMATIC SPRINKLER SYSTEMS

Automatic sprinkler systems present a paradox since they affect employee safety but are usually installed principally to protect property and to lower insurance rates. If such a system is *voluntarily* installed by the employer to protect property, should it be required to meet personal safety standards? And if an existing system does not meet current standards, should it be dismantled and withdrawn from use? This would hardly be in the interest of safety.

Sometimes a good sprinkler system is installed only to be made ineffective by incorrect usage of the space protected. One error is to allow the sprinkler heads to become fouled by materials such as paint spray residues. If a spray area is protected by an automatic sprinkler system, a good way to protect the sprinkler heads is to cover them with paper bags. If a fire does occur, the paper bags will either burn away or be washed away by the water spray so that they do not interfere with the fire suppression action of the sprinklers.

Another error with automatic sprinklers is to stack material too close to the ceiling. This interferes with the distribution of spray from the sprinkler head. At least an 18-inch clearance must be used, as shown in Figure 9.1, to permit the sprinkler spray to be well distributed.

FIXED EXTINGUISHING SYSTEMS

Technically speaking, automatic sprinkler systems are "fixed extinguishing systems," but really what is usually meant by a fixed extinguishing system is a more local system for controlling special fire hazards such as kitchen grills or tempering

Figure 9.1. Minimum vertical clearance between stacked material and sprinkler head is 18 inches to permit distribution of spray.

tanks. Again, the principal objective may be property protection and insurance rate reduction, but steps must be taken to prevent the system, which may discharge dangerous gases or other agents, from becoming a hazard to employees. Thus if the discharge of the system is not obvious, it is necessary to warn employees, perhaps with a discharge alarm, that dangerous agents are being expelled into the atmosphere. If a strategy of "total flooding" is used with a dangerous agent, an emergency action plan is necessary to assure that personnel escape. Some agents are so dangerous as to be prohibited altogether as extinguishing media; examples are chlorobromomethane and carbon tetrachloride.

Fixed extinguishing systems are somewhat like giant, fixed, automatic fire extinguishers. Many of the maintenance procedures appropriate for portable fire extinguishers are also appropriate for fixed extinguishing systems. As with portable extinguishers, fixed systems are required to be inspected annually. Corresponding to the monthly visual inspections of portable extinguishers is a somewhat more comprehensive semiannual inspection for fixed systems to determine whether containers are charged and ready for operation. If the containers are factory charged and have no gauges or indicators, they must be weighed to determine charge. A weight decrease of 5% or a pressure decrease of 10% is considered within tolerance. Discussion of specific systems follows with requirements for each.

Dry Chemical Systems

The word "chemical" in dry chemical systems should be remembered, and the question to be asked is whether the extinguishing chemical will produce any undesirable reaction with process reagents or perhaps foams and wetting agents also employed. There is more than one type of dry chemical available, and generally these chemicals are not to be mixed when filling the cylinders or containers. Mixing is permitted if the chemical to be added is "compatible" with the chemical stated on the approval nameplate of the system.

Dry chemicals used for extinguishing agents are usually not dangerous to the health or safety of personnel. However, the actual distribution of the chemical powder during an emergency may obscure vision, preventing escape. A possibility such as this calls for a pre-discharge employee alarm system as described earlier.

The biggest problem with dry chemical systems is the caking or lumping of the agent. Humid climates or moisture-producing processes subject the system to a greater risk of caking. Caking can render a dry chemical useless, so the chemical should be sampled annually to be certain moisture is not causing caking.

Other Fixed Systems Agents

Many fixed extinguishing systems employ carbon dioxide, Halon 1211, or Halon 1301 gases. These systems have the advantage of not requiring as much cleanup after the emergency as other systems, but all three of these gases can be dangerous to unsuspecting employees, especially if a total flooding strategy is employed. Standards must be followed in planning pre-discharge warning systems, employee egress routes, and maximum concentrations of gases to be released.

Water spray and foam agents are less dangerous to employees, but necessary volumes required to be effective may introduce egress hazards. Drainage must be directed away from work areas and so as not to obstruct egress paths.

SUMMARY

The many strategies for dealing with industrial fire hazards can be grouped under the general categories of prevention, suppression, and escape, or combinations of these categories. Current applicable industrial standards encompass all of these strategies.

It is necessary to keep the hazards of industrial fires in perspective. Industrial fires cause very little loss of life and injury now, compared with fire deaths elsewhere and occupational fatalities and injuries from other causes. In light of this perspective it may seem that emphasis on equipment specifications, regular inspection of extinguishers, personnel training, and written plans is somewhat misplaced. However, the excellent record of industry in controlling fire hazards on the job should not be allowed to induce complacency now that success has been achieved. There is no doubt that the adherence to strict fire codes is what has helped industry achieve such a superior control of fire hazards compared to residential and other fire exposures.

The beginning of this chapter identified fire protection as the oldest of topics in occupational safety and health. There is perhaps a correlation between the facts that industrial fire protection is an old endeavor, and it is also a very successful endeavor. Perhaps with time even the newer fields of occupational safety and health covered in other chapters of this text will achieve the same high degree of safety and health as has already been achieved by fire protection.

EXERCISES AND STUDY QUESTIONS

9.1. Should fire extinguishers be required in all industrial plants? Why or why not?

9.2. What are the three major fields of fire safety? *Prevention, Suppression, Per Protection*

9.3. What argument does industry offer against providing fire extinguishers?

9.4. How does fire rank among causes of death in America? *accidential*

9.5. How does America rank among nations of the world in numbers of fire deaths per million people? *one of worst*

9.6. What is the "occupancy category" for most fire deaths in the United States? *Residential*

9.7. What percent of the total number of fire deaths from multiple-death fires is attributed to industry? *1.7%*

9.8. Name some items to be included in fire prevention plans.

9.9. How is preventive maintenance related to fire hazards? *friction*

9.10. How can automatic audible alarm systems fail to warn employees of a fire emergency?

9.11. Are automatic smoke alarms required in industrial plants? *No*

9.12. Why does an employee sometimes notice a fire in the plant but fail to sound the alarm? *fear*

9.13. Are industrial plants required to have fire brigades? *No*

9.14. How often must fire brigade members be trained? *4 or annually*

9.15. Name some conditions that would make an employee unfit to serve in a fire brigade. *Heart, Resp*

9.16. Identify the four classes of fires. Give example extinguishing media for each. *A Wood Paper / B Oil Gas Liquid / C Elect / D F Metals*

9.17. What extinguishment medium is used for LPG fires? *Lots of water AnDep?*

9.18. Name some forbidden fire extinguishers. *Carbon Tet*

9.19. What are the advantages and disadvantages of dry chemical as a fire extinguishment medium? *Expensive, mark, Mess*

9.20. How often must fire extinguishers be inspected? *monthly*

9.21. Are fire extinguishers required to be mounted at a certain distance from the floor? *M*

9.22. How often are employees required to be trained to operate fire extinguishers? *annually*

9.23. Name some modern requirements for standpipe and hose systems. *500 gallon per minute 30 min*

9.24. Why are direct city water supplies usually unacceptable for supplying ~~standpipe and~~ hose systems for fire protection? *Not enough pressure*

9.25. Are industries required to have automatic sprinkler systems? *No*

9.26. Why are paper bags placed over sprinkler spray heads? *Keep clean*

9.27. How close to the automatic sprinkler head in a warehouse is it permissible to stack material? *14"*

9.28. If a fixed extinguishing system has no gauges or indicators, how can its charge condition be determined? *weight of bottle*

9.29. Name some gases employed in fixed extinguishing systems. *CO₂ Halon*

10

Materials Handling
and Storage

 Percent of OSHA
general industry citations
addressing this subject

6%

The usual concept of a factory is a place where things are made or materials are processed, but often the major activity in a factory consists of moving things and materials around. Lifting, a most basic material-handling activity, accounts for most back injuries, the largest workplace injury category of all. Industrial trucks, tractors, cranes, and conveyors all have the simple mission of moving materials, and they all cause injuries and fatalities every year.

The National Safety Council charges materials handling with 20 to 25% of all occupational injuries. The size of the problem is emphasized in NSC's *Accident Prevention Manual for Industrial Operations* as follows:

> As an average, industry moves about 50 tons of material for each ton of product produced. Some industries move 180 tons for each ton of product.

Before addressing specific hazards, it would be wise to examine the general nature of hazards created by materials handling.

In materials handling, masses are usually measured in tons or pallet loads instead of ounces, pounds, or kilograms. The human body is light and frail by comparison, so bulk materials' masses can easily pinch, fracture, sever, or crush. Contributing to the hazards of large masses is the reality that materials handling includes *motion* for these masses.

To illustrate the general "mass/motion" hazards of material-handling equipment, consider the following comparison of processing versus material-handling equipment. To be struck by a moving part of a processing machine may or may not cause injury, depending on the size of the machine, the motion of the moving part, and the shape or surface characteristics of the part. But to be struck by an industrial truck or conveyor is almost certain to cause injury. More indirectly, the mass/motion hazards of materials handling can affect safety by impacting facilities such as gas lines or electric lines or by overloading structural components of buildings.

Another general hazard of materials handling is its automatic or remote control nature. Materials pumps and conveyors are often started automatically on demand or from a manual switch which is located far away. Conveyor accidents are often caused by this remoteness characteristic. In another example, railroad cars often move about a yard, far from the eye of the engineer, or worse yet, they may roll almost silently, coasting independently into position from the momentum of a locomotive's momentary push, with no local control at all.

An indirect general hazard from materials handling is fire. This hazard emphasizes the storage aspect of materials handling. Warehouse fires are costly in terms of property loss but can also be dangerous to workers.

MATERIALS STORAGE

Materials-handling standards say that bags, containers, or bundles stored in tiers shall be "stacked, blocked, interlocked, and limited in height so that they are stable and secure against sliding or collapse." For general materials the standard does not explain what constitutes "blocking" or "interlocking" and does not specify height limits for stacks. Some industry practices, however, seem to defy the standard. Since the standard for materials storage is not specific in its requirements but does address results (e.g., "so that they are stable and secure against sliding or collapse"), the standard should be recognized as a performance standard.

Housekeeping is another consideration for materials in storage. Sloppy warehouse practices can lead to trip hazards or fire. Pest harborage can result from certain accumulations of materials and can constitute a hazard. Outside storage can become grown over in weeds and grass, which can be a fire hazard in dry weather.

The Safety and Health Manager should monitor the ups and downs of the company's sales and production fortunes. If there is a recent buildup in production in anticipation or in response to booming sales, there is likely to be a warehousing problem. Additional warehouse space is likely to be added later, after the upturn appears more permanent. Even after an expansion decision is made, a certain amount of lead time is necessary to plan and build add-on warehouse space. Meanwhile, the existing warehouse space becomes crowded and unsafe. Judgmental limits for material stacking are cast aside as creative ideas are explored to squeeze more material into a small space. Some of these new ideas will be good ones; it is not basically unsafe to try to conserve space. But during a period of warehouse crowding, special attention should be given to new hazards which may result from new procedures in the warehouse. Aisle and exit blockage are likely to occur during such periods of warehouse stress, as is the incidence of storage stack collapse. Any accidents that occur during such periods will point to some of the new hazards being generated, and these accidents should be analyzed to eliminate cause.

INDUSTRIAL TRUCKS

This category of material-handling equipment is typified by the forklift truck. Forklifts are typically either electric motor powered or have internal combustion engines. Besides forklifts there are tractors, platform lift trucks, motorized hand trucks, and other specialized industrial trucks. Not included are trucks powered by means other than electric motors or internal combustion engines. Also not included are farm tractors or vehicles primarily for earth moving or over-the-road hauling.

Truck Selection

"Ignorance is bliss" for most industrial managers who set out to buy a forklift truck. Little known is the fact that there are 11 different design classifications by type of power and by degree of hazard for which approved. These 11 design classifications are distributed over at least 16 different classifications of hazardous locations to which the forklift truck may be exposed.

One can easily wonder why it must be so complicated to select a forklift truck. The basis for this complication is that engines and motors can be dangerous sources of ignition for flammable vapors, dusts, and fibers. To design these engines and motors to effectively prevent ignition hazards is a costly matter, and the marketplace will not support the purchase of an explosion-safe forklift truck for use in an ordinary factory location. Thus a complicated array of classifications and regulations are set up to permit the specification of the right industrial truck for the right job—no more and no less.

Industry standards for industrial truck classifications and their corresponding hazardous location codes are a maze of abbreviations and definitions. To understand these abbreviations, keep in mind that the objective is safety from *fires and explosions*. Whether the industrial truck is diesel, gasoline, electric, or LP gas powered, the more fire-safe models are designed to prevent ignition of accidental fires, and thus they are more expensive. A simple summary is contained in Table 10.1. More details are contained in the standards, but most Safety and Health Managers will need only the general idea. Another thing to keep in mind is that it is legal to use a safer, higher-classified industrial truck than the minimum required, but of course it usually will not be economical to do so. If a firm already has an EE-approved electric forklift, however, it may be expedient to go ahead and use it instead of buying a new ES- or E-approved unit which would suffice for the given application.

Summarizing these principles and several pages of applicable regulations, Table 10.2 gives the Safety and Health Manager a perspective into the approval classes of various industrial truck designs. The "classes," "groups," and "divisions" represent varieties of hazardous locations in which the definitions correspond *roughly* to those of the _National Electrical Code_® and are covered in more detail in Chapter 13. "Class" and "Group" refer to the type of hazardous material present, and "Division" refers to the extent or degree to which the hazardous material is likely to be present in dangerous quantities.

TABLE 10.1 SUMMARY OF INDUSTRIAL TRUCK DESIGN CLASSIFICATIONS

Diesel	Electric	Gasoline	LP gas
D	E	G	LP
Standard model	Standard model	Standard model	Standard model
Cheapest	Cheapest	Cheapest	Cheapest
DS	ES	GS	LPS
Safer model	Safer model	Safer model	Safer model
More expensive	More expensive	More expensive	More expensive
Exhaust, fuel,	Spark prevention	Exhaust, fuel, and	Exhaust, fuel,
electrical	Surface-temperature	electrical systems	and electrical
systems safeguards	limitation	safeguards	systems safeguards
DY	EE		
Safest diesel	Safer still		
Most expensive	More expensive		
diesel	All motors and		
No electrical	electrical enclosed		
equipment	EX		
Temperature-	Safest of all		
limitation	Most expensive		
feature	electrical truck		
	Even electrical		
	fittings designed		
	for hazardous		
	atmospheres		

Source: Summarized from Code of Federal Regulations 29 CFR 1910.178.

TABLE 10.2 ALLOWABLE CATEGORIES FOR INDUSTRIAL TRUCKS FOR VARIOUS HAZARDOUS LOCATIONS

Class	Group	Division 1	Division 2
I	A	No industrial truck permitted	DY, EE, EX
	B	No industrial truck permitted	DY, EE, EX
	C	No industrial truck permitted	DY, EE, EX
	D	EX	DS, DY, ES, EE, EX, GS, LPS
II	E	EX	EX
	F	EX	EX
	G	EX	DY, EE, EX
III		DY, EE, EX	DS, DY, Eª, ES, EE, EX, GS, LPS

ªPermitted to continue if previously in use
Source: Summarized from Code of Federal Regulations 29 CFR 1910.178.

There are so many categories of "approvals" for industrial trucks it is easy to lose sight of the overall objective in the approval process: to prevent fires and explosions from improper use of the wrong truck in a hazardous atmosphere. The authority for approval of industrial trucks is delegated to nationally recognized testing laboratories, such as Underwriters' Laboratories, Inc., and Factory Mutual Engineering Corporation. The prudent Safety and Health Manager will leave the application-for-approval process to the manufacturer of the equipment and simply look for the UL or FM approval classification, such as DY, EX, GS, and so on. For nonhazardous locations, even an *unapproved* truck may be used if the truck *conforms* to the requirements of Type D, E, G, or LP. This meets the spirit if not the letter of the approvals process.

In these days of high energy costs and the search for alternatives, some company managements may desire to convert an industrial truck from one energy source to another. But tampering with the design or altering the truck may invalidate the approval. Industrial truck conversions can be made, but the process is a bit tricky. The enlightened Safety and Health Manager will be alert to warn the company management not to waltz blithely into a lift truck conversion scheme without doing it properly. The conversion equipment *itself* must be approved, and there is a right way and many wrong ways to carry out the conversion. The "easy way out" is often to buy a new lift truck, a characteristic of the standards more appreciated by lift truck dealers than by lift truck users.

The hazard of fires and explosions may be the most complicated factor in the selection of forklift trucks, but it certainly is not the only factor. Far more important than the fire hazard rating of the design of the truck are the operations, fueling, guarding, training of drivers, and maintenance—subjects discussed under "Operations" below.

Operations

One of the first items of interest on the part of the Safety and Health Manager should be the refueling or recharging area of forklifts. Smoking is prohibited in these areas, and this fault is found more than any other. Also a problem is the charging of forklifts in a "nondesignated area." Hazards in the area include spilled battery acid, fires, lifting of heavy batteries, damage to the equipment by the forklifts, and battery gases or fumes. All of these hazards need to be addressed by the Safety and Health Manager in some way or another.

An old chemistry laboratory adage says:

Here he lies all calm and placid;
He poured water in his acid!

Another says:

Do what you oughter:
Pour acid into water!

Federal standards require the above, but the language of the standards is not so colorful. Perhaps a sign in the area would achieve compliance with the rule. A better way to promote safety, though, would be to include in an employee training program an explanation of the violent and exothermic reaction that occurs when water is poured into a concentrated strong acid.

Everyone knows that arcs and sparks frequently fly when battery connections are made. What most do not know is that gases liberated during charging processes can reach ignitable concentrations. Fire is little or no hazard when a battery is simply ''jumped'' to another. But a forklift charging area is a different matter with large amounts of gases liberated. Adequate ventilation to remove gases is essential.

With hazards of gases and acids in battery-charging areas, the alternative of the internal combustion engine seems attractive. But each of the internal combustion engine choices—diesel, gasoline, and LP gas—emit another dangerous gas: carbon monoxide. Since forklifts usually operate indoors, carbon monoxide gas levels can be a problem. The 8-hour time-weighted average exposure limit for carbon monoxide is 50 ppm.

If the Safety and Health Manager determines that a carbon monoxide problem exists in the plant and that the forklift trucks are the culprits, several alternatives are possible. The obvious one is to switch to electric forklifts. Another solution might be to alter the building or to install adequate ventilation systems. Perhaps the cheapest solution of all would be to review procedures and operations to determine whether sources of emissions can be reduced or perhaps eliminated entirely. Key questions are:

1. Are operators leaving engines running unnecessarily?
2. Can the layout of warehouses or plant facilities be revised to reduce concentrations?
3. Are faulty or worn-out lift trucks creating more emissions than necessary?

Although there are no universal general minimum lighting requirements for industrial plants, where industrial trucks are operated, safety demands that the trucks themselves have directional lights if the plant area is too dark. Truck lights are required if the general lighting is less than 2 lumens per square foot. This is really a quite low level of light since an ordinary 100-watt incandescent bulb can produce 1700 lumens. Even in an all-black room with nonreflective walls, a 100-watt bulb could produce more than 2 lumens per square foot in an 8 foot x 12 foot x 16 foot room. Wall or other surface reflections help the overall situation, so the requirement for 2 lumens per square foot is not difficult to meet. A lighting consultant can help in this determination.

One of the greatest hazards with forklifts and other industrial trucks is the transition between dock and cargo vehicle. Figure 10.1 shows precautions to be taken. Although a highway truck is shown in the figure, the hazard also exists for loading railroad cars.

Many workers feel that because they know how to drive an automobile, they

Figure 10.1.   Preventing forklift hazards: transition from dock to cargo vehicle.

also basically know how to operate a lift truck. Unfortunately, many employers are inclined to take their word for it. But the operation of a lift truck takes a great deal more skill than the operation of an automobile. Compared to an automobile, a lift truck has a much shorter wheelbase, and when the load is lifted the center of gravity is very high. This creates stability problems to which the operator may be unaccustomed. Compounding the stability problem are the small-diameter wheels found on lift trucks, making chuckholes and obstructions more hazardous. When loaded, the center of gravity of the lift truck and load together can be shifted dangerously forward. Picking up and depositing loads requires skill in proper manipulation and safe positioning. An off-center load presents a special hazard in which the load might tip in transit even though the truck itself is in a stable position.

Besides stability problems, visibility can be a problem with lift trucks. The load itself can block the view and necessitate driving with the load trailing. Driving in worker aisles presents problems of pedestrian traffic, especially at corners where visibility is limited. Although lift trucks are not silent, they may seem so in a noisy factory environment. This increases the hazard to pedestrians and the need for greater visibility for lift truck operators.

Passengers on a lift truck can be a hazard in more than one way. For one thing, the truck is often equipped to seat only the driver, and there may be no safe place for a passenger to ride. Hitchhikers also can distract the driver, whose attention is even more important on a lift truck than in an automobile. One practice particularly frowned upon is riding on the lift *forks*. The temptation is great to use a forklift truck as a personnel elevator. Actually this practice can be safe IF:

1. A safety platform is firmly secured to the lifting fork.
2. The person on the platform can shut off power to the truck.
3. Protection from falling objects is provided if needed.

An ordinary wood pallet is not generally considered to be a "safety platform" although it is often used for lifting personnel.

Some readers may consider unreasonable the rule that the person on the platform be able to shut off power to the truck. This rule also stands in the way of usage of an ordinary pallet as a lifting platform. Workers, too, may resist this rule, and the Safety and Health Manager needs to be in a position to counter this resistance with training programs that effectively explain the reasons for the rule. As a rationale for the rule, ask the workers to consider the hazard caused by unexpected obstructions. A small obstruction could:

1. Damage the platform
2. Tip the platform, causing the worker to lose balance
3. Injure the worker on the platform
4. Knock the worker off balance

The forklift driver is really in a poor position, due to distance or angle, to see all obstructions and to judge their distance from the lifted platform. One might argue that the lift may be entirely clear of any obstructions at all, but such lifts would be unusual. There is usually no reason for employees to be lifted unless the lift is adjacent to equipment, stacks of material, or building structure. Any of these items can present obstruction hazards.

A variation of the "forklift rider" is the "carpet pole rider." In carpet warehouses the lift truck is equipped with a single pole which is guided into the spool of a roll of carpet to lift the roll and transport it around the plant. Workers have been known to ride these poles rodeo style to gain access to the top of a stack of carpet rolls. Normally, there should be no reason for a worker to ride the pole because the pole can be raised and guided into position by the driver without assistance.

When manipulating loads the forklift is often operating close to observers, supervisors, or assistants giving directions to the operator. A dangerous place to stand is beneath the elevated fork, whether it is loaded or not. Another dangerous position is between an approaching lift truck and a fixed object or bench.

The foregoing examination of operational hazards should make it clear that a measure of special training for industrial truck operators is appropriate. Equipment dealer representatives can be of assistance, and most equipment manufacturers have prepared training programs for purchasers and operators of their equipment. The Safety and Health Manager should reflect on the following question: If an inspector should visit my facility and flag down a forklift driver and ask the driver to describe the training program for forklift operators, what would the driver say? Too often the response has been disappointing, perhaps even embarrassing.

One thing to check is parked, unattended forklift trucks. The first thing to do is determine whether the lift truck is really unattended or not. If the operator cannot see the truck, it should be considered unattended. Even if the operator can see the truck, if the truck is more than 25 feet away, the truck is unattended. If the truck is unattended, the motor should be turned off. Even if the operator *is* close by, if the operator is *dismounted*, the fork must be fully lowered and the controls neutralized. The next thing to check is whether the brake is set, and if the truck is on an incline, whether the wheels are blocked. Finally, the horn should be tested.

Federal standards take very seriously the question of maintenance, inspection, and service for industrial trucks. No latitude is permitted to allow defective industrial trucks to continue to be operated until the next regular overhaul. Any condition, such as an inoperative horn, defective brake, or broken headlight in a forklift is cause to remove it from use until repaired.

Most people are amazed to learn that federal standards require industrial trucks in use to be given a safety inspection *daily*. Compare this rule with procedures for automobile safety inspections, which most states require *annually*. If the industrial truck is used on a round-the-clock basis, safety inspections are required after every *shift*. It would be wise for the Safety and Health Manager to institute some type of procedure or record to assure that this job gets done and that proof of performance will be kept on file.

A final note on the subject of industrial trucks is the provision of an overhead guard to protect the operator from objects falling from the elevated load. More and more forklifts are being equipped with these overhead guards required for protection against falling objects such as small packages, boxes, and bagged material, not for protection against the impact of a full capacity load. These overhead guards are not to be confused with the much sturdier rollover protective structures (ROPS) described in Chapter 14, although the same structure can be used for both purposes if it is constructed properly. Some loads are unitized, and objects within the load are secured from falling back on the operator. For such applications the hazard to the operator from falling objects disappears, and the overhead guard becomes unnecessary.

CRANES

An industrial truck is convenient for general material handling of palletized loads. But some material handling jobs cannot be handled by an industrial truck. Larger, heavier, more awkward loads require the versatility of a crane, especially if the path of travel is complicated.

To most people a crane is a construction industry device which typically is seen lifting heavy steel beams to lofty places. Although this image is correct, it is incomplete. Cranes are also used extensively in general industry, although they usually take on a different form. Cranes in industrial plants are generally limited in travel by a track or overhead runway structure, typified by the "overhead traveling crane" shown in Figure 10.2. Such cranes are popularly called "overhead bridge cranes" or simply "bridge cranes" by the workers who use them. Some models, such as those in Figure 10.2, are operated from a cab mounted on the crane itself. Others are operated from the floor by means of a hanging cord control called a "pendant," or from a fixed remote station called a "pulpit." "Gantry" cranes have legs which support the bridge above the railway. "Cantilever" gantry cranes have extensions on one or both ends of the bridge; these extensions extend the reach of the crane outside

Figure 10.2. Various configurations for overhead cranes. (ANSI standard B30.2.0-1976, reprinted with permission of ANSI and ASME.)

the area between rails on which the crane travels. One characteristic common to all overhead and gantry cranes is that the trolley, which carries the hoisting mechanism, rides *on top of* the rail on which it travels. Overhead cranes whose trolleys are not so mounted are called "underhung cranes" or "monorails," depending on type or application. Safety standards for overhead and gantry cranes are different from standards for monorails.

The chief concern of the Safety and Health Manager with regard to overhead cranes should be that workers will overload the crane. The rated load should be plainly marked on each side of the crane, and if the crane has more than one hoisting unit, each hoist must have its rated load clearly marked. So easily recognized are crane load markings that their absence is conspicuous.

Even if the rated load marking *is* on the crane, workers will sometimes be tempted to exceed the rated capacity. Everyone knows that engineers provide for a safety factor in their designs, so virtually every crane will support more than its rated load, even without damage to the crane. But uncertainties in the actual weight of the load, dynamic loads during transport, shock loads during lifting, variations in crane components, and unavoidable design variabilities can combine to result in a very dangerous situation even when rated load capacities are exceeded "slightly" or only "occasionally." As with running a stop sign, an alert driver can almost always avoid an accident, but once in a great while another driver will be approaching the intersection at a high rate of speed or the visibility will be obstructed, or the stop sign violator will be complacent or distracted, and a serious accident will occur. It becomes very difficult for the Safety and Health Manager or anyone else in the plant to police the use of cranes properly within their rated load limits at all times. This is where training becomes important so that workers will understand the risks and consequences of their actions. Recalling the principles set forth in Chapters 2 and 3, another way to control this problem is to take the trouble to make a big example out of any accident or near accident which occurs as a result of overloading a crane, even if no injuries occur.

Many overhead cranes operate outdoors where wind can be a hazard. Wind alone is usually not dangerous, but a wind load in combination with a working load can result in dangerous structural damage to the crane. Automatic rail clamps are required for outdoor storage bridge cranes. The idea with these clamps is to lock the bridge to the rail if the wind exceeds a certain velocity. This idea sounds good because it protects against the wind hazards described above and also protects the bridge from unintentional and uncontrolled rolling in a high wind or brake failure. But like some other devices, the safety device carries with it a hazard of its own. Think about it for a moment: In what mode of crane operation would the tripping device most likely engage the automatic rail clamp? The answer is when the bridge is traveling *at top speed* against the wind. The sudden engagement of the rail clamp when the bridge is traveling at top speed is likely to injure the operator in the cab, dangerously swing the load, and/or damage the crane. Therefore, the crane needs either a visible or audible alarm or both to warn the bridge operator *before* the rail clamp engages.

If the crane operator rides in a cab mounted on the bridge or the trolley, somehow the operator must have a way to gain access to his or her station. One idea that comes to mind is to use a portable ladder, but this is not such a good idea. Since the bridge travels, the operator, cab, trolley, and bridge may roam far from the point at which the operator mounted. The idle ladder standing far away would be an invitation for someone to remove it for some other use or perhaps just to get it out of the way. A ladder used for crane cab or bridge access should be of the fixed type. Stairs and/or a platform can also be used but must require no step over a gap exceeding 12 inches.

Overhead cranes designed or constructed in-house sometimes do not adhere to accepted principles for crane design and overlook some of the necessary safety features specified by applicable standards. Besides cab access, discussed earlier, there are specifications for footwalks for safe maintenance of the trolley and bridge. These footwalks need standard toeboards and handrails, as described in Chapter 5.

Ideally, footwalks will have at least 78 inches headroom. Sometimes, however, this is not practical because the crane may be near the ceiling of the building. The standards recognize this difficulty and permit less than 78 inches headroom. However, a footwalk becomes rather ridiculous if the headroom is less than 48 inches and would more properly be called a "hands-and-knees" walk. In these situations footwalks should be omitted, and a stationary platform or landing stage for crane maintenance workers should be installed.

An obviously serious hazard with operation of an overhead bridge crane is overtravel. Devices to control overtravel would include trolley stops, trolley bumpers, and bridge bumpers. Bumpers and stops are somewhat different in that bumpers absorb energy and reduce impact, whereas a stop simply stops travel. The stop is simpler and can consist merely of a rigid device which engages the wheel tread. For safety, such a stop needs to be at least as high as the wheel axle centerline. Bumpers soften the shock by absorbing energy but are not as positive as a true stop. For example, bridge bumpers need only be capable of stopping the bridge if it is traveling at 40% of rated load speed or less. If the crane travels only at slow speeds, shock loads are not significant, and bridge bumpers and trolley bumpers are not necessary. There may be other circumstances of operation, such as restriction of crane travel, which eliminate the need for bridge bumpers or trolley bumpers.

Related to bumpers and stops are rail sweeps. If a tool or some item of equipment happens to obstruct one of the rails on which the bridge travels, a catastrophic accident could occur. Therefore, bridge trucks are equipped with rail sweeps, projecting in front of the wheels, to eliminate this hazard. There is nothing magical about the term "sweep"; even part of the bridge truck frame itself may serve as a sweep.

One obstruction that can be encountered by an overhead crane is another overhead crane which runs adjacent to it with side rails parallel. Proper clearance should be provided between the two adjacent bridge structures. Unfortunately, such clearance may make it impossible for one crane to transfer its load to the domain of the adjacent crane. Some factories resolve this problem by using retractable extension arms on the crane. The trouble with these retractable arms is that the operator may forget and leave them extended, resulting in a collision between the two adjacent cranes.

Electric shock is a concern with overhead cranes, and the concern is principally in two areas:

- Shock from exposure to current-carrying portions of the crane's power delivery system
- Shock from a shorted connection in a hanging control pendant box (see Figure 10.3)

A third area of concern for electrical shock is the accidental contact of live, high-

Figure 10.3. Hand-held pendant control for overhead crane.

voltage overhead transmission lines. This is a hazard for mobile cranes with booms (discussed in Chapter 14) because of their wide use in the construction industry. Exposed live parts in the crane's power delivery system are usually protected by their remoteness or "guarded by location." Some older models might present hazards and need modification.

A greater hazard is the possibility of shock from a pendant control. The electrical conductors can be strained if they are the sole means of support for the pendant. The control station must be supported in some satisfactory manner to protect against such strain. If a failure does occur, it is likely to be at a connection within the box. This raises the possibility of a dangerous short to ground through the operator's body. Pendant controls take much abuse and need to be of durable construction.

Not related to electrical shock but on the subject of pendant control boxes is the requirement that boxes be clearly marked for identification of functions. Some older model or homemade cranes might have pendant control boxes without function markings. Without a great deal of trouble or expense the Safety and Health Manager can check the cranes in-house for compliance and get the control functions marked on the pendant controls.

One hazard to think about with overhead cranes is what would happen if a temporary power failure occurred. Suppose that the crane were in the process of lifting a heavy load by means of its hoist mechanism. Obviously, no one would want the crane to drop its load to the floor upon power failure, but the hazard does not end here. Suppose that the crane bridge is traveling in a horizontal direction, whether loaded or unloaded. Upon power failure the bridge would stop, and this might not be dangerous. But when power is *restored*, dangerous lurching action might occur. In fact, upon a power failure the crane operator might even leave the crane cab!

Several design alternatives can protect against these hazards. One solution is to equip the control console with spring-return controllers. Pendant boxes can be constructed with spring pushbuttons instead of toggle switches. A disconnect device can neutralize all motors and not permit a reconnect until some sort of positive "reset" action is taken. Even if the power remains on, it could be a hazard to inadvertently switch on a lever at the wrong time. Notches, latches, or detents in the "off" position can prevent such inadvertent actions.

Brakes are of obvious importance to the safe operation of a crane. But a large number of crane operators do not use brakes but instead rely on a practice called "plugging" by the industry. The operator merely reverses the control and applies power in the opposite direction, thereby stopping the load. While no OSHA standard prohibits the practice of "plugging," it should be pointed out that plugging is not as effective as applying a brake under extreme conditions, such as stopping a large, fast-moving load. Under no circumstances should a crane operator depend entirely on plugging due to the brake being inoperative. In fact, plugging would be completely useless in event of crane motor failure.

A crane has many moving parts, many of which are located far from the operator's console in the cab or from a floor operator holding a pendant. Moving machine parts are hazardous, and the remoteness characteristic amplifies the hazard. Moving parts are hazardous not only to personnel; they can be hazardous to the crane itself, which in turn can be indirectly hazardous to personnel as well. Hoisting ropes, for instance, can possibly run too close to other parts in some crane configurations and in some positions of the bridge and trolley. The result can be chafing or fouling of the hoist rope. If the configuration of the equipment will permit this situation to develop, guards must be installed to prevent chafing or fouling. Such moving parts as gears, set screws, projecting keys, chains, chain sprockets, and reciprocating components need to be checked to see if they present a hazard, and if so they must be guarded.

As with conveyors and other material-handling equipment, overhead cranes are often large and widely distributed items of equipment. Electrical power is distributed over long distances, sometimes by means of open runway conductors, and portions may be so far away as to be obscured from view from the location of the power supply switch. Picture the insecurity of the maintenance worker who is compelled to repair a crane and is in direct contact with an exposed 600-volt (but deenergized) runway conductor, but the power supply switch is so far away that it is out of sight! Therefore, such switches should be so arranged as to be *locked* in the open or "off" position.

Ropes and Sheaves

Safety standards for wire rope strength state that: "Rated load divided by the number of parts of rope shall not exceed 20% of the nominal breaking strength of the rope." A safety factor of 5 is implied as follows:

$$\frac{\text{rated load}}{\text{number of parts of rope}} \le 20\% \times (\text{nominal breaking strength}) \quad (10.1)$$

It will subsequently be explained that the "number of parts of rope" is a multiplying factor which enables a multiple-sheaved block and tackle assembly to be loaded much higher than the wire rope load. Thus

$$\text{wire rope load} = \frac{\text{rated crane load}}{\text{number of parts of rope}} \quad (10.2)$$

From equations (10.1) and (10.2),

$$\text{wire rope load} \le 20\% \times (\text{nominal breaking strength}) \quad (10.3)$$

Multiplying each side of the (in)equality by 5, we have

$$5 \times (\text{wire rope load}) \le 100\% \times (\text{nominal breaking strength}) \quad (10.4)$$

Rearranging to create a ratio yields

$$\frac{\text{nominal breaking strength}}{\text{wire rope load}} \le 5 \quad (10.5)$$

Since the ratio of the strength to the load is at least 5, the *safety factor* is 5.

The term "parts of rope" refers to the mechanical advantage provided by the block and tackle assembly. "Parts of rope" is computed by counting the number of lines supporting the load block. Of course, all the lines make up one continuous line which is reeved through several sheaves to achieve mechanical advantage. The concept is best explained by a picture; Figure 10.4 shows five different reeving combinations. Note that the mechanical advantage is numerically equivalent to the number of "parts of line."

| 1 part of line | 2 parts of line | 3 parts of line | 4 parts of line | 5 parts of line |

Figure 10.4. Five different reeving combinations. The mechanical advantage is equal to the number of "parts of line" supporting the load block.

One additional caution is in order when determining the appropriate maximum load to be applied to a given reeving setup. The weight of the load-carrying sheave must be added to the weight of the load to be picked up to arrive at the total load of the line. The weight of the load block cannot be ignored, as is emphasized by the massive load block displayed in Figure 10.5.

One way to prevent overloading the wire rope and the crane itself is to provide a hoist motor which can develop insufficient torque to overload the wire rope. The combination of such a hoist motor and correct reeving for the crane design will result in no overloading. Under this arrangement the crane simply will be unable to lift any load that would damage the crane or exceed its safety factor. Most overhead cranes today are designed this way. How fortunate it would be if the human back had this design feature.

Any time a rope is wound on a drum, the rope end anchor clamp bears very little of the load when several wraps are on the drum. The friction of the rope on the drum holds the load. But if the drum is unwound to less than two wraps, dangerous loading of the anchor clamp may result in failure, and the wire rope will break free from the drum. Usually, the overhead crane is set up so that, even if the load block is lying on the floor, several wraps remain on the hoist drum. The floor may not be the extreme low position for the crane, however. The Safety and Health Manager

Figure 10.5. Massive load block. The load block becomes part of total load on the line.

should look around for pits or floor openings into which the overhead crane might operate, resulting in dangerous unwinding of the hoisting drum.

"Don't saddle a dead horse" is a familiar safety slogan which refers to the improper mounting of wire rope clips employing U-bolts. Such a clip assembly bears a resemblance to a saddle, and the U-bolt represents the cinch strap. The U-bolt places more stress on the wire rope and has less holding power than the clip, and therefore should not be placed on the live portion of the rope when a loop is formed. The "dead" end of the rope gets the U-bolt, and the "live horse" gets the clip. Right and wrong methods are illustrated in Figure 10.6. Unfortunately, some workers in the field, unsure of the correct method, place the clips *both ways* in alternating fashion, thinking they are "playing it safe." Such an arrangement may be even more unsafe than "saddling dead horses" with every clip.

Before leaving the subject of wire rope, the hazard of rope whip action should be emphasized. Wire rope "cable" seems so heavy and inflexible that it seems unnatural that it could crack like a whip or any fiber rope. It is difficult for anyone to visualize the tremendous tension forces on a wire rope during use in a material-handling operation, until that wire rope breaks. Most workers have never seen what happens when a wire rope breaks; perhaps this explains why so many workers stand too close while the rope is drawn taut by the load. The hazard is very serious, and an injury accident is very likely to be a fatality.

(a)

(b)

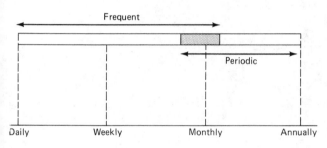

(c)

Figure 10.6. "Don't saddle a dead horse." Right and wrong ways to secure wire rope loops using U-bolt clips. (a) Incorrect—"saddle" is on dead end of rope; (b) incorrect—clips are staggered both ways; (c) correct—all clips are placed with the saddle assembly on the live portion of the rope and the U-bolt on the dead end.

Crane Inspections

Almost everyone is aware of the long, perhaps tedious checklists for inspection of an aircraft every time it flies. An aircraft must not fail in any catastrophic way during operation, and this makes frequent inspections worthwhile, even if they are repetitious and rarely uncover any defects. In some ways a crane is like an aircraft; it, too, must not fail.

With regard to crane inspections the standards use the terms "frequent" or "periodic" to specify when various items on the crane should be checked. Such usage is an attempt not to be overly specific in telling the employer *what* to do and *how often*. Some broad guidelines describing the meaning of these terms are illustrated in Figure 10.7. Note that there is some overlap, as monthly inspections can be considered either frequent or periodic.

Figure 10.7. Inspection intervals for overhead cranes.

The crane manufacturer is a good source for detailed guidance in what to look for in the frequent inspections. This type of inspection is performed by the crane operator, just as a pilot inspects his or her aircraft before a flight. This analogy between aircraft and cranes might be a starting point for a safety theme in a training program for crane operators.

The "frequent inspection" routine should include a daily visual inspection of hoist chains, plus a monthly inspection with a signed report. In the field the term "hoist chains" has been widely misinterpreted to include chain *slings* for handling the load, but there is a separate standard for slings. Figure 10.8 identifies which chain is hoist and which is sling.

Crane hooks take a great deal of abuse and are critical to the safe operation of the crane. Although they usually contain considerable overdesign, damage or wear can reduce the margin of safety. Telltale signs of an abused and dangerous hook are illustrated in Figure 10.9.

5 ton

Hoist chain

Sling chain

Figure 10.8. Hoist with sling. Hoist chain is not to be confused with the sling chain.

Normal throat opening

Excess throat opening must not be greater than 15% of normal throat opening

(a)

10° Must not be greater than 10% twist

Safety latches

Figure 10.9. Defective crane hooks: (a) bent hook; (b) twisted hook; (c) cracked hook.

(b)   (c)

A more thorough inspection of crane components is needed for "periodic" intervals. Whereas daily inspection of crane hooks is merely visual, the periodic inspection calls for a more scientific approach, such as using magnetic particle techniques for detecting cracks. More extensive checks for wear are appropriate also, such as the use of gauges on wire rope sheaves and chain sprockets.

For most kinds of plant equipment safety testing is customarily done by an independent testing laboratory such as Underwriters' or Factory Mutual. But the safety of an overhead crane is in part a function of the installation method and proper adjustments at the site. Therefore, an actual rated load test is needed prior to initial use to confirm the load rating of the crane. This is a bit tricky because if the crane is subjected to too high a load, the crane may fail, but if it is not tested with a high enough load, why conduct the test? Standards specify that the maximum load during the test should go 25% higher than the crane's load rating. This will provide some assurance that the crane will withstand its rated load. During use, however, the

rated load 100
+25% load test 125 = rated load of 100

crane should not be loaded higher than its rated load. Any extensive repair of altera-tion of the crane may subject it to a retest.

Wire Rope Wear

The two principal moving parts of an overhead crane are the wire rope and the drum and sheaves upon which they travel. In addition, the wheels for the bridge and trol-ley move as do a few other items on the crane. Whenever parts move and contact other parts, they are subject to wear. Wear on the bridge and trolley wheels may eventually lead to problems but is not likely to cause crane failure. Wear on the drum and sheaves is more dangerous because of the harm this can bring to the wire rope. But worn sheaves and drum *alone* are generally not the *direct* cause of crane failure. The wire rope remains as the most critical moving part. With continued use every wire rope will eventually wear out and fail, a hazard that generally is not toler-able. Some way has to be devised to predict the failure of wire rope and withdraw it from use before a catastrophe.

It is not a simple matter to determine when a wire rope needs replacing. A wire rope has many individual wires, and it is easy to cut or break away any one of these small wires. Almost everyone has seen broken wires in a old wire rope (or "cable" as laymen often say) which raises the question: Is such a wire rope dangerous? Safety and Health Managers do not survive long in their companies if they go around ordering wire rope replaced every time they find a broken wire. And if they did sur-vive, their companies would not survive. Before delving into this issue, some basics about wire rope need to be examined.

A wire rope should really be considered a machine because the individual wires move upon each other while the rope flexes, resulting in friction and wear. Further-more, unless the individual strands are able to move properly during flexure, tremendous tensile stresses may be placed on some of the wires, causing them to break. Rust, kinks, and other types of abuse can interfere with the movement of the wires and lead to the stresses that cause wire breakage. As some wires break, the ten-sile stresses on other wires increase, leading to their breakage as well. Eventually, the force of the crane load will be sufficient to overcome the tensile strength of all re-maining wires combined, and the wire rope fails.

Even a well-maintained wire rope will be subject to *wear* on the individual wires, especially the outside wires. As the wear causes the diameter of the wire to be-come smaller, the tensile force increases the tensile stress on that single wire due to its reduced cross-sectional area. Thus the worn wire, too, may break due to stress concentration, even if there are no kinks or rust, and even if the wires move properly to distribute the load among all wires.

For obvious purposes, a wire rope is overdesigned and will withstand more than its rated load. Equally obvious from the discussion above is that all wire rope will eventually develop broken wires with continued use. Broken wires are permis-sible to an extent, but beyond a certain point the rope becomes dangerous. A means of evaluating the *degree* of wire rope deterioration is awkward and difficult but nevertheless *necessary* to prevent catastrophic failure.

The ANSI standard recommends[1] a procedure for counting the broken wires. Figure 10.10 diagrams the components of wire rope, defining the terms "strand" and "lay." If there are more than 12 randomly distributed broken wires in a single strand in a single lay, there should be cause for questioning the continued use of the rope. A good place to look for broken wires is around end connections. Sometimes an expert can evaluate the remaining strength in a deteriorated rope after inspection. This possibility is acknowledged by the ANSI standard, which advocates the use of good judgment.

Another measure of rope condition is the amount of reduction of rope diameter below nominal. Figure 10.11 shows that when gauging a wire rope, the caliper can be placed upon a small diameter or a larger diameter. The convention for wire rope terminology is to use the large diameter for designation as the nominal diameter of the rope. Most people pay little attention to wire rope, but wire rope is no small matter considering that a quarter million tons of wire rope is sold annually.

(a)

(b)

Figure 10.10. Wire rope components. (a) This wire rope has six strands plus an inner core. Do not confuse "strands" with "wires"; (b) Wire rope lay. "Lay" is one complete wrap of a single strand around the rope. Since this rope has six strands, lay is the length from the first hump to the seventh hump, as shown in the diagram. (Courtesy of the Construction Safety Association of Ontario, Ref. 69.)

[1]ANSI B30.2–2.4.2

Correct way
This gives correct diameter

Incorrect way
This does NOT give correct diameter

Figure 10.11. Gauging wire rope. (Rotate rope to select largest diameter.) (Courtesy of Construction Safety Association of Ontario.)

Operations

The actual handling and moving of the load by the crane is a function of the skill and attitude of the crane operator and the workers who attach and secure the sling or lifting device. As with motor vehicles, the *operator* of the crane is probably the most important factor in preventing accidents.

A good deal of skill is required to attach the load safely, especially if a sling is used. Slings will be discussed later in this chapter. The hoist rope is not intended for wrapping around the load, as this kind of abuse may damage the rope and at the same time be an inadequate support for the load. Misplacing the attachment off the line of the center of gravity can cause dangerous swings when the load is lifted. After the load is lowered the tendency is to consider the hazards removed. But disengaging the load attachment can also result in dangerous shifts in material which can injure the inexperienced or unwary worker on the floor.

SLINGS

Slings are used to attach the load to the crane, helicopter, or other lifting device. Slings come in a great many varieties and are very important in the safety of material-handling operations. Components of the sling assembly are often subjected to much larger forces during lifting than the hoist rope or other material-handling equipment. Because the skill of the user is so important in the proper application of the sling, slings are often misused, resulting in much more abuse and damage than to the components of the crane.

The most important point to remember for the safe use of all slings is that the stress on a sling is greatly dependent on the way it is attached to the load. Figure 10.12 shows two different ways of applying a sling to pick up identical loads. If the angle of the supporting legs of a sling is sharp, as in Figure 10.12b, the advantage of

Figure 10.12. Comparison of sling tensile forces for two different methods of lifting identical loads: (a) tensile force on the sling is approximately 1150 lb; (b) tensile force on this sling is approximately 2000 lb.

multiple legs can be lost. Using too short a sling is the most common cause for this condition. The "rated capacity" of a sling is the working load limit under *ideal* conditions; if the sling is applied at leg angles other than specified in the rated capacity table, the capacity can be drastically reduced due to the physics of the applied forces. Therefore, "rated capacity" is an incomplete term without the accompanying leg angle.

Note the following progression in capacities of ½-inch alloy steel chain as the number of legs increase:

Single (at vertical)	11,250 pounds
Double (at 60° from horizontal)	19,500 pounds
Triple (at 60° from horizontal)	29,000 pounds
Quadruple (at 60° from horizontal)	29,000 pounds

Intuitively, the capacity of the sling increases as the number of legs or supporting members is increased. But note that no increase in capacity is shown in going from three legs to four legs. The reason for this is that, as in a four-legged chair, three legs will actually carry the load. At times the weight distribution might be equal among legs of the sling, but usually the balance is not so perfect. As the load shifts around it is entirely possible for one of the four legs to become slack and for the other three to bear the entire load.

Alloy steel chain, besides being very strong, is very durable and capable of withstanding the physical abuse that industrial slings routinely receive. Ordinary carbon steel chain as is commonly found in hardware stores should not be used for slings. Wire rope slings can be just as strong as steel chain slings, but unfortunately wire rope is more subject to wear. Individual wires in the wire rope are more easily broken, with eventual incapacitation of the sling.

Wire rope slings also have a specification for the permitted number of broken wires. The rule is no more than 10 randomly distributed broken wires in one rope lay, or five broken wires in one strand in one rope lay. Note from our earlier discussion that this is a bit more strict than the requirement for wire rope for overhead cranes.

Selection of the proper sling for a given application can consider several factors besides rated load. The nature of the item to be lifted, its surface finish, temperature, sling cost, and environmental factors must be considered. The Safety and Health Manager is usually not the person who makes this decision, but there is increasing reason for the Safety and Health Manager to have a voice in what is done in this area. Many factory superintendents, foremen, and material-handling workers are not aware of the many federal standards confronting the criteria for selection of slings. In fact, many of these personnel do not even have a clear understanding of the inherent hazard mechanisms surrounding industrial slings. Therefore, it is recommended that the Safety and Health Manager provide consultation and advice in the selection and use of industrial slings in the interest of worker safety.

For some criteria, such as load markings, repair procedures, proof testing, and operating temperatures, the requirements for various types of slings are not identical, and even vary in curious ways. Some of this is due to the physical differences in sling types, and some is due to the various origins and rationales for the requirements for various slings. Table 10.3 is intended to summarize some of the more curious differences between requirements for various types of slings. It is by no means exhaustive. For instance, nylon web slings are not permitted in the presence of acid or phenolic vapors. In the case of caustic vapors, polyester and polypropylene web slings and web slings with aluminum fittings are not permitted. The Safety and Health Manager can use Table 10.3 as a first check in an in-house inspec-

TABLE 10.3 COMPARISON OF CERTAIN REQUIREMENTS FOR SLINGS

Type of sling	Rated capacity markings required?	Repairs permitted?	Employer required to maintain records of periodic inspection?	Employer required to maintain a *certificate* of proof test?	Safe operating temperature (°F) Maximum	Safe operating temperature (°F) Minimum	Proof test of *sling* required?	Repairs records required? [a]
Alloy steel chain	Yes	Yes	Yes	Yes	1000[b]	None specified	Yes	No, except for welding or heat treating
Wire rope								
With fiber core	No	Yes	No	Yes, for welded end attachments	200	None specified		No
With nonfiber core	No	Yes	No		400[c]	−60[c]	Yes, for welded end attachments	No
Metal mesh	Yes	Yes	No	No	550[d]	20[d]	Yes	Yes
Natural fiber rope	No	No	No	No	180	20	No	N/A
Nylon fiber rope	No	No	No	No	180	20	No	N/A
Polyester fiber rope	No	No	No	No	180	20	No	N/A
Polypropylene fiber rope	No	No	No	No	180	20	No	N/A
Nylon web	Yes	Yes	No	Yes, for repaired slings	180	None specified	Yes, for repaired slings	Yes
Polyester web	Yes	Yes	No	Yes, for repaired slings	180	None specified	Yes, for repaired slings	Yes
Polypropylene web	Yes	Yes	No	Yes, for repaired slings	200	None specified	Yes, for repaired slings	Yes

[a] N/A, not applicable.
[b] If alloy steel chains are working in temperatures exceeding 600°F, load limits are reduced.
[c] Seek manufacturer's recommendations for use outside temperature range.
[d] Impregnated metal mesh slings have more restricted temperature requirements.

tion or purchase decision; then further details should be checked out in the standards.

It must be reemphasized, however, that the skill and training of the worker who uses the sling to attach the load is more important than all of the detailed specifications and sling standards combined. This is a good place for the reader to reflect on the fatality described in Chapter 3, in which two inexperienced workers attached the hoist hooks directly to the bundling wires for a 40-foot bundle of channel steel! The bundling wires were no substitute for an industrial sling, and one of the wires gave way; the falling steel killed one of the workers.

CONVEYORS

Hazards with conveyors can be quite serious, and workers seem to sense these hazards more than with some other types of machines. It seems that all of us have implanted somewhere in our imaginations the vision of the beautiful maiden tied to the conveyor in the sawmill about to be sawed in half unless she submits to the villain, or in another version, signs over the deed to her ranch. The truth is that sometimes workers *are* caught in industrial conveyors and not only killed but dismembered or even pulverized beyond recognition. The horror of this truth installs a healthy respect on the part of most workers around industrial conveyors.

By contrast, some of the worst conveyor hazards are very innocent in appearance. In-running nip points which themselves might not even seriously injure a hand or arm can start an irreversible process once the employee is caught, resulting in an employee's entire body being drawn into a machine. Loose clothing in particular can get caught, and the employee, even before being injured in the slightest way, can become doomed.

Belt Conveyors

On a belt conveyor, in-running nip points are seemingly everywhere. Pulleys are required to drive the belt, change direction of the belt, support the belt, and tighten the belt. One side of every pulley is always an in-running nip point. Defense against this hazard generally consists of one of three means: isolating the nip points, installing guards, and installing emergency tripping devices.

The best method of protection is to isolate the in-running nip point so that an employee would not or could not come into the danger area. If isolation is impractical, a guard can sometimes be installed to keep out the worker's body or extremities. The design of the guard must vary with the application, and sometimes it is difficult to make the guard practical because it may interfere with the operation of the conveyor. Because of body geometry, distance from the danger zone is a design factor in building the guard, a principle that is covered in more detail in Chapter 11.

If both isolation and guarding are impossible or infeasible for the application, the workers can be protected by some type of emergency tripping mechanism. A wire or rope can be run along the length of the conveyor so that a worker, falling into the conveyor, can grab the trip wire and stop the machine. Unfortunately, this method of protection requires the overt action of an alert worker or coworker.

Overhead Conveyors

Large appliance parts or vehicle assemblies are often handled by overhead conveyors. Hooks attached to a moving chain support each item as it is moved. This type of conveyor is particularly suited for products that have delicate or finished surfaces because so little of the conveyor actually contacts the product. For the same reason, overhead conveyors are very useful also for paint spray or finishing operations.

Overhead conveyors avoid many of the risks of belt conveyors by eliminating many of the in-running nip points and by removing the moving parts from worker access. But overhead conveyors have hazards of their own, such as dropping conveyored materials to the plant floor or upon worker stations. Screens or guards can protect against this hazard but not entirely, because the moving parts must be accessible to work stations for processing. A good rule is to place screens or shields under the conveyor whenever it passes over an aisle or other area where personnel are likely to gather. Another good place for screens is where the conveyor chain moves up or down an incline. Such movements cause the loads to shift on their hangers and increase the possibility of a falling load.

Figure 10.13 shows three different orientations for the hangers or hooks which support the work held by an overhead conveyor. Note how much safer is the orientation in which the work is held in front of the hook. If the work encounters an obstruction, it is more likely to catch and stop the conveyor if the work is in front of the hook. If the work trails the hook, an obstruction may lift and knock off the load.

Figure 10.13. Three different orientations for conveyor hooks: (a) hook trails the load; (b) side orientation; (c) load trails the hook (dangerous).

Screw Conveyors

Screw conveyors can be very dangerous. The very principle of their operation is an ingoing nip point at the intake. Complicating the hazard is the fact that in order to operate at full capacity the intake must be completely submerged in the material to be transported. Submerged usually also means hidden, so an unseen serious hazard exists at the intake. Finally, there may be a need for the worker to be fairly close to the screw conveyor for many applications, in order to shovel or distribute material into the intake.

A simple and often good way to protect workers from the hazards of the screw intake is to box the intake area in a small screen enclosure which allows passage of the material but keeps out fingers, hands, and feet. If even a coarse screen mesh is too fine to permit passage of the material, an enclosure with larger openings may be necessary, perhaps with openings large enough to admit a finger or hand. This type of enclosure can be made safe, too, by making the box large enough that the worker's *reach* will not permit entry of hands or fingers into the danger zone even if the openings are large enough for hands or fingers. This follows principles of machine guarding which are discussed in more detail in Chapter 11.

LIFTING

Before closing this chapter on materials handling we return to the subject of lifting. At the beginning it was stated that back injuries, mostly from lifting, are the biggest compensable injury category of all. Lifting injuries are very complex and very difficult to control. Naturally, the amount of weight lifted is important, but many other factors determine whether an injury occurs. Even a lightweight lift such as 5 to 10 pounds can cause serious back injuries if conditions are just right (rather, wrong). The physical condition of the person doing the lifting is also important.

Much emphasis has been placed on technique, with the most frequently heard saying being:

Lift with your legs, not with your back.

Unfortunately, the rule is rather difficult to follow because almost everyone is able to lift a heavier weight with the back than with the legs. Lifting with the legs requires squatting down and then lifting both the load and the lifter's body. This requires a great deal of leg strength for heavy lifts and is especially difficult when the worker is unaccustomed to lifting with the legs. Training and exercise with light loads can help in developing the technique, but there are other disadvantages of lifting with the legs. Park and Chaffin have shown that if the shape of the load is such that it must be brought out in front of the knees, lifting with the legs *increases* the compression force on the lower back! Also overlooked by the often-quoted rule is the fact that lifting with the legs takes as much as 50% more energy as lifting with the back, especially when the load is light and the frequency of lifting higher, as can be concluded from Figure 10.14.

The capacity to lift varies greatly with the horizontal position of the load, which is determined largely by the shape of the object being lifted. Various independent studies of this relationship have been analyzed by NIOSH, resulting in a proposed specification for maximum weight lifted versus the horizontal distance of the load from the center of gravity of the body. This specification is summarized in Figure 10.15, but it must be remembered that the graph represents merely a NIOSH recommendation—not an established standard.

SUMMARY

This chapter has brought recognition to the fact that the handling of material in a manufacturing plant can be as hazardous as the industrial process itself. The basic nature of materials-handling hazards were examined; then hazards of specific machines and equipment were discussed.

The Safety and Health Manager needs to be aware of not only the proper safety features to seek in new equipment, but also the inspection, service, and maintenance of equipment in-house. However, for industrial trucks, cranes, slings, and perhaps *all* material-handling equipment, the operator skill, attitude, and

Figure 10.14. Comparison of the number of lifts that can be made using two different lift postures while consuming identical amounts of energy. (From NIOSH, Ref. 51; adapted from Ref. 28.)

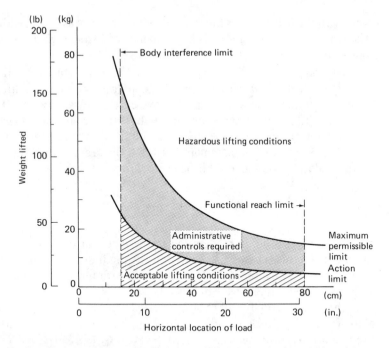

Figure 10.15. NIOSH-recommended specification for maximum lift weights at various horizontal distances for infrequent lifts from floor to knuckle height. (From NIOSH, Ref. 51.)

awareness of hazards are probably more important to the safety of the worker than are the safety features of the equipment itself.

EXERCISES AND STUDY QUESTIONS

10.1. Why are four-legged slings rated no higher than three-legged slings?

10.2. An often-heard safety slogan is "Don't saddle a dead horse!" What does this slogan mean?

10.3. What *safety* design feature do most modern overhead cranes have that is unfortunately not characteristic of the human back?

10.4. Name a part of the plant the Safety and Health Manager should give special attention to when production and sales are booming.

10.5. In the diagram, which orientation of the 4000-pound load will place less stress on the sling that handles it?

(a) (4 0 0 0 lb)   (b)   (4 0 0 0 lb)

10.6. Suppose that a company was able to save some money by trading in its Type LPS forklift truck for a Type DY. Would this introduce some safety problems? What if the trade were from DY to LPS? $p - 197$

10.7. Why might spring pushbuttons be a better crane control than toggle switches?

10.8. In the diagram in Figure 10.16, what is the mechanical advantage? If the nominal breaking strength of the rope is 5000 pounds and the load block weighs 200 pounds, what is the maximum rated load of the hoist?

10.9. What is a rail sweep, and why is it needed?

10.10. Name in order of preference three methods of protection for in-running nip point hazards on conveyor belts.

10.11. Why are overhead crane access ladders specified to be of the fixed type?

10.12. Because of rising gasoline costs, a company desires to convert existing gasoline-powered forklifts to LPG powered. What implications would such a decision have?

10.13. Identify a "performance" standard in the standards for materials handling. Explain why it is a performance-type standard.

10.14. Explain the following terms as applied to industrial cranes: bridge, trolley, pendant, pulpit, gantry, cantilever gantry.

10.15. Name at least four characteristics of forklift trucks that require more skill for safe operation than is required for operation of an automobile.

10.16. At least four general characteristics of materials handling contribute to its intrinsic hazard potential. Name and explain four such characteristics.

Figure 10.16. Blocks and pulleys for Question 10.8.

11

Machine Guarding

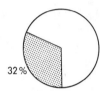

Percent of the OSHA
general industry citations
addressing this subject

32%

Most people think of a machine guard when industrial safety is mentioned, and there is good reason. More efforts and resources have been expended to guard machines than in any other industrial safety and health endeavor. To modify or guard a single machine is generally not a major project when compared with installing a ventilation system or a noise abatement system. But although each machine guarding modification is usually small, the aggregate becomes a major undertaking involving plant maintenance, operations, purchasing, scheduling, and of course the Safety and Health Manager. The Safety and Health Manager should take a leadership role in the implementation of machine guards—enumerating problem areas, setting priorities, selecting safeguarding alternatives, and assuring compliance with standards.

GENERAL MACHINE GUARDING

If Safety and Health Managers are to be able to "enumerate problem areas" and "set priorities" as suggested above, they need to be knowledgeable about what makes a machine dangerous. Despite wide differences between machines, some mechanical hazards seem to be shared by machines in general, and these mechanical hazards will be discussed first.

Mechanical Hazards

Following are general machine mechanical hazards listed in approximate order of importance:

226

1. Point of operation — *most accidents*
2. Power transmission
3. In-running nip points
4. Rotating or reciprocating machine parts
5. Flying chips, sparks, or parts

In addition to the mechanical hazards listed above are others, such as electrical, noise, and burn hazards. These hazards, though, are usually controlled by other methods and are covered elsewhere in this book. It is predominantly the mechanical hazards which are controlled by machine guards, the subject of this chapter.

Although the sequence of priority in the list above is only approximate, there is no question about which hazard should be at the top of the list. By far the largest number of injuries from machines occur at the "point of operation" where the tool engages the work. This machine hazard is so important that it is discussed in a separate section later in this chapter, detailing various strategies and guarding devices in controlling the hazard.

Power transmission apparatus of the machine, typically belts and pulleys, are the second most important general machine hazard. Belts and pulleys are usually easier to guard than is the point of operation. Access to the belts and pulleys is usually necessary only for machine maintenance, whereas the point of operation must be accessible, at least to the workpiece, every time the machine is used. Although belts and pulleys are easier to guard, they are easily overlooked by the Safety and Health Manager. A section in this chapter is devoted to belts and pulleys, due to their overall importance to safety.

Machines that feed themselves from continuous stock generate a hazard where the moving material passes adjacent to or in contact with machine parts. This hazard is called an "in-running nip point" or an "ingoing nip point." Even on machines not equipped with automatic feed, in-running nip points occur where belts contact pulleys and gears mesh. Figure 11.1 shows examples of in-running nip points. In-running nip points are not only direct hazards but can cause injury indirectly by catching loose clothing and drawing the worker into the machine.

Rotating or reciprocating moving parts can present hazards similar to those of belts and pulleys and in-running nip points; in truth there is overlap between these categories. But rotating or reciprocating moving parts bring to mind other parts of the machine that might need guarding. Particularly dangerous is a part of the machine that moves *intermittently*. During the motionless part of the cycle workers might forget that the machine will later move. Material-handling apparatus, clamps, and positioners are in this category. The most intermittent motion of all is *accidental* motion. It pays to consider what would happen in event of a hydraulic failure, broken cotter pin, loosened nut, or some other accidental occurrence. Would the guard on the machine protect workers in these circumstances? Is the risk of occurrence of enough significance to warrant installing a guard?

The fifth item listed earlier, "flying chips, sparks, or parts," is not necessarily the least significant; it merely stands alone as a somewhat different category. Many

Figure 11.1. In-running nip points. (From OSHA, Ref. 11.)

machines obviously throw off chips or sparks from the area of the point of operation. Flying objects should also be included because sometimes the product being
manufactured actually breaks, and pieces may be hurled at the operator. It is also

possible for parts of the machine to break and fall on or be thrown at the operator. One means of protecting workers from flying chips and sparks is by personal protective equipment. But this is not nearly as effective as using machine guards to protect the operator and other workers in the vicinity. Often, the term "shield" or "shield guard" is used to describe guards that protect the operator from flying chips, sparks, or parts.

Guarding by Location or Distance

The easiest and cleverest way to guard a machine is not to use any physical guard at all but to so design the machine or operation so as to position the dangerous parts where no one will be exposed to the danger. This is usually in the realm of machine design, and more and more attention is being given to safety in modern machine designs. But even without altering a machine, the machine can be turned and backed into a corner so that its belts, pulleys, and drive motor are impossible to reach during normal operation. A good example would be a portable concrete mixer. Admittedly, this strategy makes the motor and drive difficult to reach for maintenance, but on the other hand, so do ordinary bolt-on guards.

Guarding "by distance" refers to the protection of the operator from the danger zone by setting up the operation sequence such that the operator does not need to get close to the danger. For some difficult-to-guard machines, such as press brakes, the distance method of guarding the point of operation (see Figure 11.2) is expressly permissible. Press brakes are used to bend sheet metal, and their long beds

Figure 11.2. Press brake guarding "by distance."

make it difficult to guard their points of operation. When the workpiece is a large piece of sheet metal, the operator of necessity stands well back from the point of operation and is thus safeguarded by "distance."

Tagouts and Lockouts

A surprising number of industrial machine accidents occur not when the machine is in operation but when it is down for repair. A worker simply turns a machine back on, not realizing that it is down for repair and that a maintenance worker is still close to or inside the machine!

Such accidents sound like freak occurrences, but that is because most of us are more accustomed to small machines around the home where only a few persons, usually family members, may be in the area. But in a factory the machines may be large, and their repair status may not be obvious. Large numbers of personnel may have access to the machine, and a miscommunication between operating supervisors and maintenance crews can easily occur. Human beings have been literally mangled and digested by large industrial machines.

Two simple safety procedures for preventing accidents of this type are the "tagout" system and the "lockout" system. In the tagout system the maintenance worker places a tag on the on–off switch or control box, so that anyone who might have occasion to turn the machine back on will be warned by the tag to leave it alone. The lockout system (see Figure 11.3) provides positive protection to the maintenance worker because he or she is the only person who has the key to the lock-

Figure 11.3. Lockout system for protection of maintenance workers while the machine is being repaired.

out. Note in Figure 11.3 that the lockout has several positions for separate padlocks for each maintenance worker exposed. Each maintenance worker therefore has an element of personal responsibility and control for his or her safety while working on the machine.

The tagout system is simpler, but the lockout system is gaining in favor. It would seem that a lock would not be necessary, because after all, who would turn on a machine when a maintenance worker's tag warned not to? But factories are operated by humans, and various mistakes can lead to a tagout accident. For instance, the maintenance worker can forget to remove the tag after the repair work is complete. Operating personnel may think that the maintenance personnel have forgotten to remove the tag and ignore it. The reader can no doubt think of other scenarios which would lead to an accident with tagouts but which would have been prevented with the lockout. When the maintenance worker has the only key to the lock, there is no way for an operator to turn the machine back on, that is, with the assumption that the maintenance worker was careful enough to actually use the lock.

Interlocks

open door to dryer; it stops

Contrasted with the lockout is a safety device called an "interlock." Modern clothes dryers stop rotation as soon as the door is opened, and thus comply with industry safety standards for revolving drums, barrels, and containers. Even if the drum itself is closed, its rotation can present a hazard unless it is guarded by an enclosure. An interlock between the enclosure and the drive mechanism is specified to prevent rotation whenever the guard enclosure is not in place.

A tumbling machine is a popular industrial machine which uses a revolving drum to rotate metal parts in the presence of an abrasive tumbling medium to improve surface characteristics of the metal parts. Many tumbling machines found in industry do not have interlocked enclosure guards, and Safety and Health Managers should check for such deficiencies.

Trip Bars

turn off

Large machinery layouts are often difficult to guard but can be provided with trip bars that stop the machine if the operator falls into or trespasses into the danger zone. The operator's hand or body deflects the bar that trips a switch. Figure 11.4 illustrates an emergency trip bar on a rubber mill, a very dangerous machine.

Sometimes it is impractical to place the trip bar such that anytime the worker falls into the danger zone the bar will trip automatically. An alternative in these situations is to provide a triprod or tripwire for the worker to grab to switch off the machine. Examples are seen in Figure 11.5. Such devices deserve some careful study and experimentation to be sure workers are able to reach the triprod or tripwire if they get into trouble.

Figure 11.4. Pressure-sensitive body bar on a rubber mill. (From OSHA, Ref. 11.)

Figure 11.5. Safety tripods and tripwires. (From OSHA, Ref. 11.) *(continued)*

Figure 11.5. *(continued)*

Fan Blade Guards

A very prominent safety standard is the one that requires fan blades to have guards whose openings do not exceed 1/2 inch. Literally millions of ventilation fans are distributed throughout industries all over the country, and many of them have guards with openings larger than 1/2 inch. New fan designs are generally built according to the latest standard, but a principal problem is the retrofit of old fans to meet the 1/2-inch-opening-size standard.

An enterprising safety equipment manufacturer got the clever idea of marketing a nylon mesh which could be wrapped around the existing noncomplying guard and then drawn up tightly with a drawstring in the rear (see Figure 11.6). The nylon mesh was 1/2-inch gauge, and if it was drawn tight, it kept workers free from danger. Before rushing out to purchase these inexpensive nylon mesh guards, the Safety and Health Manager should realize that they are not a panacea. First of all, the nylon mesh, or any guard for that matter, cuts the efficiency of the fan. Furthermore, any fan will gradually accumulate oil and lint on its blade surfaces and on the guard. The nylon mesh seems even more susceptible to oil and lint buildup than does the metal guard, generating a nuisance to maintenance in keeping the guard clean or else resulting in a gradual deterioration in the effectiveness of the fan.

To deal with the fan blade guard dilemma, many innovative guard and fan blade designs are beginning to appear to the market. Some of the new guards are removable for easy washing, and some small plastic-bladed fans have no guards at all. A very lightweight fan with plastic blades, powered by a small motor, presents no hazard to personnel. With no guard at all, the increased efficiency can go a long way toward making up for the low-power motor on such an unguarded fan.

Figure 11.6. Nylon mesh guard for ventilation fan blades.

Anchoring Machines

Another troublesome machine guarding standard is the rule for anchoring fixed machinery to the floor to keep it from "walking" or moving. Such anchoring is required for all machines "designed for a fixed location." Machines that have reciprocating motions such as presses have a tendency to "walk" unless securely anchored. Drill presses and grinding machines can also be hazardous unless anchored.

One interpretation of the phrase "designed for a fixed location" is any machine that has mounting holes in the feet or bases of the legs. It is true that these holes are for the purpose of anchoring machines, but the mere presence of the holes is not proof that the machine must be anchored. The mounting holes might simply be a convenience feature for ease in shipping or to permit the machine to be mounted at the discretion of the user for whatever reason, such as security purposes instead of safety.

SAFEGUARDING THE POINT OF OPERATION

Injury statistics bear witness to the fact that the point of operation is the most dangerous part of machines in general. On some machines the point of operation is so dangerous that some type of safeguarding is required for every setup; mechanical power presses is one example. Taking guidance from the specific rules for mechanical power presses, the Safety and Health Manager can extend the principles to other machines because most of the safeguarding methods specified will work also for other machines.

A general classification of methods of safeguarding the point of operation is into guards and devices as follows:

1. Guards
   a. Die enclosures
   b. Fixed barriers
   c. Interlocked barriers
   d. Adjustable barriers
2. Devices
   a. Gates
   b. Presence-sensing devices
   c. Pull-outs
   d. Sweeps (no longer accepted for mechanical power presses)
   e. Hold-outs
   f. Two-hand controls
   g. Two-hand trips

The reader will note that nowhere in the list above do hand-feeding tools such as tongs appear. Hand-feeding tools (see Figure 11.7) are helpful in eliminating the *need* for operators to place their hands in the danger zone, but it must be emphasized that these tools do *not* qualify as guards or devices for safeguarding the point of operation.

Guards

The function of a guard is obviously to keep the worker out of the danger area, but many guards do not succeed in this function. Some guards merely screen part of the danger area around the point of operation, but this can be hazardous. Too many workers will defeat the purpose of the guard by reaching through, over, under, or around the guard, exposing themselves to perhaps a greater hazard than if the guard were not present. Of course, every guard has to be removable between jobs for maintenance or setup purposes, but even this must be kept inconvenient, or operators will take it upon themselves to unfasten the guard to get it out of their way. Wing nuts or quick-release devices should not be used to secure guards. Nuts and bolts are better, but even better than ordinary nuts and bolts are recessed head fasteners such as Allen screws.

Most guards are metal, and popular design utilizes expanded metal, sheet metal, perforated metal, or wire mesh as filler material. A secure frame is needed to maintain the structural integrity of the guard. When a guard panel becomes larger

Duckbill pliers

Pliers with curved handles
for two-handed usage

Feeding tongs

Right angle jaw tongs for handling
flanged or cup-shaped work pieces

Lightweight tweezer made of steel spring

Vacu-tongs for feeding, positioning,
and retrieving heavy
formed parts

Magnetic lifter—"twist-off"

Tong devised to
fit tube or cup

Double magnet with release lever

Double-cup lifter with release button

Push stick

Stock

Push block

Figure 11.7. Hand-feeding tools help but do not replace the function of a point of operation guard. (From OSHA, Ref. 11.)

than 12 square feet, the rigidity of the guard is in jeopardy, and additional frame components are needed. Many types of ordinary wire mesh are unsuitable because the wires are not secure at the cross points. Ordinary window screen would be in this category. Galvanized screen is better as are some types that are welded or soldered at the cross points.

One principle of machine guarding borrowed from the power press standard is the maximum permissible opening size. A contemplation of human anatomy results in the conclusion that the farther away from the danger zone, the larger can be the openings in the guard without creating a hazard. If the guard is at arm's length from the danger zone, an opening of several inches still might not be dangerous. But if the guard is immediately adjacent to the danger zone, no opening should be large enough to permit a finger to reach through. Standard guard opening sizes are speci-fied in Table 11.1. The principle behind the standard openings is illustrated in Figure 11.8. Some companies have made available a simple go/nogo guard gauge (see Figure 11.9): Insert the point of the gauge through the guard. If the gauge reaches the danger zone, the guard opening is too large.

TABLE 11.1 OSHA'S SPECIFICATION FOR MAXIMUM PERMISSIBLE GUARD OPENING SIZE VERSUS DISTANCE FROM THE POINT OF OPERATION

Distance of opening from point of opera-tion hazard (in.)	Maximum width of opening (in.)
½–1½	¼
1½–2½	⅜
2½–3½	½
3½–5½	⅝
5½–6½	¾
6½–7½	⅞
7½–12½	1¼
12½–15½	1½
15½–17½	1⅞
17½–31½	2⅛

Source: OSHA Standard 1910.217, Table O-10.

Figure 11.8. Maximum permissible guard opening should depend on distance to the danger zone.

Figure 11.9. Guard opening size gauge. (Courtesy of Rockford Safety Equipment.)

Visibility through the guard is a problem with some guards. An old practice was to paint all guards orange. But the bright orange color makes it difficult to see through the guard to the point of operation of the machine; black is by far the superior color for point of operation guards.

Die Enclosures

Punch presses and similar machines have mating dies that close on each other to act as the point of operation. The space between the upper and lower dies is the danger area, and the die enclosure guard is designed to enclose only this small area. The advantage over other guard types is that the die enclosure guard is small, but still it is not the most popular guard. Since dies vary widely in size and shape, the die enclosure guard must be essentially custom-made to fit the die or at least the die "shoe" which acts as a base to hold the die. Another disadvantage is that the die enclosure guard is essentially right at the point of operation, which permits no latitude for the guard mesh size or grillwork spacing. The maximum permissible opening size at this point is ¼ inch, and this may limit visibility. Figure 11.10 illustrates a die enclosure guard.

Die stop

Handle

Die slide

Figure 11.10. Die enclosure guard used with sliding die for feeding. (From OSHA, Ref. 11.)

Fixed Barriers

Fixed barrier guard is a general term for a wide variety of guards which can be attached to the frame of the machine. Figure 11.11 shows one example of a fixed barrier guard, but remember that there is no set style or shape for fixed barrier guards. Even the mesh or spacing of the bars is variable, dependent on the distance of the guard to the point of operation (refer back to Table 11.1). Large fixed barrier guards can permit large distances between the guard and the point of operation and a coarser mesh for the guard material.

Figure 11.11.   Fixed barrier guard. (From OSHA, Ref. 11.)

Interlocked Barriers

More sophisticated is the interlocked barrier guard shown in Figure 11.12. An interlock, typically electrical, disables the actuating mechanism whenever the guard is opened. But the interlock is not required to stop the machine if it has already been tripped, and thus it usually affords inadequate protection to an operator who is at-

Figure 11.12.   Interlocked barrier guard.

tempting to hand-feed the machine. If the barrier is so easy to open and close that the operator can open it and reach in while the machine is still moving, the interlocked barrier is not doing its job. Instead of a guard, such an arrangement would be more properly labeled a gate, a device that will be discussed later.

Adjustable Barriers

Guard manufacturers have devised some clever ways for guards to be adjusted to individual applications during the setup. Unlike the fixed barrier guard, the adjustment is temporary, and the same guard can be reshaped later for a different setup. The trick with adjustable barriers is to make them easy enough to adjust to be practical but not so easy as to tempt an unauthorized person to tamper with them or gain access to the danger zone. Figure 11.13 shows one type of adjustable barrier guard.

Awareness Barriers

Some people confuse the term "adjustable barrier guard" with the term "awareness barrier." An awareness barrier (see Figure 11.14) is not recognized as a guard and does not meet the guarding criteria of keeping the operator's hands or fingers out of the danger zone. Although the awareness barrier is not a guard, it does provide a reminder that the hands are in danger. In the style pictured in the figure, metal rings or cylinders lie on the table and are lifted by the operator's fingers when the fingers are too close to danger. The operator at that point could go ahead reaching further into the machine, resulting in injury, but training and good judgment should inhibit him or her from taking this action. Contact with the awareness barrier should be a

Adjustable bars

Figure 11.13. Adjustable barrier guard. (From OSHA, Ref. 11.)

Figure 11.14. Awareness barrier installed on a shear.

learned cue to immediately withdraw the hands. The effectiveness of awareness barriers remains in doubt, as some feel that a mere deterrent is not enough to protect the operator. An added complication is that the awareness barrier may obscure from view the real danger. Many operators believe that it is not only a matter of convenience, but also one of safety to be able to see the actual point of operation.

Sometimes the term ''awareness barrier'' is also used to describe a simple rope or chain suspended in front of a danger area with perhaps a sign hanging on it to warn personnel to keep out. An example is the back side of a metal shear, as shown in Figure 11.15. The chain will not ensure that personnel will stay out of the point of operation or danger zone, but it will warn employees of the danger.

Jig Guards

The design of a jig guard is integral to the engineering of the manufacturing operation. The guard has the function of both protecting the operator and facilitating the operation to increase productivity. There is nothing standard about a jig guard be-

Awareness barrier

Figure 11.15.   Rear view of power squaring shear. (From OSHA, Ref. 11.)

cause it is designed to fit the individual workpiece and hold it in place while the machine performs the cut or other operation. Jig guards usually move with the work while the operation is performed. The jig guard depicted in Figure 11.16 is being used to notch cross members in the manufacture of four-way pallets. This clever guard keeps the circular table saw blade covered at all times, either by the jig guard between cuts or by the workpiece itself during the cut.

Point of operation guards are excellent when the machine can be fed effectively by automatic means or through a guard window without the operator having to reach into the danger zone. But the only feasible way to feed some machines is manually, and some of these machines are very dangerous. This class of machine is typified by the mechanical power press to be discussed below. The only way to assure protection of the operator while hand feeding these very dangerous machines is to use some kind of *device* that keeps the operator's hands out of the danger zone while the machine cycles and does its work. Some ingenious devices have been developed for this purpose and will be discussed in the section on power presses that follows.

POWER PRESSES

Punch presses are at the same time the most inherently dangerous and one of the most useful production machines in industry. The epitome of mass production machines, the press excels when huge volumes of identically produced products are required. Mass production depends on interchangeable manufacture, which in turn requires machines that successively produce parts which are essentially identical. The power press is superbly qualified for this task.

Figure 11.16. Jig guard for notching cross members of a four-way shipping pallet. (Idea credited to the Occupational Center of Central Kansas, Inc.)

Figure 11.17 illustrates two popular model power presses. The term "power press" is really a general one that encompasses hydraulic-powered models and forging presses in addition to the popular mechanical punch press. The outstanding feature of a power press is the set of mating dies that close on each other to cut, shape, or assemble material, or to do combinations of these operations in one or more strokes. The more subtle features of presses, such as methods of power transmission and control of the stroke, are important determinants of permissible safeguarding. It is at the mechanical models, powered by flywheels, that the most restrictive press standards are aimed.

Press Hazards

There are, of course, reasons why so much importance is attached to safeguarding the point of operation of mechanical power presses. The injury record for power presses is not a good one. When feeding the press by hand, the operator is close to danger every time the dies close, and this happens thousands upon thousands of times in a press operator's career. One slip and in a fraction of a second a finger or a hand is amputated. Such accidents were commonplace in the first half of the twentieth century. Shortly before World War II, there developed an awareness that

Figure 11.17. Typical power presses: (a) open-back inclinable (OBI) model; (b) straight-side model.

even the careful operator could become a victim of the power press, and efforts were initiated to eliminate the hazard.

To understand the nature and significance of the power press hazard it is necessary to study the interaction between human being and machine. In a hand-feed setup the press and the operator are alternating actions in a rhythm that cycles every few seconds and in some bench press operations may cycle in a fraction of a second. Figure 11.18 depicts the sequence of actions in a typical press cycle employing hand feeding without safeguarding. Production incentives motivate the operator to higher and higher speeds as skill develops. The operator learns to develop rhythm with the press motion which becomes very regular. The sound of the press tripping mechanism, the closing dies, and other press motions can become cues to the operator to make a hand or foot motion. The process involves eye–hand–foot coordination every cycle. Operating a power press is not as difficult as simultaneously patting your head and rubbing your stomach, but it is easy to visualize the hazards involved in hundreds of thousands of repetitive cycles.

One of the biggest causes of accidents with power presses is an attempt on the part of the operator to readjust a misaligned workpiece in the die. The motivation is

Figure 11.18.   Press cycle operation sequence (no safeguarding).

a very powerful one to reach back in to correct the error even after the ram has been tripped. If the operator lets the error go, the misaligned workpiece can wreak havoc when the dies close. At the very least, the workpiece will be ruined. More likely is that the expensive dies will be broken or ruined, and it is very possible that the press frame itself will be damaged or broken. One misaligned workpiece can result in many thousands of dollars damage to the dies and the press itself. But even worse, the misaligned workpiece or the dies can be fragmented when the dies close, subjecting the operator to injuries from flying metal pieces. It is no wonder, then, that the operator has a powerful urge to reach back in to correct a misaligned workpiece. The operator's eye will see the error, and the hand will reach in—even though the foot may have already depressed the pedal to actuate the press.

The somewhat insidious nature of power-press hazards motivated the standards writers to initiate a policy of "no hands in the dies." The theory was that tongs or other tools and devices could be used to feed workpieces, eliminating the need for the worker to put his or her hands into the danger zone. In addition to the tongs or other feeding tools, press guards or safety devices would prevent operators from placing their hands or fingers into the danger zone, even if they tried, while the dies were closing.

The theory worked for most jobs, but some applications presented awkward feeding problems which defied solution with the "no hands in the dies" approach. It became clear that the metal-stamping industry would not be able to comply with a rigid "no hands in the dies rule" in all situations, and support for the theory began to crumble. In its place new rules were established for assuring the reliability of safeguarding devices for protecting operators at the point of operation.

Press Designs

To understand the somewhat complicated rules for safeguarding a power press, it is necessary to examine the basics of how power presses work. Most power presses are mechanically powered, although there is a large number of hydraulically powered models, generally recognizable by the presence of a large hydraulic cylinder above the ram. The cylinder usually resembles the large cylinder in an automobile service station lift. Hydraulic power presses are expressly excluded from coverage by standards for mechanical power presses, as are pneumatic presses, fastener applicators, and presses working with hot metal, *even if they are mechanically powered*. Also excluded is the press brake, a power press with a long bed (refer back to Figure 11.2) which is used to bend sheet metal. Shears, because they employ blades instead of dies, are not considered within the definition of mechanical power presses.

The easiest way to distinguish a mechanical power press from other types is by the presence of a large heavy flywheel which by its rotation carries energy which is imparted to the ram when the press is tripped. The flywheel is *usually* mounted on one side of the press near the top, as was shown in the typical press in Figure 11.17.

One of the most important features is whether the press is "full revolution" or "part revolution." This refers to the method of engaging and disengaging the flywheel to deliver power to the ram. Full revolution types make a positive engagement that cannot be broken until the crankshaft and flywheel make one complete revolution together. During this revolution the press ram goes down, the dies close and then reopen, and the ram returns to its full-up position to await another cycle. At the end of the revolution the flywheel is disengaged and rotates freely under the power of the motor.

The part revolution machine typically has a friction clutch which can be disengaged at any time during the press cycle. Compressed air is used to engage or disengage the clutch instantly at the discretion of the operator. Upon disengagement of the clutch a brake is applied which immediately or almost immediately stops the ram. It is easy to see the advantage of being able to interrupt a press stroke at any point in the cycle, but the part revolution advantage does not stop there. The instantaneous engagement is also valuable in causing the press to cycle quickly once engaged, giving less time for the operator to get into trouble by making an afterthought reach into the point of operation.

The Safety and Health Manager must know which presses are full revolution and which are part revolution in order to know how to equip the press with the proper safety equipment. Thousands of dollars have been foolishly spent by purchasing the wrong safety equipment for a power press. One way to get a rough idea whether the press is full revolution is by the age of the press. Most presses are full revolution, and invariably the older presses are full revolution unless they have gone through a renovation process. Not incidentally, the average age of mechanical power presses in the United States is steadily increasing, and at the time of publication of this book statistics suggest that about half of all presses in the United States are more than 20 years old!

Since part revolution presses employ a friction clutch, the housing on the flywheel often has a bulge to accommodate the clutch, as shown in Figure 11.19. The

Figure 11.19. Full-revolution and part-revolution machines: (a) full revolution; (b) part revolution.

clutch is actuated pneumatically, and this generally means that an extra line can be seen on the outside of the flywheel cover running to the center of the clutch, as shown in the figure. Full-revolution clutch machines may also have a small line running outside the flywheel housing, but this is an oil line for lubricating purposes. None of these criteria for distinguishing machines is 100% reliable, so they should be used only as a means of preliminary screening for possible troublespots. Press engineers or the equipment manufacturer's representative can be consulted for an authoritative determination.

Point-of-Operation Safeguarding

Once having determined whether a press is full or part revolution, the Safety and Health Manager or engineer can proceed to a determination of the most effective means of safeguarding its most dangerous zone, the point of operation. There are at least 10 recognized methods of power press safeguarding, but acceptability of each depends on the press configuration and method of feeding. The various safeguarding methods can be divided into the following four categories, ranked according to degree of security:

1. Methods that prohibit the operator from reaching into the danger zone altogether
2. Methods that prohibit the operator from reaching into the danger zone any time the ram is in motion
3. Methods that prohibit the operator from reaching into the danger zone only while the dies are closing
4. Methods that do not prohibit the operator from reaching into the danger zone but which stop the ram before the operator can reach in

Only categories 1 and 2 are to be trusted for a full-revolution press. Some category 3 methods are permissible for full-revolution presses, but with category 3 there is exposure to the hazard that the press will "repeat." So far, no antirepeat mechanism has been found to assure absolutely that a full-revolution press will not repeat with an extra unwanted stroke. These repeat strokes are a terrifying possibility, but fortunately that possibility has been significantly reduced in recent years. Recalling that it is likely that about half of all presses are more than 20 years old, the threat of repeats is still something to consider.

Category 4 methods of safeguarding are permitted by OSHA only for part-revolution presses, and this is where many Safety and Health Managers go wrong. It should be obvious that any method that depends on stopping the ram to protect the operator must be installed only on part-revolution presses. Full-revolution presses, by definition, cannot be stopped. It is surprising, though, how many category 4 devices are seen in industry installed on full-revolution presses. It is true that these devices can afford some protection on full-revolution presses by locking out the tripping mechanism while the operator is in the danger zone. Once the press is tripped, however, these devices are powerless to stop the ram.

Press Guards

The section on guarding earlier in this chapter described types of guards for use on machines in general. Four of these types—die enclosure, fixed barriers, interlocked barriers, and adjustable barriers—are acceptable on mechanical power presses. Indeed, the die enclosure guard is almost exclusively used on mechanical power presses, although it is not as popular as some of the other types of guards. The fixed barrier guard is a very popular method of guarding power presses which employ automatic feeding of coils of strip stock and automatic ejection of the finished parts. The interlocked press barrier guard is not permitted for hand feeding, but a gate *device* (not a guard) can be used. Indeed, *none* of the four types of guards are permitted for hand feeding the press (putting the hands in the dies) because by definition of a press guard "it shall prevent the entry of hands or fingers into the point of operation by reaching through, over, under, or around the guard."

    With only one exception a guard or some type of point of operation safeguarding device must be installed on *every* mechanical power press. That one exception is where the full-open position of the ram results in a gap between the mating dies of less than ¼ inch, too small to permit entry of the fingers (see Table 11.1), and thus too small to be a hazard. But for all other mechanical power presses, even the automatic-feed models or the robot-fed setups, point-of-operation safeguarding is required. An accident can occur even with automatic feeding if a worker tending the automatic setup attempts to adjust a workpiece during operation.

    We have already established that guards cannot be used where the operator feeds the press by putting his or her hand in the die. We have also stated that virtually every mechanical power press is required to have point of operation safeguarding. Then does this make hand feeding illegal? The answer is "no," and the key is the difference between the terms "guarding" and "safeguarding." Safeguarding is a more general term and encompasses a variety of mechanical or electromechanical devices which protect the operator even when hand feeding is used. The standards are specific about which devices can be used with which types of machines and appropriate setups. Each of these configurations for mechanical power press devices will now be considered.

Gates

Gates look somewhat like a guard (see Figure 11.20) but are different in that they open and close with every machine cycle. Contrasting with interlocked barrier guards, gates *can* be used for manual feeding. Gates are used almost exclusively with mechanical power presses.

    There are two types of gates: Type A and Type B. The Type A gate is the safer of the two because it closes before the press stroke is initiated, and it *stays closed* until all motion of the ram has ceased. Type B gates are the same except that they remain closed only long enough to prevent the operator from reaching in during the more dangerous downward stroke. Although the upward stroke is less dangerous, there is still the hazard of repeats when an operator reaches in on the upward stroke.

CAUTION

Gate

Figure 11.20. Gate devices. (From
OSHA, Ref. 11.)

Type B gates are not forbidden for full-revolution clutch machines, but the occasional tendency of these machines to repeat is a sobering thought, and Type B gates are not recommended for presses with full-revolution clutches.

In defense of the Type B gate, however, is its efficiency over the Type A gate. A substantial percentage of the press cycle time can be saved if the operator can start reaching in as soon as the ram starts back upward. The savings may be only a fraction of a second per cycle, but over hundreds of thousands of cycles, the difference can be significant. The labor and overhead cost associated with the operator's time is saved, and also conserved are productive capacity of the press and the plant floorspace in which it is housed. It is possible for these combined costs to represent over $50 per hour. This represents an incentive to modernize the power press equipment so that it qualifies for the most efficient safeguarding systems.

Presence-Sensing Devices

Modern electronic detection devices have made their way into the machine guarding industry, and there are several devices of this type available for protection of the point of operation. One type utilizes a bank of photoelectric cells to set up a light screen, the penetration of which will immediately stop the ram. Figure 11.21 illustrates this type of device.

It becomes a game for workers to defeat these devices intended for their protection. It is obvious that if a worker can reach over or around the light screen, the

Figure 11.21. Photoelectric presence-sensing screen. (From OSHA, Ref. 11.)

machine will not stop. Alternate points of entry not covered by the sensing device should be guarded so that the operator cannot reach into the point of operation without tripping the device.

Another way to defeat a photoelectric device is by flashlights or other sources of light aimed to maintain the sensors in an energized condition at all times even if the worker's hand has broken the field. Even when deliberate attempts to defeat the device are not a factor, ordinary ambient light in the factory may keep sensors energized so that the screening device will not function effectively. A device that cannot be trusted to function when the outcome is so critical has little or no safe-guarding value.

To compensate for the ambient light problem, some presence-sensing devices function in the infrared frequencies rather than in the visible light spectrum. This makes the "light screen" invisible, and this feature may have some advantages, too.

Another way to trick a light screen is to somehow squeeze between the beams. If the sources and sensors are tightly spaced, squeezing between the beams becomes impossible for a part of a human body. The concept can be demonstrated, however, by carefully slipping a paper or cardboard edgewise between adjacent beams. Some commercially available screens have a sophisticated programmed scan which crisscrosses the field in such a complicated fashion as to prevent defeating the device. Such crisscrossing may reduce the number of sensors required to protect a given area. The concept is illustrated in Figure 11.22.

Another type of presence-sensing device uses a conductor to set up an electromagnetic field in its vicinity. Many variables affect the tripping threshold of these devices, sometimes called radio-frequency sensors, and this has damaged their reputation. For instance, one person's body, due to its mass or conductivity characteristics, might trip the mechanism from a distance of 2 feet from the point of operation. Another person might not trip the device until he or she has actually entered the danger zone. If this type of device is used, it should be "tuned" to have the proper sensitivity for a given operator and setup. Figure 11.23 illustrates the electromagnetic field type of presence-sensing device. Late model versions of this device have been found to be very effective.

Presence-sensing devices are quite practical for hand feeding in conjunction with a foot switch. Thus if the operator's rhythm is off and the foot switch is depressed too soon (while the operator's hand is still in the sensing field), the machine will not operate. Even more important, if the operator sees a misaligned workpiece and attempts to reach in after the ram has started its downward motion, the sensing field will detect this action and stop the ram. Thus the device not only prevents operator injury, it arrests a costly smash-up of the dies and possible damage to the press itself.

From what the reader has already learned, it should be obvious that this advantage of the presence-sensing device is feasible only for part-revolution presses.

Figure 11.22. Programmed crisscross scan for sensing plane.

Figure 11.23. Electromagnetic field type of presence-sensing device. (From OSHA, Ref. 11.)

Indeed, the use of presence-sensing devices for safeguarding the point of operation on full-revolution mechanical power presses is prohibited.

Although the presence-sensing device is designed to stop the ram before the operator can reach the danger zone, the worker should not tempt the machine. The author knows of a case in which a new machine[1] equipped with a presence-sensing device was being demonstrated by a proud worker to his family during an open house. The worker repeatedly thrust his hand into the machine to prove that the photoelectric cell was quicker than his hand. Finally, he succeeded in beating the machine and lost the ends of his fingers. This "accident" actually happened.

Presence-sensing devices are like gates in being subject to the issue of when to return access to the point of operation to the operator. Since presence-sensing devices are permitted only for part-revolution presses, it seems sensible to afford them the same production efficiencies permitted the Type B gate. Therefore, it is permissible to deactivate the sensing field on the upstroke. This process of bypassing the protective system is called "muting." As soon as the ram reaches its home position, the system is reset to its protective mode. Muting permits the same production efficiencies that the Type B gate holds over the Type A gate.

While we are discussing efficiency, why not go one step further and eliminate the foot switch? It would be feasible for the press to be restrained by the control system for as long as the operator's hand or arm breaks the sensing field during hand feeding. Then, as soon as the operator withdraws his or her hand from the danger zone, the press would *automatically* cycle without a signal from the operator! Feasible? Yes. Legal? No. The presence-sensing device is prohibited as a press tripping mechanism, although many European factories do employ this highly productive mode of operation for power presses.

---

[1]A printing press, not a punch press, in this instance.

There is a way that a presence-sensing device *can* be used to trip the press—when the presence-sensing device is used in conjunction with another safeguarding device such as a gate. Thus the gate represents the safety device, and the presence-sensing field acts as a tripping device. This would be a complicated and expensive system and should be considered rare.

Presence-sensing devices must be designed to adhere to the General Fail-Safe Principle stated in Chapter 3. Thus if the presence-sensing device itself fails, the system must remain in a protective mode. A failure in the device must prevent the press from operating additional cycles until the failure is corrected. But such a failure must not deactivate the clutch and brake mechanisms which are essential in stopping the press. If a failure in the device causes an interruption of the main power supply to the machine, the clutch must automatically engage. Of course, the clutch and brake system should have this characteristic, regardless of the choice of safeguarding devices; the clutch and brake simply represent additional examples of systems which should be designed in accordance with the General Fail-Safe Principle.

One final note about failures in the presence-sensing system should be mentioned. It is not enough that the failure inhibit the operation of the press; the system must also indicate that a failure has occurred. This is usually done with a trouble light on the panel.

Pullbacks

A very popular method of safeguarding a power press is by means of cables mechanically linked to the travel of the ram. These cables are attached to wristlets which pull the operator's hands out of the danger area as the ram makes it downward stroke. Figure 11.24 shows one example setup of pullbacks or "pull-outs" as they are sometimes called.

One reason for the popularity of pullbacks is their versatility. They can be used with virtually any power press, regardless of power source or type of clutch. However, the pullbacks do have their disadvantages.

Proper adjustment is very important to the effectiveness of pullbacks, especially with respect to close work done on bench model machines. Even the method of attachment to the wrists is important because ordinary wristlets permit too much variation in reach. Figure 11.25 is a close-up of a wristlet assembly which minimizes variations in an operator's restricted reach. Even with properly designed wristlets, proper adjustment is critical. Differences in operators' hand sizes can be a factor, but much more important are variations in die setups. A large die set will of course have a danger zone that extends closer to the operator, requiring an adjustment in the pullback limit of reach.

Recognizing the hazard of improper adjustment of pullbacks, safety standards require an inspection for proper adjustment at the start of each operator shift, following a new die setup, and when operators are changed. This inspection requirement, and especially the *frequency* of the inspections, decidedly cools the interest of

Pullback
mechanism

Pullback straps

Wristbands

Figure 11.24. Pullback devices for
safeguarding the point of operation.
(From OSHA, Ref. 11.)

employers in the use of pullback devices. Were it not for this problem, pullbacks
would be extremely popular. This drawback notwithstanding, the pullback device
remains one of the most popular press safeguarding devices.

It is usually the diesetter or operating supervisor who makes the actual check
of the pullbacks for proper adjustment. The concern of the Safety and Health
Manager should be that the job is done effectively and that a record is made of the
inspections. Simplicity is the key to the inspection record system for pullback
devices. Simplicity will help to assure that the job gets done and will also minimize
the impact on production efficiency. One convenient method is to use a tag attached
to the pullback device itself, with blank lines for indicating "date inspected" and
"initials" by the responsible party. In multiple-shift operations, in new die setups,
or in operator changes, more than one line would be initialed in one day on the tag,
but the tag could be easily designed to accommodate multiple entries on the same
day.

Figure 11.25. Close-ups of wristlet assembly for pullback device.

Sweeps

Very popular in the past, and still seen on some power presses, are devices that sweep away the operator's hands or arms as the dies close. These devices, illustrated in Figure 11.26, have fallen into disfavor as a means of protecting the operator. Operators even fear injury from the sweep device itself, as it comes swinging down in front of the machine. The design dilemma of these devices is that they must be powerful enough to inflict injury themselves in order to be effective in positively removing the operator's hands from the point of operation. But the overriding reason for their disfavor is that the design and construction of sweeps are inherently inadequate as a press safeguarding device. Sweeps are no longer recognized as adequate point-of-operation safeguarding devices on mechanical power presses.

Hold-Outs

A simplification of pullbacks is the hold-out (sometimes called "restraint") device, which is feasible only for setups in which it is unnecessary for the operator to reach into the danger area. Figure 11.27 shows that hold-outs look almost exactly like pullbacks, but the big difference is that the hold-out reach is fixed and does not permit the operator to reach in at all, even between strokes of the machine. If tongs, suction cups, or other gripping devices can be used to feed a machine manually, it is feasible to use hold-outs instead of pull-outs to protect the operator. Even without these gripping devices, long workpieces can be hand-fed into the machine without actually

Figure 11.26. Sweep device. This device no longer qualifies as acceptable safeguarding for mechanical power presses. (From NIOSH.)

DANGER AREA

WRISTBANDS

Figure 11.27. Hold-outs or restraints for restraining the operator's hands from reaching into the danger zone at *all* times (compare with pullbacks). (From OSHA, Ref. 11.)

placing the hands in danger. For these applications, hold-outs are appropriate. If the operator's hands must enter the zone between the dies, however, hold-outs are infeasible as safeguarding devices.

It would seem that protection for the operator would be unnecessary in applications where tongs or other feeding devices are used instead of the hand to feed a machine. However, this is a delusion that does not recognize the strong tendency for the worker to reach in when something goes wrong. Therefore, although hand-feeding tools are somewhat useful in promoting safety, they are not recognized as point-of-operation safeguarding devices. Another means must be used to *assure* safety, and hold-outs are a good method when hand tools are used.

Two-Hand Controls

Since people have only two hands, neither hand will be injured by the point of operation if the machine can require both of them to operate the controls, or so the theory goes. The theory is a good one, but some sophistication is necessary to assure that the device achieves its goal. Workers take pride in "beating the system" or in tricking the machine to make it operate without their using the controls. One trick is to use a board or rope to tie down one control so that the operator can operate the machine with one hand and feed it with the other. Another trick is to use your head, nose, or even toe to depress one of the controls. Workers have been known to try almost anything to defeat the safety features of a machine in order to achieve a production breakthrough and receive a higher production incentive payment. This points to the power of money over personal safety in motivating the worker. It may also reveal some inefficiencies in safety devices as currently designed—inefficiencies that unduly compromise productivity in the name of safety. As was pointed out in Chapter 3, industry workers and managers will tolerate some slowdown of an operation for the cause of safety, but not much.

Figure 11.28 illustrates a two-hand control device and some of the features that attempt to prevent the worker from defeating the device. Note the smooth, rounded surface of the palm button, which makes it convenient for the palm but not for tying down. Also note the cups around the button which are intended to foul attempts at tying down the button. Control circuitry can also be used to detect foul play and stop the machine if the buttons are not depressed concurrently and released between cycles.

Controls versus Trips

The term "control" implies a more sophisticated device than a mere "trip" to actuate the machine. Within the context of safeguarding the point of operation, a two-hand control means a device that not only requires both hands concurrently to actuate the machine but also stops the machine by interrupting its cycle if the controls are released prematurely. The nature of some machines does not permit such a degree of control over the machine cycle. Two-hand controls are infeasible for these machines.

Control buttons

Figure 11.28. Two-hand control. (From OSHA, Ref. 11.)

For full-revolution presses and other machines that cannot be stopped once their cycle has begun, two-hand trips are used instead of two-hand controls. Two-hand trips require both hands to start the machine cycle, but once started there is no protection for the operator. If the machine cycle is fast, and if the operator is far enough away to stay out of danger, two-hand trips effectively protect the operator. But if the machine is slow or the trip station is close enough, the operator can reach into the machine *after* it has been tripped. This gets into the subject of safety distances, covered in the next section. After studying the section on safety distances, it will be easy to see that controls are superior to trips.

Safety Distances

In reviewing the safeguarding devices for protection of operators from the points of operation of machines, we see that most of them protect the operator by making it impossible to reach into the danger zone after the machine cycle has begun. However, two of these devices—the presence-sensing device and the two-hand control—rely on a capability of interrupting a machine *in midcycle*. Every mechanical machine has inertia, so some time must elapse between the signal to stop and the complete cessation of motion of the machine in the point of operation area. If the inertia is great, an operator might be able to reach quickly into the danger zone before the protective device is able to completely stop the machine. Therefore, the operator station must be moved back from the point of operation a sufficient distance so that a reach into the danger zone before the machine stops is impossible.

In addition to the presence-sensing device and two-hand control, the two-hand trip device must also be located at a sufficient distance, as was mentioned earlier. Although the two-hand trip is not capable of stopping the machine, protection is incomplete unless the distance to the danger zone is great enough to prevent the operator from reaching in after releasing the palm buttons.

To compute a "safe distance" for a presence-sensing device of a two-hand control, it is necessary first to compute the stopping time of the machine. Figure 11.29 shows one model of a stop-time measurement system which connects one sensor to a palm button and the other to the machine motion. Upon a signal that the palm button has been depressed, the system starts counting in fractions of a second until motion of the machine ceases. Keep in mind that we are speaking here of the motion of the machine's ram, die, cutter, or other part *at the point of operation* which could cause injury. Flywheel motion or motor rotation may continue and indeed due to their inertia would usually be impossible to stop in time to be of any benefit. The resulting stopping time is generally displayed as is shown in the model in Figure 11.29. The reader should take care not to confuse the brake stop-time measurement device in Figure 11.29 with a "brake monitor" to be discussed later.

Once the stopping time has been determined, this time must be multiplied by the maximum speed the hand can move toward the point of operation, as in the following formula:

safety distance = (stopping time) × (hand speed constant)   (11.1)

OSHA relies on a maximum hand movement speed of 63 inches per second, some-

Figure 11.29. Brake-stop-time measurement device. (Courtesy of the Rockford Safety Equipment Company.)

times referred to as the hand speed constant; this constant is credited to L. Lobl of Sweden. Consider the following example:

Example 11.1

A power press is protected by an infrared-beam-type presence-sensing device. A stop-time measurement system is used to compute the time between breaking of the infrared beam and stoppage of the press ram. This stop time is found to be 0.294 second. The safety distance then is computed to be

$$\text{safety distance} = 0.294 \text{ second} \times 63 \text{ inches/second}$$
$$= 18.5 \text{ inches}$$

Thus all points in the plane of the infrared sensing field must be at least 18½ inches away from the point of operation for this power press.

The calculation would have been identical if the device in the example had been a two-hand control (but not a two-hand trip!). The palm buttons of the two-hand control would have to be placed 18½ inches away from the point of operation.

The safety distance calculation for the two-hand *trip* uses an entirely different logic from that used for the two-hand *control*. True, the same hand speed constant of 63 inches per second is used. But with two-hand trips the idea is for the machine to complete the dangerous part of its cycle before the operator can reach in after releasing the palm buttons. Thus the *slower* the machine, the more dangerous it is when using trips. It is paradoxical that the *faster* the machine, the harder it is to stop and the greater must be the safety distance for two-hand *controls*, whereas for *trips* it is just the opposite: the *slower* the machine, the greater must be the safety distance.

With many machines, hydraulic presses for example, the commencement of the cycle is virtually instantaneous with the depressing of the palm buttons. But with machines that commence the cycle with the mechanical engagement of a flywheel, there is an inevitable delay as the engagement mechanism waits for its mating engagement point on the flywheel. The delay can be considerable for a slow-rpm machine, especially if it has only one place on the flywheel suitable for making the mechanical engagement. The process is best explained by a diagram, as in Figure 11.30. Figure 11.30a shows a very unlucky happenstance in the position of the fly-

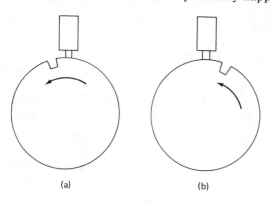

(a)   (b)

Figure 11.30. Two possible positions of the flywheel when a machine is tripped: (a) unlucky position—the engagement point has just passed the tripping mechanism; (b) lucky position—the engagement point is approaching and is very near the tripping mechanism.

wheel with its engaging point just past the tripping mechanism at the moment of actuation. This follows the principle that the *worst state* of the machine should be used to determine how to make the operation safe. Since the position of the flywheel at the moment the machine is tripped is purely a matter of chance, it is equally likely that the flywheel will be in the lucky position indicated in Figure 11.30b. But since this position cannot be counted on, the safety distance is computed assuming the flywheel is in the position indicated in Figure 11.30a.

One complete rotation of the flywheel completes the machine cycle for most machines. One-half of this rotation of the flywheel is the dangerous portion of the stroke, the closure motion of the machine. Thus adding a complete revolution for engagement plus one-half revolution for closure motion, the resultant danger period is one and one-half revolutions of the flywheel for a machine whose flywheel has only one engagement point. For machines that have several engagement points evenly spaced around the flywheel the danger period is shorter, depending on how many engagement points exist. To see this, consider an example of a machine with four engagement points. In the worst case, the farthest the engagement point could be from the tripping mechanism at the moment of tripping would be 90° or one-quarter revolution. Add this to the one-half revolution during machine closure, and the total danger period becomes three-quarters revolution.

Summarizing the calculation of safety distances for two-hand *trip* devices is the following formula:

$$\text{safety distance} = \frac{60}{\text{rpm}} \left(\frac{1}{2} + \frac{1}{N} \right) \times 63 \qquad (11.2)$$

where   rpm = flywheel speed in revolutions per minute while engaged

   N = number of engaging points on the flywheel

In equation (11.2) the factor 60/rpm is used to determine the time in seconds required for the flywheel to complete one revolution. The computation in parentheses is the worst-case number of flywheel rotations until closure, and the factor 63 is the hand speed constant in inches per second. The computed safety distance is in inches.

It is important to remember that equation (11.2) is to be used for two-hand *trips*, not two-hand controls. It is easy to determine that two-hand controls are superior to two-hand trips. From a production efficiency standpoint, two-hand controls can normally be placed much closer to the machine, which facilitates hand feeding. Two-hand trips can be more dangerous unless moved to a greater safety distance, which in turn impairs productivity.

Figure 11.28 was captioned "two-hand control," but the figure could also represent a two-hand trip. The device in Figure 11.28 could have been mounted on a pedestal so that it could be moved closer or farther from the point of operation, dependent on calculation of the formula for the given setup. Two-hand devices can also be mounted directly on the machine if the safety distance is short enough to permit this.

Neither equation (11.1) nor (11.2) are constant for a given machine; both are dependent on a given setup because of variations in die dimensions and die weight, which in turn affect flywheel speed, stopping time, and dimension of the danger zone on the machine. The pedestal is movable, but only a supervisor or safety engineer should be capable of moving it. This can be accomplished by locking the pedestal in position with a key or by bolting it to the floor and forbidding operators to unbolt it, or some other means of fixing the position. Failure to fix the position results in a temptation to the operator to move the pedestal closer to the machine to speed up production.

Because of the superiority of two-hand controls over two-hand trips, both in terms of safety and productivity, many industrial plants are converting their older machines equipped with trips to the more modern two-hand controls. Such a conversion is not a mere change in the two-hand device, but also generally represents a major modification to the machine itself and its power transmission apparatus, so that the machine can be classified as part revolution instead of full revolution. Brake monitoring and control reliability features are also required on part-revolution presses and will be explained next.

Brake Monitoring

It can be seen from the preceding discussion that the stopping time is very important in computation of permissible safety distance for part-revolution machines. But stopping time is dependent on the brake, and unfortunately brakes are subject to wear. Stopping time is also dependent on die setup, which can change from lot to lot. Therefore, it is naive to apply the principle of safety distances and then to walk away and trust the press to always respond in the same way as it did the day it was tested for safety distances. So for every part-revolution press whose safeguarding device depends on the brake, a brake monitoring system is needed to monitor the brake *every stroke*. Note how different this is from the brake stop-time measurement system of Figure 11.29, which would be set up only occasionally to check or set safety distances. In contrast, the brake monitor is a permanent installation on the press, monitoring the overtravel of the ram past the top stop every stroke.

Since the system is mechanical, there will be some overtravel, and a tolerance must be set which permits this overtravel. Employers can set this tolerance as high as they desire, but it will not be to their advantage to set the tolerance very high because a larger overtravel means a longer stopping time, which means a greater safety distance, which means reduced efficiency. There is no safety distance for Type B gates, but the employer is still allowed to reasonably establish a "normal limit" for ram overtravel.

The brake monitor can be engineered to measure either stopping time or overtravel distance. The most popular style is electromechanical with a pair of limit switches triggered by a cam linked to the press crankshaft. This type is commonly known as a "Top-Stop" monitor (see Figure 11.31). The first switch signals the application of the brake, and the second signals overrun. The cam must not trigger the overrun switch until a new cycle is initiated. Sooner or later the brake will deteriorate, and the overrun switch will be tripped which means that brake-stopping-time

Top dead center

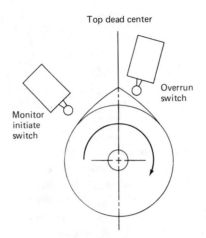

Monitor
initiate
switch

Overrun
switch

Figure 11.31. "Top stop" brake monitor (From *Manufacturing Engineering,* February 1976; courtesy of the Society of Manufacturing Engineers.)

tolerance has been exceeded. At that point the monitor system must provide an indication to that effect.

In addition to the brake monitor, a control system is required to assure that the press will cease to operate after a failure in the point of operation safety system occurs, *but* the brake system will *not* be shut down due to the system failure. This requirement is a direct application of the General Fail-Safe Principle stated in Chapter 3.

The reader should be reminded that brake system monitoring and control reliability is not required on all power presses. The logic is that on many press setups, the brake monitor and control system would have marginal benefit, whereas on others, they would be of critical importance due to the selection of safeguarding methods, the mode of operation, and the construction of the press itself. Table 11.2 summarizes the various options for guarding or safeguarding presses, the need for brake monitors and control systems, and alternate permissible configurations.

TABLE 11.2 POWER PRESS SAFEGUARDING SUMMARY[a]

Guard type or device	Full revolution		Part revolution	
	Hands in	Hands out	Hands in	Hands out
Guards Barrier guards (fixed, adjustable or die enclosure)	Illegal	Inspect weekly	Illegal	Inspect weekly or Brake monitor and control system
Interlocked barrier guards	Illegal	Inspect weekly	Illegal	Inspect weekly or Brake monitor and control system

(continued)

TABLE 11.2 (CONTINUED)

Guard type or device	Full revolution		Part revolution	
	Hands in	Hands out	Hands in	Hands out
Movable barrier devices Type A gate	Inspect weekly	Inspect weekly	Inspect weekly or Brake monitor and control system	Inspect weekly or Brake monitor and control system
Type B gate	Inspect weekly	Inspect weekly	Brake monitor and control system must detect top-stop overrun beyond limits	Inspect weekly or Brake monitor and control system
Devices Presence sensing	Illegal	Illegal	Safety distance and Brake monitor and control system	Safety distance and Inspect weekly or Brake monitor and control system
Pull-out	Inspect: each shift each die set-up, each operator	Inspect: each shift each die set-up, each operator	Inspect: each shift, each die set-up, each operator	Inspect: each shift each die set-up, each operator
Sweep	Does not qualify as safeguard	Does not qualify as safeguard	Does not qualify as safeguard	Does not qualify as safeguard
Hold-out (restraint)	Illegal	Inspect weekly	Illegal	Inspect weekly or Brake monitor and control system
Two-hand control	see Two-hand trips	see Two-hand trips	Safety distance and Fixed control position and Brake monitor and control system	Safety distance and Fixed control position
Two-hand trips	Safety distance and Fixed trip position and Inspect weekly	Safety distance and Fixed trip position and Inspect weekly	Inspect weekly or Brake monitor and control system	Inspect weekly or Brake monitor and control system

[a]This summary compares inspections requirements, brake monitoring and control system requirements, safety distances, and legal and illegal arrangements. The summary does not include detailed specifications, requirements for multiple operators, and other details too numerous to include in this summary. For details, see OSHA Standard 1910.217.

GRINDING MACHINES

Grinding machines are in almost every manufacturing plant—on the production line, or in a toolroom, or maintenance shop.

There are two or three items which create the most trouble as follows (shown in Figure 11.32):

- Failure to keep the workrest in close adjustment (within ⅛ inch) to the wheel on offhand grinding machines
- Failure to keep the tongue guard adjusted to within ¼ inch
- Failure to guard the wheel sufficiently

These rules may seem "nitpicking," but there is a grave hazard with grinding machines that most people do not know about—the breakup of the wheel while rotating at high speed. It does not happen very often, but when it does, the injuries to the operator can be fatal. It is against this hazard that all three of the items above are aimed, even the workrest adjustment requirement.

A severe stress can be placed on the grinding wheel if the workpiece becomes wedged between the workrest and the wheel. A large gap invites the workpiece to become pinched and then drawn down by the wheel resulting in a severe wedging action, as illustrated in Figure 11.33. The forces of this wedging action threaten the integrity of the bonded abrasive wheel, possibly causing it to break up and hurl stone fragments at the operator at near-tangential velocities. The only protection to the operator from these flying stone fragments are a good wheel guard, a properly adjusted tongue guard, and any personal protective clothing the operator might be wearing.

Upper gap must be less than 1/4 inch

2

Workrest gap must be less than 1/8 inch

1

Guard must cover spindle end, nut, and flange

3

Figure 11.32. The three biggest troublespots on ordinary grinding machines.

Figure 11.33. Severe wedging action due to a large gap between workrest and wheel.

It is a simple matter for anyone to check the gap adjustment on the grinding machine workrest—easy to check, but not so easy to *keep* in adjustment. Since the wear of the grinding wheel causes the gap to gradually widen, constant surveillance is needed to assure that the gap is maintained less than ⅛ inch. There is little tolerance for overcompensating for the wear because ⅛ inch leaves little margin for setting the workrest closer than required. Since there is no way to avoid frequent adjustment, some convenient means need to be initiated to make adjustments quickly and efficiently. To this end, a workrest gauge is recommended as illustrated in Figure 11.34. The go/nogo character of this gauge aids in a quick and sure decision as to whether to adjust the workrest every time it is checked.

Another easy check is the maximum spindle speed; it should not exceed the maximum speed marked directly on the wheel. Operation of an abrasive wheel above its design speed subjects the wheel to dangerous centrifugal forces which also could lead to wheel breakup.

Sometimes an abrasive wheel has manufacturing imperfections or transportation damage which make it dangerous. Before the wheel is mounted, it should be visually inspected for such damage or imperfections. Invisible imperfections can sometimes be detected by tapping the wheel gently with a nonmetallic implement such as the plastic handle of a screwdriver or a wooden mallet. A good wheel typically produces a ringing sound, whereas a cracked wheel will sound dead. The reason nonmetallic objects are used to make the "ring test" is that metallic objects might themselves ring, giving the impression that a defective wheel is good.

1/8″

1/8-in. hook on end of gauge
should not be able to hook
over the edge of the workrest;
if it does hook over, the
workrest needs to be
adjusted closer to the wheel.

Figure 11.34. Grinding machine workrest gauge.

SAWS

Saws have some obvious hazards and some not-so-obvious ones. Almost everyone respects the danger of a power saw, but serious injuries continue to occur, and acceptable means of guarding both obvious and subtle hazards from saws need to be considered.

Radial Saws

If the Safety and Health Manager learns about no other saw, he or she should become familiar with this one. Radial saws or "radial arm saws" can be quite dangerous, and in addition they are difficult to guard. Figure 11.35 depicts a typical radial arm saw, but although the blade can be seen to be partially guarded on the top half, the lower portion of the blade is exposed. Figure 11.36 illustrates one type of lower blade guard for radial arm saws. The guards are very unpopular and are often removed by employees.

Another problem with radial saws is the return-to-home position. The saw should be mounted such that "the front end of the unit will be slightly higher than the rear, so as to cause the cutting head to return gently to the starting position when released by the operator." A radial saw which creeps out while running is an obvious

Figure 11.35. Typical radial saw, not properly guarded, and in violation of OSHA standard. (From NIOSH, Ref. 42.)

Figure 11.36. Radial arm saw equipped with lower blade guard.

hazard, but one that is adjusted too much at an angle will "bounce" on the home position stop—another hazard. A radial saw also will have a stop that prevents the operator from pulling the saw off the radial arm or beyond the edge of the saw table. This stop should be adjustable for limiting travel of the saw head only as far as necessary.

A radial saw is usually used as a cutoff saw and is mounted as shown in Figure 11.35. It is possible, though, to reorient the cutting head 90° so that the blade is parallel to the table and thus to convert the saw to a ripsaw for long pieces of stock. In this mode of operation the saw head is locked into position, and the material is pushed into the saw. There is a hazard, though, if the material is fed into the saw in the wrong direction. To be safe, the material must be fed *against* the rotation of the blade. If the material is fed *with* the rotation of the blade, especially if the feed rate is fast, the saw teeth are likely to grab the workpiece and pull it into the machine at high speeds, possibly pulling the operator's hands into the blade with the material being sawed. It is not uncommon that a radial saw fed incorrectly in this manner will hurl the workpiece through the machine and toss it across the room.

Table Saws

A "table saw" is an everyday term for a hand-fed saw with a circular blade mounted in a table. Unlike the radial saw, the table saw head always remains stationary during a cut while the work is fed into it. With table saws, the three biggest problem areas are hood guards, spreaders, and nonkickback fingers (see Figure 11.37). Antikickback protection is more important for ripsaws than for crosscut saws.

The hood guards present the most problems because the obstructed view makes the saw operator's job more difficult and awkward. Although most of the

Push stick

Spreader (splitter)

Antikickback fingers

Figure 11.37. Hood guard, spreader, and antikickback fingers on a table saw.

hood guards in the field are metal, most new machines come with see-through plastic guards. But the rapidly rotating saw blade can cause a static charge to develop on the nonconducting plastic guard, causing it to become covered with sawdust so that the blade cannot be seen. Also, the plastic guard is easily scratched, further reducing visibility through the guard.

It is true that hood guards will not absolutely prohibit contact of the operator's hands with the saw blade. The hood guard, with its spring action, acts more as an "awareness barrier," as was discussed earlier in the section on safeguarding the point of operation on general machines. But there is another reason for using a hood guard: to protect the operator from flying objects. The saw blade rotates at 3000 rpm, and this produces large centrifugal forces and high tangential velocities. Consider the following calculation for a popular make of a table saw:

Blade speed: 3450 rpm
Blade diameter: 10 inches
$$\begin{aligned}
\text{Tangential velocity} &= (\text{rpm}) \times (\text{blade perimeter}) \\
&= 3450 \text{ rev/min} \times 10 \text{ in.} \times \pi \\
&= 108,385 \text{ in./min} \times \frac{60 \text{ min}}{1 \text{ hr}} \times \frac{1 \text{ mile}}{5280 \text{ ft}} \times \frac{1 \text{ ft}}{12 \text{ in.}} \\
&= 102.63 \text{ miles per hour}
\end{aligned}$$

This means that if a tooth breaks off the saw or if a small chip or block of wood breaks off and is carried on the blade for nearly a full revolution as in Figure 11.38,

Block of wood leaves blade at tangential velocity and flies toward operator

Tangential velocity 103 mph

Block of wood breaks off and is carried around blade

Figure 11.38.   Broken saw tooth or chips of wood are a hazard to the saw operator.

it will be propelled at the face of the operator at a velocity of over 100 miles per hour. It is little wonder that eye protection is considered necessary for operating a table saw.

Kickback

Kickback is the word used to describe the situation in which the entire workpiece is picked up and thrown back at the saw operator. The energy for the kickback comes from the sawblade itself. The rotation of the blade is toward the operator. At the front of the blade where the saw first meets the work the direction of blade motion is toward the operator and *down*. But at the back of the blade the direction of motion is toward the operator and *up*. Because the saw teeth are slightly wider than the blade thickness, a correctly aligned workpiece will contact the blade only at the point at which the cut is taking place. But if the workpiece should shift slightly, the exiting portion of the cut at the back of the blade will become misaligned, causing the edge of the material adjacent to the cut to contact the blade as it emerges from the table. This contact can result in a sudden and powerful upward thrust which causes the material to break contact with the table surface. Further misalignment becomes almost inevitable at this point, and the workpiece is grabbed firmly by the blade. If the workpiece is extremely thin or fragile, it will break up at this point, with thin portions or chips continuing with the blade down under the table. But a much more likely result is that the rigid material cannot follow the blade, and it is hurled at tangential velocity directly at the operator.

Both the spreader and the nonkickback fingers are intended to help prevent kickback. The spreader keeps the saw kerf open or spread apart in the completed portion of the cut so that the material will not contact the blade. The nonkickback fingers or "dogs" are designed to arrest the kickback motion should it start to occur. The shape of the dog permits easy motion in the direction of the feed. A backward motion, however, causes the dog to grip the material and prevent a kickback.

Band Saws

It is essentially impossible to guard the point of operation in most band saw operations. However, the unused portion of the blade can feasibly be guarded. A sliding guard is used for this and is moved up or down to accommodate larger or smaller workpieces, respectively. This sliding guard, together with the wheel guards, permit the entire blade except for the working portion to be guarded.

Other Saws

Hand-held circular saws are subjected to a variation of the hazard of kickback, except that it is the *saw* that is kicked back instead of the material! Proper training and operator respect for the saw are important, as are a clean sharp blade, a "deadman'' control, and a retractable guard for the lower portion of the blade. A dead-man control is simply a spring-loaded switch (button or trigger) which will immediately cut off power to the saw if the operator releases the switch.

The retractable guard for the lower portion of the blade on a hand-held circular saw is perhaps analogous to the lower blade guard on a radial saw and the hood guard on a table saw. However, on a hand-held circular saw the retractable guard is much more important. If the operator locks the retractable guard open with a small wedge as operators sometimes will do, the saw becomes very hazardous both before and after the cut. The blade is exposed, and direct damage or injury can result if the saw is dropped or set down on a surface.

Cutting aluminum with a hand-held circular saw can be a real problem. Hand-held circular saws are widely used to cut aluminum extrusions to length to manufacture windows, storm doors, shutters, and other architectural aluminum extrusion products. The problem is that the saw blade gets very hot, reaching the melting point of aluminum at around 1200 °F; drops of molten aluminum start flying off the blade onto the guard, causing quite a mess. The aluminum later solidifies, causing the guard to stick and malfunction.

One prominent manufacturer of architectural aluminum extrusions wrestled with this problem for five years before turning to alternate means of protecting the operator. Reverse polarity brakes were installed on the saws, which caused the blade to come to an immediate halt as soon as the operator released the trigger. This removed most of the hazard, because it is while the blade coasts to a stop after use that the lower blade guard is most important. For added protection, the workers were provided with protective gloves and pads for the hands and wrists.

Power hacksaws are difficult to guard, more so than their cousins, the horizontal band saws. The band saw can feasibly be guarded with an adjustable guard along all portions of the blade except the portion actually performing the cut. Because of the reciprocating action of the power hacksaw, however, a much more complicated guard would be required to adjust back and forth *during* the cut every stroke. Guards, quardrails, or guarding by location would be appropriate for this hazard. Some modern power hacksaws are equipped with enclosures which contain the entire stroke of the reciprocating blade.

BELTS AND PULLEYS

Almost every industry has a wide variety of belts and pulleys and other driving systems for transmitting power from motors to machines. The hazards are well understood, and the technology is simple. Belts and pulleys represent a good target area for the Safety and Health Manager to institute a low-cost program of in-house safety improvement.

Before getting carried away, it should be recognized that not all belts and pulleys are hazardous, and the standards acknowledge this fact by excluding certain small, slow-moving belts. The exclusions are intricate and are best represented by a decision diagram, as in Figure 11.39. A height of less than 7 feet off the floor or working platform is generally considered a working zone where personnel need protection from belts and other machine hazards.

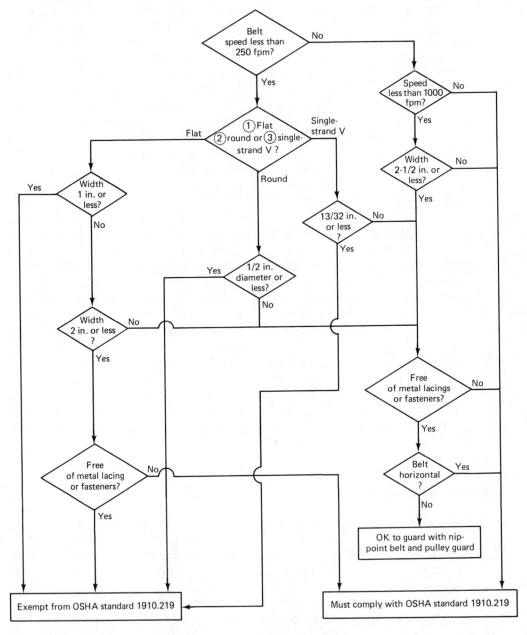

Figure 11.39. Decision diagram for OSHA belt guarding standard.

Related to belts and pulleys are shaft couplings, such as are typically found between a pump and the motor that drives it. The preferred method of eliminating hazards with these couplings is to design them such that any bolts, nuts, and setscrews are used *parallel* to the shafting and are countersunk, as shown in Figure 11.40. If these fasteners do not extend beyond the edge of the flange as shown in the figure, it is unlikely that they will cause injury. The greatest hazard with exposed setscrew heads is that they will catch parts of loose clothing and then draw the worker into the machine. The setscrews and other projections are typically invisible on the rapidly spinning shaft or flange, adding to the hazard. The installation of U-type guards, wherever needed throughout the plant, is an objective which needs to be set by the Safety and Health Manager. Once aware of the hazard, maintenance person-

(a)   (b)

(c)

Figure 11.40. Safety with shaft couplings: (a) screws or bolts are mounted parallel to the shafting and are countersunk; (b) setscrew is mounted in the periphery of the flange, but is countersunk and does not extend beyond the flange; (c) exposed setscrews present no hazard because the shaft coupling is covered by a U-shaped guard.

nel can go about systematically taking measurements, having U-type guards fabricated in the sheet metal shop, and then installing these guards.

The possibility of safeguarding belts and pulleys by *location* should not be overlooked. Some belts and pulleys are located in a part of the machine that is protected from worker exposure. Some people believe that location is the answer to safeguarding the belt and pulley on motor-drive air compressors which operate intermittently. But because this equipment starts automatically, safeguarding by location may not be sufficient to eliminate the hazard.

One approach to guarding large air compressors is to place them in a room by themselves. The door should be kept locked. It is best that the room be kept small if heat dissipation methods permit so that it will not be used for storage or other purposes which will result in worker exposure. The safety of maintenance personnel who must enter the room to service the compressor should not be overlooked. Administrative procedures and training can be used to reduce hazards to these personnel.

It is important to call attention to the hazards of compressed air hoses used for cleaning. Compressed air hoses with nozzles are often used to spray chips away from machine areas. Safety standards specify that the air pressure used for such purposes must not exceed 30 psi. Most industrial compressed air systems operate at working pressures greater than 30 psi. Therefore, a reducer at the nozzle or a reducing nozzle is used to bring the air pressure within the specified maximum of 30 psi. Figure 11.41b shows one type of nozzle for this purpose.

Excessive air pressure from these nozzles can make flying chips hazardous. Even with proper air pressure, chip guarding and personal protective equipment are needed to protect the worker. If alternate means can be utilized to remove chips, it is usually in the interest of safety to discontinue the use of air nozzles. Unfortunately, metal chips are very sharp and hazardous to handle, making the cleaning process somewhat of a problem.

(a)   (b)

Figure 11.41. Two ways to comply with OSHA's requirement to reduce compressed air for cleaning to pressures less than 30 psi (a) pressure reducer in line; (b) special ventilating nozzle reduces pressure to less than 30 psi. (From NIOSH.)

The hazard of flying chips is fairly obvious, but most people do not realize that compressed air hoses used for cleaning can even present hazards of *fatalities*. There have been a few recorded cases of fatalities where horseplay has failed to consider the hazards of compressed air. There is no way that the human body can contain without serious damage internal overpressures of even 30 psi. Unfortunately, workers have rarely been trained to respect the lethal pressures presented by the ordinary and seemingly harmless compressed air nozzle used for cleaning.

Jacks

A frequent weekend accident is the familiar fatality when a "shade tree mechanic" is killed under an automobile which has fallen from the jack supporting it. A jack is essential for lifting, but the supported load is usually much more stable if transferred to secure blocks, removing the load from the jack. What is unsafe for the shade tree mechanic is also unsafe for the industrial worker as far as jacks are concerned.

As with a crane or a hoist chain, the temptation is to use a jack until failure, but the result of this policy will eventually be a catastrophic failure, an unacceptable alternative for the end of life for a jack. Therefore, the only other alternative is to inspect the jack at intervals throughout its life to watch for signs that the jack either needs repair or is worn to the danger point.

SUMMARY

Machine guarding is a term almost synonymous with industrial safety and is a high-priority item with Safety and Health Managers. Although passed off by some health and safety professionals as old fashioned and nontechnical, machine guarding is actually a challenging task with new guarding technologies pressing the state of the art.

The most dangerous part of most machines is the point of operation, where the tool meets the workpiece. Unfortunately, it is also the most difficult part of the machine to guard in most cases. When guarding becomes impractical or infeasible, electromechanical devices have been devised for protecting the operator from the point of operation. Not to be overlooked are the indirect hazards at the point of operation such as flying chips or sparks.

Next in importance to the point of operation are belts, pulleys, and other power transmission apparatus. Usually more easily guarded than the point of operation, belts, pulleys, gears, shafts, and chains should receive plant-wide attention. Guards can often be fabricated in-house with supervision of the Safety and Health Manager. Remember to consider the possibility of guarding by location or by distance.

One of the most important and dangerous of production machines is the mechanical power press. Safety requirements for guarding the point of operation of presses can be quite technical and complicated. Many of the guarding methods specified for presses can be used as principles for guarding machines in general.

The hazards of radial saws are exemplary of those of other types of saws. Kickback is a hazard with most saws, and various mechanical devices in addition to training workers in the mechanism of kickback are viable remedies.

OSHA's abrasive wheel machinery standard is quite complicated, but the principal problems are short and simple—guarding the wheel, nut, and flange; adjustment of the workrests on grinding machines; and adjustment of the tongue guards. The Safety and Health Manager should also give attention to such miscellaneous equipment as hand-held power tools, compressed air equipment, and jacks. Significant among these is the requirement to reduce air pressure to 30 psi or less when the air is used for cleaning.

EXERCISES AND STUDY QUESTIONS

11.1. Explain the term "point of operation."

11.2. Name several types of mechanical hazards on machines in general. Which is the most important from a safety standpoint?

11.3. Identify several examples of in-running nip points.

11.4. Name two ways of safeguarding a machine that require no physical guard or device at all. What is the difference between these two methods of safeguarding?

11.5. What is a lockout? How does it differ from an interlock?

11.6. What are the disadvantages of nylon mesh guards for fans?

11.7. Name some reasons a machine might have bolt holes in its feet.

11.8. Name several types of safeguarding devices for the point of operation.

11.9. What is the difference between an interlocked barrier guard and a gate?

11.10. What is the difference between two-hand controls and two-hand trips?

11.11. What is the difference between Type A and B gates?

11.12. What is the advantage of Allen-head screws over wing nuts for machine guards?

11.13. A guard has a maximum opening size of ¾ inch. The guard openings are 6 inches from the danger zone. Does the opening size meet requirements?

11.14. What is an "awareness barrier?" What is a "jig guard?"

11.15. Explain the terms "full revolution" and "part revolution" as applied to press clutches. Which is safer?

11.16. What is "muting" of safeguarding devices, and when is it permitted?

11.17. What is the difference between pullbacks and restraints?

11.18. What is the biggest disadvantage of pullbacks?

11.19. Explain the difference between a "brake monitor" and a "brake stop-time measurement device."

11.20. A part-revolution clutch press has a brake stop time of 0.37 second. At what minimum distance should two-hand controls be placed?

11.21. Where should a presence-sensing device be placed on the press of Question 11.20?

11.22. If the press of Question 11.20 had been a full-revolution press operating at 60 rpm and having four engagement points, at what distance should a two-hand trip device be placed?

11.23. Would a two-hand control offer any improvement in the press of Question 11.22? What about a presence-sensing device?

11.24. What are the three biggest problems with grinding machines? Why are they so important from a safety standpoint?

11.25. What is the "ring" test?

11.26. Compare hazards for table saws which are used as ripsaws versus those used as cross-cut saws.

11.27. When do shaft couplings need no guards?

11.28. How can compressed air used for cleaning be dangerous? Under what circumstances is it permitted?

12

Welding

3% Percent of OSHA
general industry citations
addressing this subject

After as broad a subject as machine guarding it may seem ridiculously specific to address a field as narrow as welding. However, it may surprise some to learn that welding processes present some of the greatest hazards to both safety and health. In terms of breadth of hazard, welding encompasses even more than the chapter on machine guarding, and for that matter, even more than any other chapter of this book.

Although this chapter is entitled "Welding," this term is to be taken in a very broad sense to include gas welding, electric-arc welding, resistance welding, and even related processes such as soldering and brazing, which technically are not really welding processes. Welding processes are so diverse that before addressing the hazards of these processes it is necessary to name them and to provide the Safety and Health Manager with the necessary background in welding process terminology.

PROCESS TERMINOLOGY

The key to understanding welding hazards is to know how the process itself works, and unless Safety and Health Managers have this knowledge, their credibility with their manufacturing and operations counterparts will be minimal. Everyone knows that welding requires that material melt or fuse to form a rigid joint. The first question to ask to determine the process is "What material melts?" If the melted material is of the parts to be joined themselves or of like filler material, the process is welding. If the material is some other material of lower melting temperature, the process is brazing or soldering. The breakpoint between brazing and soldering is 800°F (427°C), with brazing above 800°F and soldering below.

280

Since welding requires materials to be melted, heat is required, usually applied intensely, to meet the demands of high melting points of welding materials. The method of applying this intense heat usually identifies the process. Excluding the unusual and exotic processes such as thermit and laser processes, there are three basic categories of conventional welding as follows:

- Gas welding
- Electric-arc welding
- Resistance welding

Gas welding is typified by the familiar oxyacetylene torch process, in which the very hot burning acetylene gas is made to burn even hotter by supplying pure oxygen to the flame. For welding lower-melting-point materials, alternate and safer gases such as natural gas, propane, or MAPP gas can be used. These alternate gases are often used for brazing and soldering.

The Safety and Health Manager should be careful not to confuse "gas welding" with some types of electric-arc welding which use an inert gas to facilitate the process. Indeed, some of these processes have names such as "gas metal arc welding" or "gas tungsten arc welding," but they are not gas welding. The telling feature of gas welding is that the gas must be used as a fuel for the process, not as an inert gas.

Even more diverse than the types of gas welding are the various types of electric-arc welding. Arc welding requires a small gap between electrodes, one of which is usually the workpiece itself. The intense heat is provided by the electric arc that forms between the electrodes. The process that typifies electric arc is "stick electrode" or shielded metal arc welding (SMAW),[1] shown in Figure 12.1. This highly portable and most popular operation is seen in welding structural steel for buildings, in repair of steel components, and in a wide variety of manufacturing processes. The "stick" is a piece of "welding rod" which is held by a gripper and is consumed by the process. The stick consists of a filler metal surrounded by a "flux," a term to be

Figure 12.1. "Stick electrode" or SMAW (shielded metal arc welding).

[1]American Welding Society recommended abbreviation.

explained later. Similar to SMAW welding is "flux-cored" arc welding (FCAW), in which the flux is on the inside of the rod, which is reminiscent of acid- or resin-core solder. Sometimes the welding process does not consume the electrode, a good example being the process commonly called TIG (tungsten inert gas) or GTAW (gas tungsten arc welding) shown in Figure 12.2. In a related process, GMAW (gas metal arc welding), the electric arc consumes a *flexible* electrode which is spooled on a reel and is continuously fed to the arc during welding.

Figure 12.2. TIG (tungsten inert gas) welding or GTAW (gas tungsten arc welding).

The terms "flux" and "inert gas" have been used earlier and need some explanation. The extremely hot melting temperatures of steel and other metals make these metals very vulnerable to oxidation, which is harmful to the weld. Flux is typically a chemical compound which combines with impurities and with oxygen to prevent harmful oxidation of the hot metals. After combining with impurities while the flux is in the molten state the resultant molten liquid is called "slag," which later solidifies and must be removed from the finished weld. With some processes one of the inert gases such as argon or helium is used for the same purpose. The inert gas displaces the air ambient to the weld and thus keeps harmful oxygen away from the hot metals. Unfortunately, this inert gas also sometimes keeps the oxygen away from the *welder*, an obviously undesirable characteristic that will be explored subsequently in the section on hazards mechanisms.

One other electric arc process should be mentioned at this point in the text: the "submerged arc" process or SAW (submerged arc welding). In this process the flux is granular and, as can be seen in Figure 12.3, the electric arc is hidden under the pile and puddle of granular and melted flux, respectively. This has great safety and health advantages, and submerged arc is growing in popularity. Automatic welding machines are often programmed to apply flux automatically and move the electrode over a long, straight path in the manufacture of large structural steel beams or

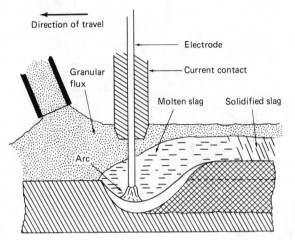

Direction of travel

Electrode

Granular flux

Current contact

Molten slag   Solidified slag

Arc

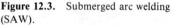

Figure 12.3. Submerged arc welding (SAW).

"plate girders." However, SAW welding in overhead positions is a problem because the granular flux will fall by gravity instead of covering the weld.

Resistance welding (Figure 12.4) is one of the least hazardous of welding processes. It is widely used in manufacturing of the mass production variety. However, resistance is generally restricted to relatively thin sheets of material. The concept of resistance welding is to pass electrical current *through* the material to be welded, enabling the heat so generated to melt the material. Physical pressure is also applied at the point of the weld. The attractive thing about resistance welding is that the melting generally occurs only where the mating surfaces meet. The outside and adjacent surfaces which are exposed to atmospheric contaminants harmful to the weld do not reach melting point, and damage to the material is minimal. This fact also precludes the need for a flux or inert gas to complicate both the production process and the safety and health aspects.

Resistance spot welding (RSW) is widely used to join sheet metal coverings, housings, guards, and shields on products as varied as space heaters and grain bins. Another important industry for spot welding is the automobile industry. Seam welding (RSEW) is generally preferred over spot welding for watertight seals because a pair of rollers apply continuous pressure and a series of electrical pulses make in effect a seam of overlapping spot welds.

Some of the more unusual or exotic processes should be mentioned because, although they may rarely be used or seen by the Safety and Health Manager, the nature of their hazards may be entirely different. The "thermit" method of welding (TW) employs a chemical reaction to produce the welding heat. Thermit is good for awkward applications or perhaps where the welding to be done is remote from convenient electrical power or gas sources. Laser beam welding (LBW) uses a concentrated laser light beam to generate the welding heat. Lasers pinpoint the weld so precisely that they are used for almost microscopic applications on tiny parts.

Figure 12.4. Resistance spot welder.

GAS WELDING HAZARDS

The myriad different processes for welding should give some clue as to why it was stated earlier in this chapter that the breadth of welding hazards is greater than that of any of the subjects of any of the other chapters of this book. We now examine these hazards, beginning with gas welding because it has given Safety and Health Managers the most problems.

Acetylene Hazards

Gas welding cylinders are so familiar that it is difficult to keep in mind their devastating destructive power. Acetylene gas, the fuel gas for most gas welding, is so unstable that its pressurization in manifolds to pressures greater than 15 psig (30 psia) is prohibited. Contrast this low pressure with the familiar oxygen, nitrogen, or other ordinary compressed gas cylinder, which contains pressure greater than 2000 psi.

There are tricks to avoid the hazards of instability of acetylene gas; the most popular one is to dissolve the gas in a suitable solvent, usually acetone. Then the pressure can be raised to around 200 psi. As the acetylene gas is used from the acetylene cylinder, the pressure is lowered slightly, permitting a greater quantity of gas to bubble out of the solution, resulting in equilibrium at a pressure suitable for welding. In this fashion a relatively large quantity of acetylene can be stored in a reasonably portable cylinder.

Figure 12.5 shows the inside of an acetylene cylinder. Most people do not know that the cylinder contains a solid absorbent filler material for the acetone–acetylene solution. The contents are more liquid than gas, and this gives rise to a hazard mechanism. Acetylene cylinders should be kept valve end up both while in storage and while in use. There is no harm in tilting cylinders slightly, and in fact this is common practice in the use of hand trucks to handle cylinders connected for use. Good advice, however, is not to tip acetylene cylinders to an angle more than 45° from vertical.

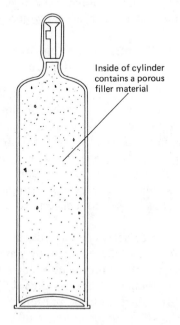

Inside of cylinder
contains a porous
filler material

Figure 12.5. Acetylene cylinder.

If the cylinders are stored horizontally or valve end down, liquid acetone could enter the valve passages instead of the intended acetylene gas. Then later when the valve is opened, the welder could get an unexpected flow of highly flammable liquid acetone instead of acetylene gas. Since the purpose of opening the valve is usually to ignite the torch with a sparking source, it is obvious that it would be easy to ignite the liquid acetone by accident. A burning quantity of spilled acetone is difficult to control and is quite dangerous. The reader is referred back to Chapter 7 to consider the flammability characteristics of acetone.

Another way to get liquid acetone through the cylinder valve is to use the cylinder when it is nearly empty. Acetone coming through the valve or leaking elsewhere is easy to detect. The principal active ingredient in nail polish remover is acetone, and the odor of nail polish remover is familiar to everyone.

Acetylene cylinders have been known to leak around the valve stem, causing what welders call "stem fires." Another place to check for leaks is from the plug in the bottom of the cylinder. In one incident, fortunately not a serious accident, the welder was perplexed by occasionally hearing a small explosion, audible but causing no damage. The mystery was finally solved when it was discovered that the explosions were occurring in the small concave area beneath the bottom of the cylinder. The little explosions were harmless, but imagine the hazard that could have accumulated if there had been no ignition, and the defective cylinder had been allowed to slowly release acetylene, as in overnight storage.

It is important to be able to turn off the fuel flow quickly in an emergency, especially when the fuel is acetylene. Some acetylene cylinder valves are designed to accept a special wrench, but ordinary hammers and wrenches are in general inappropriate for opening cylinder valves. The special wrench should be kept available for immediate use.

A carbide "miner's lamp" burns on acetylene by dropping lumps of calcium carbide into water. The resulting chemical reaction is as follows:

$$CaC_2 + 2H_2O \longrightarrow Ca(OH)_2 + C_2H_2 \uparrow$$

Thus acetylene gas slowly bubbles out of the water solution and fuels the lamp. The same chemical reaction can be used for generating acetylene gas for welding by means of an acetylene generator. This process avoids the hazards of large-scale storage of acetylene cylinders. But the process of storing acetylene in cylinders has been so perfected, made relatively safe, and so commercialized that acetylene generators are rarely seen anymore. The cylinder supplier has a large acetylene generator somewhere, but this is of no concern to most Safety and Health Managers.

Before leaving the subject of acetylene hazards, alternative fuel gases should be considered. If a Safety and Health Manager really wants to win the approval of top management, he or she will offer a production innovation which will make the workplace safer, reduce the company's legal vulnerability, and cut production costs—*all at the same time*! It is difficult to do but not impossible, and selection of welding fuel gas presents a possible opportunity. It takes some "homework" and particular diligence to pursue the issue, but the rewards to the company and to everyone involved can be dramatic.

MAPP gas, natural gas, and propane were mentioned earlier as possible alternatives to acetylene. The first complaint the Safety and Health Manager will hear using these gases is that they do not burn hot enough. It is true that acetylene excels when a very hot flame is needed, but many industrial applications do not need temperatures as high as welders want for *some* applications. Alternate gases are certainly hot enough for brazing and soldering, but they are also hot enough for some welding applications.

One reason halfhearted attempts to switch to alternate gases do not work is that welders attempt to switch gases while using the same torch tips they used with acetylene. Special torch tips may be the secret to making the new idea a success. Welders also may need to be taught how to adjust their torches to achieve a "neutral flame" with the proper burning characteristics.

Natural gas is the cheapest alternative welding fuel of all in many cases because it is piped into the plant. On the negative side, welders will say that public utility natural gas pressure levels of 4 oz/in^2 are too low to use for welding. But the author knows of one company that took this problem to the utility and negotiated a special arrangement in which the utility agreed to supply the welding manifold with natural gas at high pressures. The results were cost savings, a safer workplace, decreased exposure to acetylene regulations which no longer applied because the acetylene was gone, a happy plant manager, and a *delighted* Safety and Health Manager.

Oxygen Cylinders

We have seen the hazards of the highly unstable acetylene, and by comparison oxygen is much more stable; in fact it is almost completely inert if it is kept away from fuel sources. But ironically, oxygen cylinders are more dangerous than acetylene cylinders. The reason for this danger is the extremely high pressure contained by the oxygen cylinder. Figure 12.6 depicts the familiar oxygen cylinder; note its resemblance to the shape of a bomb or a rocket. It is difficult to comprehend the energy that can be released by the sudden rupture of the valve on an oxygen cylinder containing 2000 psi pressure. There have been numerous accounts of cylinders becoming airborne and crashing into brick walls, demolishing the walls. If a cylinder valve ruptures while the cylinder is confined in a relatively small room, it may ricochet off walls until it kills anyone unfortunate enough to happen to be in the same room when the valve breaks. Consider the hazard of a heavy cylinder flying wildly around the room like a rapidly deflating balloon.

Workers often unwittingly drop oxygen cylinders onto the ground or bang them violently together. Oxygen cylinders are often seen standing alone and unsupported. Even though they are quite heavy, their small bases make them easy to tip over, with the resultant danger of knocking off the valve. The temptation to leave oxygen cylinders standing alone and unsecured needs to be dealt with in safety training sessions.

Figure 12.6.   Oxygen cylinder.

Another temptation with oxygen cylinders is that they seem to make perfect rollers for supporting and moving heavy items about. No matter whether the cylinders are full or "empty" (even a spent cylinder is not completely empty), use of cylinders as rollers or support can damage the cylinder and perhaps the valve. Furthermore, the large, heavy cylinders used as rollers present a problem of *control*. Once a heavy load begins to roll on a set of welding cylinders it can even *run over* an unsuspecting victim.

A perennial problem is keeping track of the valve protection cap, which must be removed to use the cylinder but which also must be screwed back into place when the cylinder is in storage. This cap protects the important valve from damage, and if the cylinder is ever moved about without it, there is a risk of the cylinder falling and knocking off the valve, with disastrous results.

When an oxygen cylinder is strapped to a wheeled cart or hand truck along with its companion acetylene cylinder and the regulator assembly is in place for welding, the cylinder is generally considered to be in operational status, not in storage. Therefore, industry practice says that the valve protection caps do not have to be in place in these situations.

Looking carefully at the valve protection cap in Figure 12.6, a somewhat odd-looking vertical slot can be seen; actually, there are two, but one is hidden. These slots have a definite engineering purpose, but not the purpose most people think. If the valve comes off while the protection cap is screwed into place, the escaping gas will impact at high velocity on the closed top portion of the cap, tending to counter the force of the gas escaping at the valve. The slotted parts on the sides of the cap permit the gas to escape, but in directions exactly opposite to each other, balancing forces and leaving the cylinder relatively at rest.

Unfortunately, the existence of the slotted openings on the cap are an invitation to their misuse by the worker who is attempting to handle the cylinder. Cylinders are heavy and unwieldy, especially to the worker who has had to handle a lot of them in one day. Furthermore, in cold weather these cylinders have a tendency to become frozen to the ground, slab, or even to each other. The worker will want to find a means to break them apart from each other or from the slab to which they are frozen. The slotted opening on the cap seems to be an ideal place to insert a pry bar to obtain some leverage. But this is *not* the purpose of the slots, and misuse in this fashion is what can lead to a broken or damaged valve.

As if the extreme pressure hazards were not enough, oxygen presents additional hazards due to its chemical properties. As stated before, it is relatively stable in the absence of fuel sources, but the fire hazard of pure oxygen under pressure in the presence of a combustible substance is extraordinary. A substance as benign as ordinary grease can suddenly become explosively combustible in the presence of pure oxygen under pressure. A worker will often hold his or her hand over the valve opening when first opening the valve to test the cylinder. If grease is on the hands or if the worker is wearing greasy gloves, a hand can easily be lost in the explosive combustion that can follow.

We have examined the separate hazards of acetylene and oxygen, but when oxygen and acetylene cylinders are stored together the hazards are multiplied. There is always the possibility that one or more cylinders will leak. The reader will perhaps recall the account earlier in this chapter describing small explosions from a leaking plug in the bottom of an acetylene cylinder. Acetylene is already highly flammable, and the presence of pure oxygen makes the situation about five times as serious. (Ordinary air is only about 20% oxygen; pure oxygen is 100%.) A noncombustible barrier at least 5 feet high must separate oxygen and acetylene cylinders, or they must be moved apart at least 20 feet.

TORCHES AND APPARATUS

Because of their vital role in safety, torches, manifolds, regulators, and related apparatus must be "approved," usually taken to mean by a recognized testing laboratory, such as Underwriter's or Factory Mutual.

The familiar torch, illustrated in Figure 12.7, is a more sophisticated piece of engineering than most people realize. The torch is often considered a handy double tube and valve assembly for delivering both oxygen and fuel gas to the weld flame, but it is more than that. Note in the figure that it has a *mixing chamber*. The torch is designed so that the mixing takes place at the right time and in the correct *total* volume. The welder controls the *proportion* of volumes of the mix by adjusting the torch valves for oxygen and/or acetylene, an oxidizing flame being oxygen-rich and a reducing flame being fuel-rich, but the total volume of the mixture is determined by the torch itself. The mixing chamber is mated to the various correct apertures for

Figure 12.7.   Oxyacetylene torch.

the approved torch tips, and this also is an important balance. If the balance is disturbed, flow rates may also be disturbed, and the flame may begin to travel back up the mixture stream and begin to burn *inside* the torch! This is really not all that uncommon, and all welders know this phenomenon and commonly call it "flash-back." A popping or snapping sound is a warning that flashback is about to occur. Once the flashback begins, a distinctive humming sound can be heard. The heat being generated inside the torch will soon ruin it, besides being a safety hazard. The dangerous situation is alleviated by turning off both torch valves quickly.

Even with approved torches and tips, flashback can occur because of the deterioration of the equipment, especially the tips. The tips are close to the heat and naturally become brittle or burned and crack, or pieces may break off. If the tip is not replaced, flashback is likely.

Despite the importance and sophistication of the torch and tip, it is not un-common to see the torch used as a hammer or chisel! The temptation arises due to the formation of slag, a waste product of the flux mentioned earlier. This hard coating of slag generally covers the weld and sticks to it. To finish the job, the welder or helper must chip the brittle slag off the finished weld. The torch and tip are so handy for this purpose (take note of shape in Figure 12.7) that welders will often take the shortcut of using the torch as a chipping tool. This is a good way to ruin an expensive piece of apparatus and at the same time increase the likelihood of flashback due to a damaged torch or tip.

The torch assembly is expensive and may be owned personally by the welder. Even when the company owns the torch, the individual welder to whom it is assigned may rightfully be very possessive because of the importance of care of the ap-paratus. This can cause another safety problem. Welders may want to keep their torches in their locked toolboxes, but the danger here is that almost all toolboxes contain at least some grease or oily materials. Grease or oil on the torch is dangerous because of the oxygen hazards discussed earlier.

Another dangerous procedure is to lock the torch in a personal locker whose door is designed to allow the torch to remain connected to the hose. This might be all right if the welder is careful to close the valves at the regulator or tanks, in addition to the convenient torch valves, but if the regulator and tank valves are not closed, a tiny leak from a torch valve may allow an explosive mixture to gradually develop overnight inside the confined locker space. In one accident a welder opened his locker door one morning and was decapitated as the explosive mixture was apparently ignited by his cigarette.

The next most abused piece of welding apparatus is the hose for delivering the gas to the torch. To be practical, this hose must be flexible, and it is thus subject to the physical hazards of wear and tear and deterioration to which such materials are generally susceptible. Multiple hoses can easily get tangled, and because of this, welders commonly wrap the acetylene and oxygen hoses together with tape simply to keep them more orderly, but such taping practices may hide defects in the hose. It is a good idea to keep at least 8 of every 12 inches uncovered.

Manifolds are the rigid tubing networks that enable one or more cylinders to supply one or more torches. Manifold setups are sometimes found when a regular production operation requires long-term, regular use of welding gas. The principal purpose of welding manifolds is to increase gas welding volumes, but safety is usually enhanced by the more permanent arrangement.

SERVICE PIPING

''Service piping'' is another matter altogether. Some plants use so much welding gas that it becomes practical to pipe the gas to the work station, which can present some problems. The piping for acetylene must be either steel or wrought iron because copper might react with the acetylene to produce copper acetylide, a dangerous explosive. A danger with oxygen piping is again the possibility of contact with oil or grease. Fittings and pipe must be checked before assembly and thoroughly cleaned if necessary. A hot water solution of caustic soda or trisodium phosphate is suggested for this purpose. A modern solvent such as chlorothane (1,1,1-trichloroethylene) is recommended by some engineers today. Oxygen piping systems should be purged after assembly by blowing out with oil-free air, oil-free nitrogen, or oil-free carbon dioxide. Chlorothane should be used to be sure that every trace of oil is removed.

Flashback can occur in service pipe systems, too, and flashback protection devices are specified, as are check valves in appropriate positions. One design for flashback protection devices involves a simple water-lock. But if the water freezes, the system will not work, so antifreeze protection is needed.

The things that can go wrong with a service piping system for welding are numerous and sometimes subtle. These problems involve inadequate emergency venting, improper joints, mistakes in installation in tunnels, and other items too numerous to mention here. The purpose of this text is to alert the Safety and Health Manager to the potential problems accompanying these systems so that he or she can be sure that personnel obtain and follow the appropriate standards for installing these systems.

ARC WELDING HAZARDS

Gas welding may have a more stormy safety record, but arc welding is a more popular process and is in many ways even more hazardous. This is one of the ironies of the subject. The major hazards of arc welding are health hazards, fires and explosions, eye (radiation) hazards, and confined space hazards. Although these hazards may be major for arc welding, they appear perhaps to a lesser degree in gas welding and other welding. Therefore, these subjects are addressed in sections of their own later in this chapter. Before these arc welding hazards are covered, the equipment used in arc welding needs to be understood.

Equipment Design

Industrial arc welding manufacturers do all they can, by means of federal standards they have helped to write and in other ways, to promote their equipment to the exclusion of lesser and cheaper models that might compete. But there is a logic to their efforts other than the profit motive. Small and relatively inexpensive models of arc welding machines are available which operate off ordinary household 110-volt current. But there are physical disadvantages with welding off ordinary household current. Welding requires large amounts of electrical power in the form of low-voltage, high-amperage circuits. Since household circuits are rated for amperages insufficient for effective welding, the small household machines make up in voltage what they lack in amperage. The high-voltage hazards of these small welding machines are admittedly difficult to understand because the industrial welding machines are supplied by much higher voltages—typically 240 or 480 volts! The key to this paradox is that the industrial machines step down the high voltage to less than 80 volts while raising the amperage to effective levels. Chapter 13 will return to the subject of voltage and amperage and will perhaps add clarity to the welding machine problem.   .

Grounding

Even with machines operating at proper voltages, the welder or other personnel can receive electrical shock from contact with the machine if something goes wrong. The protection for this is to be sure that the frame of the welding machine is properly grounded. Thus if there is a dangerous short to the frame of the machine, the overcurrent protection mechanism on the circuit will be tripped, protecting personnel. Welding machine grounding needs to be strong, both physically and electrically, to meet the demands of the current that may be applied to it. This is an especially important consideration for portable machines.

Operation

In the case of welding equipment, safety and health training will pay dividends in longer service lives of the equipment, a point sometimes overlooked. Welding cable carries so much electrical current that it can overheat and damage the insulation.

Coiling the cable, although convenient, contributes to this hazard. Coiled cable should be spread out before welding. Splices in the cable are not permitted within 10 feet of the electrode holder. The splices themselves must be properly insulated. Some judgment is required to determine when welding cables should be replaced. Certainly, damage to the extent that some conductors have bare spots is cause for replacement.

Care must be taken by the welder in preventing the wrong items from becoming a part of the welding circuit, either during welding or while the electrode holders are not in use. Voltage is not so much the danger as is heat produced by the potentially high amperages. Compressed gas tanks or cylinders must not be part of the electrical circuit, regardless of the flammability of their contents. The heat buildup caused by a high-amperage current through the conducting metal cylinder can cause pressure buildup in the cylinder which can exceed its design limits.

Some types of arc welding are safer than others, and they are gaining in popularity. This is especially true with respect to the hazards of fume generation and radiation, which will be discussed later.

RESISTANCE WELDING HAZARDS

The cleanest, most healthful, and probably safest form of welding is resistance welding. There are still hazards from electrical shock, but more important are the mechanical hazards surrounding the point of operation.

Shock Hazards

As in the spark coil of an automobile, many resistance welding machines build up electrical energy in a bank of capacitors for sudden release when the weld is made. The voltage may reach hundreds or even thousands of volts at peak. These voltages are not the empty variety seen in discharges of static electricity collected by walking on a thick carpet. The voltages may be at the same level, but the welding machine voltages carry with them the capacity to deliver a burning current. The capacitors that store this electrical energy should have interlocked doors and access panels. Not only must the interlock stop the power to the machine, it must also short-circuit all capacitors. Without this short circuiting, the capacitors could deliver a lethal shock even with the power *disconnected*. Parenthetically, the same is true for the inside back of an ordinary television set.

Guarding

Spot and seam welding machines apply pressure to the materials when the weld is made. For spot welding machines this pressure makes the machine analogous to a power press, and the operator can be injured from the mechanical hazards alone.

Seam welding machines are not like power presses, but they, too, have point of operation hazards. The nature of the hazard for seam welders is that the opposing rotation of the rollers produces a pair of in-running nip points both above and below the material being welded. The reader may want to refer back to Figure 12.4 to study the operation of the seam welder. The hazards of seam welders do not seem to be as pressing as those of spot welders because seam welding is more often a part of an automatic or mechanized production operation than spot welding, and thus operator exposure is not as great.

FIRES AND EXPLOSIONS

Welding is one of the principal causes of industrial fires. Perhaps even more than for any other welding hazards, the Safety and Health Manager can have an impact upon this particular hazard because preventing welding fires is more of a procedural matter than anything else. This means that training becomes a very important element in the hazard prevention strategy. Fortunately for the Safety and Health Manager, there is a wealth of audiovisual aids, literature materials, and case studies on the subject of welding fires.

To use just one case history as an example, one of the most devastating and tragic industrial accidents in the nation's history occurred in Arkansas in the 1960s. A welder's spark started a fire in a missile silo, and 53 workers trapped inside the silo were killed. This accident dramatically illustrates how welding adds to other hazards, such as confined workspaces, flammable and combustible materials, and lack of ventilation. Welding on old oil drums or pipes that have contained asphalt or other petroleum products has resulted in a large number of explosions and senseless fatalities.

People do not seem to realize the ignition potentials of welding operations. Welding is not safe to watch directly because of the eye hazards, so unfortunately, except for welders themselves or their helpers, few people realize what kind of a fireworks display is really taking place. Some industrial movies are good for illustration of these fireworks. Sparks are flying everywhere—not just the benign variety as seen flying from a typical bench grinder, but visible chunks and spatters of red hot molten metal which can sit and burn a hole completely through heavy fabric, plastic containers, and cracks in floors. Welding is often a short repair operation, and the temptation is to take a few chances because of the short duration of the hazard.

Welding Permits

To counter the temptations to ignore the short-run hazards of welding repair operations, the Safety and Health Manager might do well to institute a procedure for mandatory permits to weld in general plant areas and in warehouses. There are so many special precautions to take that a signed checklist is a good idea. The responsible party is usually the supervisor of the area in which the welding is to take place, but in some instances the welder can make the necessary checks and sign the form. The Safety and Health Manager's responsibility is to set up the permit system and ensure that it is executed properly by actually checking permits occasionally when welding is seen to be taking place in areas of potential hazard. Any responsible party will give the matter some conscientious attention before signing a welding permit form, especially if personnel have received safety training which exposes them to the devastating hazards of welding fires such as the one that killed 53 in Arkansas. Such training will certainly make the supervisor or welder think twice before signing the form.

Judgment should be exercised to attempt to set up a permit system which will be considered reasonable by welders and plant personnel alike. The key to this reasonableness will be to set up blanket permits or exemptions from the permit system in those areas of the plant in which fire hazards are minimal. Welding done in the welder's own welding shop, for instance, should be his or her responsibility, and for the Safety and Health Manager to attempt to impose a permit system in the welding shop would obviously constitute unwise interference. For some plants the entire plant might be reasonably safe from welding fires, and no permit system may be needed at all. Hazard identification surveys and advance planning for exactly when and where the permit system is really needed will go a long way toward the establishment of a reasonable system.

EYE PROTECTION

Eye protection comes under the topic of personal protective equipment (Chapter 8), but eye protection for welding operations is so important that this chapter on welding would not be complete without a section on this subject.

Note the careful reference above to welding opera*tions,* not welding opera*tors.* The welders themselves need protection, of course, but it is easy to forget about welding helpers and others in the area. Almost every welder has already received an eyeburn some time during his or her career and knows to be careful to wear eye protection as much to prevent pain and discomfort as to prevent long-range eye injury. But less experienced personnel may need more supervision and administrative controls to assure protection.

Remember when referring to welding lens shade numbers that the higher the shade number, the darker the shade (i.e., the more rays are shielded). The various arc welding methods produce much more intense radiation and thus require higher shade numbers than does gas welding. The radiation from arc welding, except submerged arc methods, is so intense that a helmet is necessary to protect the entire face area from painful burns. Gas welding is typically done while wearing goggles.

PROTECTIVE CLOTHING

Proper clothing is serious business to the professional welder. Virtually every experienced arc welder has sometime in his or her career sustained a "sunburn" from the ultraviolet rays produced by the welding arc. Such a welder does not need to be told to use protective clothing to cover all skin areas that would otherwise be exposed to the arc rays. But ultraviolet burn is only one of several hazards against which protective clothing is intended. Hot burning sparks or small pieces of molten metal can fall into openings around shoes or between pieces of clothing. Even when the clothing is seamless or when openings are well shielded, a hot piece of molten metal can be trapped in a fold and burn its way through to the welder.

Leather is the traditional favorite material for welding gloves, aprons, and leggings because of its superior thermal protective qualities. Wool is also very durable. But Nomex and other synthetic materials are also becoming popular for welders' protective clothing. Cotton fabric is attacked by the radiation and soon disintegrates even if it escapes ignition by the sparks.

Hazards to the welder from falling sparks and weld metal are greatly increased when the welding must be done overhead. Here the welder is virtually taking a shower in welding sparks. All clothing openings must be carefully shielded, and even under the helmet the welder's head must be protected from such misfortunes as a welding spark in the ear.

GASES AND FUMES

There are two extremes in degree of concern over welder breathing hazards. One extreme, call it position A, is taken generally by welders themselves, who often have no concern at all over chronic exposure to welding "smoke." Some welders even enjoy the smell of welding fumes in the air. The other extreme, position B, is the occasionally overzealous industrial hygienist who can find a hazard somewhere in almost all welding fume situations. Both extremes are only partially correct and can lead to dangerous errors in safety and health strategies.

The principal error in position A is that persons taking this extreme position are usually overlooking the long-time effects of chronic exposure. These persons tend to believe that if the welding smoke does not make them nauseated, dizzy, or give rise to some other acute symptom, the fumes are safe. Recalling the principles covered in Chapters 1 and 6, the chronic exposures can actually be the most dangerous because of their adverse effects on worker health.

Position B exaggerates the effects of tiny exposures to dangerous contaminants. It is terrifying to realize that some welding releases phosgene gas, the same gas that has been used in chemical warfare. But the exposures are generally very low and can be controlled by appropriate procedures. In the final analysis, no epidemiological studies have shown welding to be an unusually dangerous occupation. From a health standpoint, welders do not have significantly shorter life spans than workers in general. Having considered this perspective, let us categorize the hazards of welding atmospheres and examine rationally what should be done about them.

Contaminant Categories

Figure 12.8 diagrams the major kinds of welding atmosphere contaminants: particulates and gases. The particulates are dust particles or even tinier smoke particles. Metal fumes in the welding atmosphere are tiny particles of metal which have been vaporized by the arc and then resolidified into particles as they cool. The gases may be either already present, as in inert-shielding gases, or they may be chemical reaction products of the process.

    The term "pneumoconioses" in Figure 12.8 was explained in Chapter 6; it is

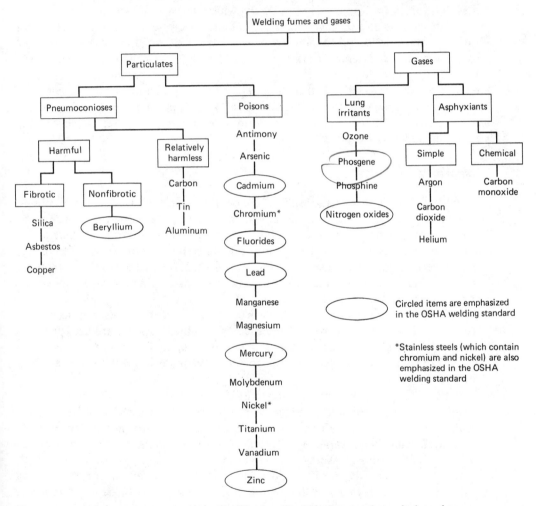

Figure 12.8. Classification of welding fumes and gases by hazard.

merely a general term which literally means "reactions to dust in the lungs." Everyone's lungs must deal with dust to some extent, and some welders' pneumoconioses are no more hazardous than would be caused by sweeping the floor. Some welding dusts are more hazardous, however, because they cause "fibrosis" or the building up of useless fibrous tissue in the lungs. The most harmful dusts are those in which the microscopic particles have the shape of fibers instead of more rounded particles. Asbestos and silica are good examples.

The "pulmonary irritants" are simpler in that they attack the lungs directly, whether the irritants be particulate or gas. The more insidious hazards, though, are from those particulates or gases that do not irritate the lungs directly, but through the lungs gain access to the rest of the body, where they act as systemic poisons.

It would be nice to have quantified tables of expected contaminant levels for various atmospheric constituents for various types of welding. There have been some experimental attempts at this (see Ref. 26 and 76), but there are so many variables to be controlled that it is almost impossible to obtain reliable predictions of weld fume content. The best strategy is to be aware of potential hazardous contaminants and to know what conditions are most likely to produce these contaminants. Atmospheric sampling can then be used in suspect situations to establish whether contaminant levels are indeed excessive.

Hazard Potentials

The biggest contributor to atmospheric contaminants around welding is the coating or condition of the surfaces to be joined. It is true that welding on clean iron or ordinary construction steel produces fairly large concentrations of iron oxide fume, but fortunately "siderosis," the pneumoconiosis resulting from iron oxide, is not really a very dangerous disease when it occurs alone. If the surface of the metal has a coating of material containing asbestos, however, this coating must be removed to prevent asbestos contamination of the air.

Even the act of cleaning the metal surfaces to be welded can result in secondary hazards. If chlorinated hydrocarbons, such as trichloroethylene, are used to clean the metal, these solvents must also be removed thoroughly before the welding takes place. The energy of the welding arc can cause decomposition of the solvent into dangerous phosgene gas.

"Galvanized" is a term referring to a zinc coating on the metal to prevent rust. Welding on galvanized steel needs extra caution and good ventilation because the welding arc can produce a fume of zinc or of zinc oxide. Zinc is not as dangerous as its relative, lead, but it can cause a brief but uncomfortable "metal fume fever." Daily exposure results in a sort of immunity, but this immunity is lost in a few days, even in just a weekend away from exposure. The next Monday morning the nausea and chills are back again, causing this disease to be known as "Monday morning sickness," although admittedly there are other things about Mondays that often make workers "sick."

Plating metals are usually much more dangerous to weld than the iron or steel upon which the plating is used. Cadmium is a plating metal whose welding fumes are considered very dangerous. This is one fume that has been known to be fatal in a single acute exposure. Even worse, acute exposures to cadmium usually do not display warning symptoms. Chronic exposures have been associated with emphysema and kidney impairment.

Stainless steel is one of the most dangerous materials to weld because of its high chromium content. Chromium trioxide is formed by the oxidation triggered by the welding heat, and chromium trioxide reacts with water to produce chromic acid. Ample sources of water can be found on human skin and the mucous membranes, resulting in chromic acid ulceration of these surfaces. Other chromic acid hazards were discussed in Chapter 8, where the phenomenon of "chrome holes" was discussed.

Welding in confined spaces complicates the atmospheric contamination problem. In confined spaces the hazards of gases increase dramatically. Nitrogen and argon are inerting agents for the protection of the weld, but they are also simple asphyxiants to the welder. Another simple asphyxiant in welding atmospheres is carbon dioxide. Contrasted with the simple asphyxiants is the chemical asphyxiant carbon monoxide, also present to some extent in welding atmospheres, especially for gas welding.

Nitrogen is not as inert as argon or helium, mentioned earlier as inerting agents, although it is true that nitrogen is a relatively stable element. But nitrogen can be oxidized, especially in the extremes of welding temperatures, creating oxides that can be harmful. Nitrogen dioxide (NO_2) is considered more dangerous than nitric oxide (NO); in fact, nitric oxide was formerly used as a dental anesthetic. But both oxides have been shown to have harmful effects.

Lead and mercury are well-known systemic poisons, and airborne fumes are the prime avenues for entry of these poisons into the body. Most welding does not involve these two metals. Soldering is used widely with lead alloys, but the low temperatures of soldering render the lead fumes relatively harmless.

Beryllium is a very useful alloy metal used in steel, copper, and aluminum. Unfortunately, the presence of the beryllium alloy in the material makes the metal very dangerous to weld. Since beryllium fume (particulate) hazards are both acute and chronic, most welders are wary of beryllium dangers.

Fluorine and fluorine compounds, usually fluorides, enter the welding atmosphere via welding flux or coverings. The popular shielded metal arc welding (SMAW) process is subject to hazards of fluorine compounds. The principal hazard is chronic exposure, not acute, and long-term exposures cause abnormalities in the victim's bones. Other cleaning compounds and fluxes may also be hazardous, and personnel should check ingredients and heed manufacturers' instructions.

Before leaving the subject of welding gases and fumes, an important point is to be emphasized. None of the toxic materials or hazardous conditions described in this section are so dangerous as to prohibit welding. Welding atmospheres can be made safe by local or general exhaust ventilation or by personal protective equipment. The key is to recognize the potentially hazardous conditions, test atmospheres for excessive contaminant levels, and take corrective action if required.

SUMMARY

Welding represents a microcosm for the study of the entire field of occupational safety and health. There are mechanical hazards, fire hazards, air contamination hazards, personal protective equipment considerations, and almost every other subject addressed in this book. Welding processes are many and varied, and most Safety and Health Managers know little about the technical aspects and terminology. A little study of the basics of welding, however, can open up opportunities for revision or substitution of processes which can enhance health and safety and at the same time improve efficiency and cut production costs. No other subject area seems to offer so much opportunity for Safety and Health Managers.

EXERCISES AND STUDY QUESTIONS

12.1. What are the three basic categories of resistance welding? Which of the three is the cleanest and most healthful?

12.2. What distinguishes soldering and brazing from welding?

12.3. What distinguishes soldering from brazing?

12.4. What is the commonly used name for the most popular process of arc welding? What is the official AWS designation for this process?

12.5. Identify the following welding processes:

GTAW

GMAW

SAW

RSEW

RSW

12.6. Why should acetylene cylinders be stored valve end up?

12.7. Why are oxygen cylinders charged at so much higher pressures than acetylene cylinders?

12.8. In what way are greasy gloves of particular hazard to the oxyacetylene welder?

12.9. Describe a way in which Safety and Health Managers might cut production costs and at the same time avoid hazards.

12.10. Why are valve protection caps important to safety? Explain the purpose of the slots in the caps. How are the slots often misused?

12.11. Explain the phenomenon of "flashback" in welding operations.

12.12. How can taping welding hoses together to keep them orderly result in a hazard?

12.13. How can small arc welding machines that operate off ordinary household current be more dangerous than industrial arc welding machines?

12.14. Why should welding cables be uncoiled before using?

12.15. What is the principal hazard of allowing metal tanks to become a part of a welding circuit?

12.16. What arc welding process is gaining in popularity because it is more healthful than other arc welding processes? What big disadvantage does it have?

12.17. What is the principal mechanical hazard of spot welders?

12.18. Give at least two reasons why people are psychologically inclined to risk the chance of welding fires.

12.19. Against what principal hazard are welding permits aimed?

12.20. Which of the following welding operations require the most eye protection? SMAW, SAW, RSEW

12.21. What is the best natural material for welders' protective aprons, gloves, and leggings?

12.22. Name the pneumoconiosis resulting from exposure to iron oxide fume. Is it a severe hazard?

12.23. Why is it difficult to provide accurate tables of welding fume content?

12.24. What distinguishes welding fumes from toxic gases produced by the welding process?

12.25. Describe some conditions which when present make welding fumes and gases more dangerous to the welder.

13

Electrical Hazards

Percent of OSHA
general industry citations
addressing this subject

Year after year the National Center for Health Statistics reports approximately 1000 accidental electrocutions annually in the United States, with about 1 in 4 being industry and farm related. Everyone knows that electrical shock can be fatal, but the mechanism of the hazard is a mystery to most people. The mystery is due in large part to the fact that electricity is invisible. The use of electricity throughout our homes has led to a degree of complacency which is a factor in most electrocutions.

ELECTROCUTION HAZARDS

The first step toward safety from electrocution is to overcome the myth that "ordinary 110-volt circuits are safe." The truth is that ordinary 110-volt circuits can easily kill and actually do kill many more people than do 220-volt or 440-volt circuits, which nearly everyone respects. But the myth about 110 volts persists because almost everyone has sustained without serious injury an electrical shock around the home or on the job. An accident like this leads victims to the false conclusion that, although a 110-volt shock can be startling, it probably will not be fatal. Although they know that others have been killed by such shocks, they somehow feel resistant or too strong to be seriously injured.

It is true that some persons are more resistant to electrocution hazards than others, but a far more important factor is the set of conditions surrounding the accident. Wet or damp locations are known to be hazardous, but even body perspiration can provide the dampness which can make electrical contact fatal. Another important condition is the point of contact. If current flow enters the body through the fingers and passes out through a contact at the elbow, no vital organs receive direct exposure. But if the flow is from a hand through the body to the feet, vital organs such as the heart and lung diaphragm are affected with possibly fatal results. Contact by the body torso to complete a circuit can also produce vital exposure to electrical current. Another factor can be the presence of wounds in the skin, which can result in a much higher current flow if contact is made where the skin is broken.

Physiological Effects

The central nervous system of our bodies is the conduit of signals between our brains and our muscles, including muscles of vital organs such as the heart and lung diaphragm. These signals are tiny electrical voltages which tell our muscles when to contract and when to relax. An external electric shock can send through the body currents which are many times greater than the tiny natural currents within our nervous systems. These larger currents can cramp or freeze muscles into a violent contraction—one that will not allow the victim to let go of the object contacted, or one that will stop breathing or stop the heart.

The heart is obviously our most important muscle. Its function is a rhythmic contraction/relaxation which is timed by natural electrical pulses. The heart is thus very vulnerable to any pulsating electrical current. Common electric utility power supplies alternating current which cycles at a frequency of 60 hertz. It is ironic that 60 hertz is one of the most dangerous frequencies to which the heart can be exposed. This frequency tends to cause the heart to convulse weakly and irregularly at a rate too rapid to accomplish anything, a phenomenon known as fibrillation. Once fibrillation starts, death is almost a certainty, except that fibrillation has sometimes been stopped by controlled electric shocks to the heart muscle. The controlled electric shocks reestablish the heart's natural rhythms. Unfortunately, a defibrillation device is rarely available soon enough to save the life of an electrocution victim.

Stopped breathing from electric shock is due to cramped muscles responsible for respiration, such as the diaphragm and those controlling rib cage expansion. The first aid remedy is artificial respiration, the same as for near-drowning or other respiratory crises.

Just how much electrical current is fatal? There is no set answer to this question, but Figure 13.1 summarizes the opinions of several experts. The horizontal scale is logarithmic and is in units of milliamps or thousandths of an ampere. To put the chart in perspective, an ordinary table lamp with a 60-watt bulb draws about 500 milliamperes of current, far more than is needed to be fatal. An ordinary house circuit of 20 or 30 amperes will not trip the circuit breaker until there is a current flow of 20,000 to 30,000[1] milliamperes, respectively, about 100 to 1000 times as much as the lethal dose.

With such lethal potential available from an ordinary 110-volt house circuit it

---

[1] 1 ampere = 1000 milliamperes.

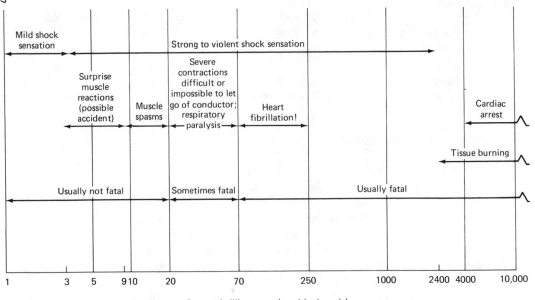

Figure 13.1.   Effect of alternating electric current on the human body.

would appear that almost no one could survive an electrical shock from such a circuit. But the body, especially the skin, has resistance which limits the flow of electric current when exposed to 110-volt potential. To understand this resistance, some fundamentals of electricity may need review.

Ohm's Law

The basic law of electric circuits is Ohm's Law, stated as:

$$I = \frac{V}{R} \tag{13.1}$$

$$where\ I\ =\ \text{current in amperes}$$
$$R\ =\ \text{resistance in ohms}$$
$$V\ =\ \text{voltage in volts}$$

The law can be rewritten as

$$V = IR \quad \text{or} \quad R = \frac{V}{I}$$

Wattage is a measure of power and can be computed from known quantities of current and voltage or resistance as follows:

$$W = V \times I \quad \text{and} \quad W = I^2R \tag{13.2}$$

where W is the power in watts.

Of principal concern are alternating current (AC) circuits, which are the predominant type in both domestic and industrial use. Standard AC circuits cycle 60 times per second (in the United States and Canada), as shown in Figure 13.2. Alternating currents are more convenient to generate and distribute than are direct currents. But computations of current, resistance, and voltage, using Ohm's Law are somewhat awkward for AC circuits because the voltage varies from zero to positive, back to zero, to negative, and back to zero again every cycle. For convenience an "effective" current for an AC circuit is computed as a value somewhat less than the current peaks. A direct current operating through a given load is found to generate as much heat as an alternating current which has peak currents 41.4% higher than the direct current. Thus the ratio of effective current to peak current is computed as:

$$\frac{\text{effective current}}{\text{peak current}} = \frac{100\%}{100\% + 41.4\%} = 0.707 = 70.7\%$$

Effective voltages are computed by the same ratios as effective currents since they are related by Ohm's Law. An ordinary 110-volt circuit then has an effective voltage of 110 volts even though peaks of voltage over 150 volts occur every cycle.

The current drawn by an ordinary 60-watt lamp bulb can be computed by arranging equation (13.2) as

$$I = \frac{W}{V} = \frac{60 \text{ watts}}{110 \text{ volts}} = 0.55 \text{ ampere}$$

Since the wire and other parts of the circuit would consume some power, a fair approximation to the current flow in the 60-watt table lamp is ½ ampere or 500 milliamperes as stated earlier.

Returning now to the question of why more people are not killed by ordinary 110-volt circuits, we can use Ohm's Law to determine how much the skin can limit the flow of electric current through our bodies. Human skin, if it is dry enough, is a

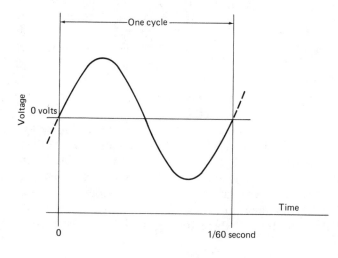

Figure 13.2. Alternating current voltage.

good insulator and may have a resistance of 100,000 ohms or more. Using Ohm's Law, a 110-volt exposure then would result in only a tiny current:

$$I = \frac{V}{R} = \frac{110 \text{ volts}}{100,000 \text{ ohms}} = 0.0011 \text{ ampere}$$
$$= \text{approximately 1 milliampere}$$

Referring to Figure 13.1, it can be seen that such a tiny current will probably not even be noticed. But add any perspiration or other moisture to the skin, and the resistance drops sharply. Due to perspiration alone the skin resistance can be reduced 200 times, to a level of about 500 ohms with good contact with the electrical conductor. Once inside the body the electrical resistance is very low and the current flows almost unimpeded. If the total resistance in the circuit is only 500 ohms, the current is calculated as

$$I = \frac{V}{R} = \frac{110 \text{ volts}}{500 \text{ ohms}} = 0.22 \text{ ampere}$$
$$= 220 \text{ milliamperes}$$

From Figure 13.1 it can be seen that an alternating current at this level passing through the body, including the heart, will most likely be fatal. Therefore, readers, if you have ever received an electrical shock, and most of us have, you can be glad that you were not perspiring enough, or that you did not have a good enough contact, or the path of the current bypassed the trunk of your body, or you were poorly grounded, or some other resistance impeded the current. Otherwise you would have been killed by the ordinary 110-volt circuit, no matter how "tough" you are or how resistant you think you are to electrical shock.

Grounding

In the discussion above the term "grounded" was used. Just what does this electrical term mean? A requirement for electrical current to flow is that its path make a complete loop from the source of electrical power through the circuit and back again to the power source. We understand this loop as we connect a lantern bulb to the posts of a lantern battery, as shown in Figure 13.3. Disconnection of the circuit at any point in the complete loop stops the flow of current. This means that there must always be two conductors: one to carry the current to the device (usually called "load") which uses it and another to carry the current from the load back to the electrical source. However, a trick makes the long trip back to the electrical source very simple in most applications of electrical power.

The earth for the most part is a fairly good conductor of electricity. Besides this, it is so massive that it is difficult for a human-made source of electricity to affect it much one way or another. Thus no matter what we do on the surface of the earth, the earth maintains a relatively even potential or charge. This means that if we drive two stakes firmly into the earth, even at great distances from each other, we may consider the resistance between them to be nil. The current flow may not be directly from one stake to the other because there are millions of electrical contacts

Figure 13.3.　An electrical circuit makes a complete loop.

to the earth at all times. Some of these contacts are positive, some are negative, but the total result is zero or earth potential. Thus any electrical conductor driven into the earth immediately assumes the zero reference potential of the earth. This is a very convenient characteristic of the earth because it enables us to use it as one great common conductor back to the source of power. Figure 13.4 illustrates the use of the ground as a return conductor.

A careful examination of Figure 13.4 reveals that the power company provides a separate neutral conductor for the completion of the circuit back to the source. There are conditions that make dependence on the common potential of the earth somewhat unreliable. A very dry season, for instance, may make the surface of the earth lose its conductivity. This is especially a problem if the area is dry around the grounding conductor stake which has been driven into the ground. The neutral conductor then assures the completion of the circuit regardless of conditions.

The use of ground in electrical circuits is so advantageous as to be considered indispensable. However, the very convenience and proximity of the ground everywhere present a hazard. If a person contacts an energized conductor and at the same time is in contact with the ground or some other object which has a conductive path to ground, that person completes the electrical circuit loop by passing electric current through his or her body. A major portion of the *National Electrical Code*® [2] is devoted to prevention of this hazard.

The principal way in which persons are protected from becoming a part of the path to ground is by insulation of conductors. In addition, exposed conductive surfaces are given a good connection to the ground, usually by means of the ground wire, so that opportunity for a person's body to be the path to ground is minimal. Paradoxically, in some rare instances the *National Electrical Code*® takes exactly

²The *National Electrical Code*®, commonly abbreviated NEC, is published regularly by the National Fire Protection Association (NFPA), 470 Atlantic Avenue, Boston MA 02201.

Figure 13.4. Alternate conductive paths to source via earth.

the opposite approach. For some systems it makes more sense to *isolate* the entire structure from ground. If the structure is isolated, workers are protected by not being in contact with conductors that could connect them with ground.

Wiring

A typical 110-volt circuit has three wires: "hot," "neutral," and "ground." Sometimes the neutral is called the "ground*ed*" conductor, in which case the ground is called the "ground*ing*" conductor. The purpose of the hot wire (usually a black insulated wire) is to provide contact between the power source and the device (load) that uses it. The neutral (usually a white insulated wire) completes the circuit by connecting the load with ground. Both the hot and the neutral normally carry the same amount of current, but the hot is at an effective voltage of 110 volts with respect to ground, whereas the neutral is at a voltage of nearly zero with respect to the ground.

The third wire is the ground wire and is usually either green or is simply a bare wire. The purpose of the ground wire is safety. If something goes wrong so that the hot wire makes contact with the equipment case or some other conductive part of the equipment, the current in its path to ground can bypass the load and take a shortcut, commonly called a "short." Since the load is bypassed, the short is a very low resistance path to ground and by Ohm's Law draws a very high current. This high current in a properly protected circuit will almost immediately either "blow" a fuse or "trip" a circuit breaker, depending on the type of overcurrent protection provided in the circuit, and stop all flow of current in the circuit.

It is possible of course to have a short without a ground wire. The equipment may be naturally grounded by its location or installation, or the hot wire can somehow contact the neutral. Sometimes the short to ground is only partial because there is considerable resistance in the short path to ground. Such a short may go undetected because the short flow of current is of insufficient amperage to cause the total circuit current to trip the overcurrent protection in the circuit. In this case current will continue to flow, and equipment loads will continue to operate in the presence of these shorts or "ground faults," as this type of short is sometimes called. Such ground faults can be especially dangerous on construction sites. This hazard is the basis for <u>ground fault circuit interruptor (</u>GFCI) devices on construction sites. The GFCI protection is in addition to overcurrent protection such as circuit breakers or fuses.

Figure 13.5 explains how a GFCI works. Whenever the current flow in the neutral is less than the current flow in the hot wire, a ground fault is indicated, and the current flow is stopped by a switch that breaks the entire circuit. One difficulty with GFCIs is that some leakages to ground are almost impossible to prevent, especially when conditions are wet or extension cords are extremely long. This

Figure 13.5. Ground fault circuit interruptor (GFCI). The 0.5-ampere fault to ground causes a current flow imbalance between the hot and neutral. This imbalance triggers the GFCI to break the circuit.

causes the GFCI to trip even when no hazard exists, a condition known in the construction industry as "nuisance tripping." An alternative to GFCIs is for the employer to test, inspect, and keep records of the condition of equipment grounding conductors.

One misconception about shorts is the idea that a good fuse or circuit breaker is sufficient to stop the flow of a dangerous short through a person's body. A reexamination of Figure 13.1 shows that a person will almost certainly be killed by a bodily exposure to a current that would not blow even the smallest popular household fuses (i.e., 15- or 20-ampere fuses). A fuse or breaker rated at 15 amperes will handle up to 15,000 milliamperes before blowing, several times as great as the fatal current shown in Figure 13.1. The beauty of the third wire or grounding wire is to provide a very low-resistance, high-current short to ground which will go ahead and trip the fuse or breaker immediately, before *other* short paths to ground (e.g., through a person's body) can do their damage.

Double Insulation

Unfortunately, less than half of the electric hand tools in actual use are properly grounded. Studies of equipment returned to the factory for repair have shown that a large number of units have been altered so that the grounding system is no longer intact. A common alteration is to cut off the third prong of the plug so that it can be plugged into an old two-wire receptacle. To counter this practice the use of "double-insulated" tools is permitted in lieu of equipment grounding. A second covering of insulation gives an extra measure of protection to the operator of double-insulated tools in case of a short to the equipment case.

Most double-insulated tools have a plastic, nonconductive housing, but this is not a fully reliable indication that the tool is double-insulated. The second covering of insulation must be applied according to precise specifications before the tool can qualify to receive the designation "double-insulated." Qualifying tools have the manufacturer's mark "double-insulated" or a square within a square ▢ to indicate double insulation.

Miswiring Dangers

In the original wiring job sometimes the electrician makes mistakes or uses slipshod practices which increase hazards. One of these practices is to "jump" (connect) the ground wire to the neutral wire. Actually, this is a trick that will work, and usually no one will be the wiser, but the practice does cause hazards. Figure 13.6 shows how a circuit is correctly wired, revealing that both the neutral and grounding wire are connected directly to ground. So in Figure 13.7, where the ground is jumped to the neutral, a third wire is not used in the wiring system.

The main hazard in jumping the ground to neutral is that it can create low voltages on exposed parts of equipment. The equipment case or housing is connected to the ground wire. Since normally no current flows through the ground wire, it serves as an excellent method of keeping the voltage on the equipment case close to

Figure 13.6. Correctly wired 110-volt circuit.

Figure 13.7. Ground jumped to neutral.

zero with respect to ground. But the neutral does carry considerable current. Using Ohm's Law it can be determined that this large current on the neutral may cause the neutral terminal to have a low voltage with respect to ground, especially if the neutral wire must travel a long distance back to the ground at the meter. If the circuit is carrying a current of 20 amperes and the resistance of the neutral wire is ½ ohm, the voltage on the equipment case is calculated to be

$$V = IR = 20 \times \tfrac{1}{2} = 10 \text{ volts}$$

This is a low voltage but is theoretically capable of producing a fatal current through a person's body if conditions are just right (rather, just wrong). The real hazard, though, is that a loose or corroded connection somewhere in the neutral circuit would increase its resistance, perhaps to 4 or 5 ohms, causing the voltage to increase several times.

Another common wiring mistake is "reversed polarity," which simply means that the hot and neutral wires are reversed. This is another subtle problem because most equipment will operate perfectly well with reversed polarity. One hazard of reversed polarity is that the designated leads (black lead hot, white lead neutral) become reversed, and the confusion could bring on an accident to an unsuspecting technician. Another hazard is that a short to ground between the switch and the load could cause the equipment to run indefinitely, independent of whether the switch is on or off (see Figure 13.8). Finally, bulb sockets can become hazardous when the polarity is reversed. In Figure 13.9a a correctly wired socket shows the screw threads to be neutral. But in a reversed polarity socket, as shown in Figure 13.9b, the exposed screw threads become hot and the button, which is naturally more protected at the bottom of the socket, becomes neutral.

Perhaps the most common wiring mistake of all is to fail to connect the ground terminal to a ground wire, a condition known as "open ground" or "ground not continuous." This is another error which can easily go unnoticed because equipment attached to circuits miswired in this way will usually operate normally. But if an accidental short to the equipment case occurs, the worker is in danger of electrocution.

The three cases of miswiring discussed in this chapter are not the only mistakes

Figure 13.8. Reversed polarity. A short in the position indicated in the circuit for this drill will cause the drill to operate continuously, independent of the switch.

Screw threads
are neutral   Button is hot

Screw threads are hot!

(a)   (b)

Figure 13.9. Hazards of reversed polarity in lamp socket: (a) correctly wired socket; (b) reversed polarity socket.

that can be made in wiring electrical circuits; they are not even the most hazardous. But because they permit electrical circuits to "work normally," they go unnoticed by uninformed users of equipment attached to such circuits. Because the errors are not immediately crippling to the function, they are frequently committed. Some simple checks with inexpensive testers can easily demonstrate the problems. Later in this chapter these testers will be examined.

FIRE HAZARDS

Most people think of electrocution when they think of electrical safety, but electrical codes have as much to do with fire hazards as they do with electrocution. Many systems, such as fuses or circuit breakers, protect against both fire and electrocution, but their primary function is fire prevention.

Wire Fires

One of the most common causes of fires of electrical origin are wires that become overheated because they conduct too much current. Wire diameters (gauges) must be properly sized to handle the current load expected, and overcurrent protection (fuses or breakers) must assure that these loads are not exceeded. Substitution of fuses with copper pennies is a common method of defeating the overcurrent protection so that the circuit will handle larger loads. If no fuse is present to burn in two, the wire itself may act as the next weakest link. If the wire becomes hot enough to burn in two, any contact with combustible material along the wire run is likely to produce a fire.

Arcs and Sparks

Whenever two conductors make a physical contact to complete a circuit, a tiny (or not so tiny) electric arc jumps the air gap just prior to contact. This arc may be so small as to be undetectable, but it is hot enough to ignite explosive vapors or dusts within their dangerous concentration ranges.

When the electric arc is an instantaneous discharge of a statically charged object, it is sometimes called a "spark." Such sparks are capable of igniting an explosive mixture, the spark plug of an automobile engine being ample testimony. Prevention of sparks is accomplished by electrically connecting or "bonding" two objects that may be of different static charge. This is especially important when pouring flammable liquids from one container to another.

The arc that occurs when an ordinary electrical circuit is completed is virtually impossible to prevent. This means that switches, lights, receptacles, motors, and almost any electrical device, even telephones, are a source of ignition to hazardous concentrations of explosive vapors or dusts. Chapter 7 discussed the ranges of explosive vapors and defined the LEL and UEL explosive limits. Since the arc is impossible to prevent, some means must be used to separate the arc from the hazardous concentrations in the air. This is done by using wire, conduit, or equipment which is either vapor tight or strong enough to contain and prevent the propagation of an explosion inside the conduit or equipment. This is an expensive undertaking, and it is tempting to take shortcuts. The *National Electrical Code* ® has a strict code for electrical wiring and equipment designed for hazardous locations. The Safety and Health Manager should be able to identify the operations or hazardous locations within the plant that require special wiring and electrical equipment. Accordingly, this identification scheme is discussed next.

Hazardous Locations

One of the most difficult tasks in the field of industrial safety is the definition of various industrial locations which require special wiring and equipment to prevent explosions. Industrial processes are so diverse as to defy a general definition. In addition, the ignition mechanisms are different for different materials. For instance, the hazard of heat buildup on electric equipment housings and bearings coated with ignitable dusts is altogether different from the hazard of spark ignition of explosive vapors derived from flammable liquids. Difficult as the problem is, it must be dealt with because some industrial locations are just too dangerous to allow exposure to electric ignition sources.

The *National Electrical Code*® meticulously defines various conditions to classify hazardous locations roughly into six categories. Within these classifications are various groups which identify the substance group that is causing the hazard.

The major classification is according to the physical type of dangerous material present in the air and is designated "Class." The next classification is called "Division" and relates to the extent of the hazard by considering the relative frequency with which the process releases hazardous materials into the air. The criteria for "Division" is subjective, not quantitative, except around paint spray areas. This subjectivity introduces gray areas which cause problems.

Figure 13.10 is a decision chart that attempts to simplify the complicated process of classification of hazardous locations. The chart is approximate only

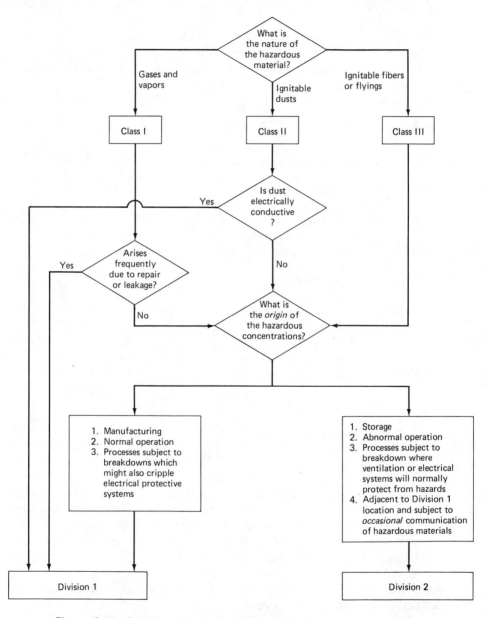

Figure 13.10. Decision chart for classifying "hazardous locations" which are dangerous from the standpoint of ignition of materials in the air.

because strict definition would require enumeration of pages of exceptions and conditions, many of which would overlap. The thing to remember is that the "Class" is the material and the "Division" is the extent of the hazard. Thus one can state that Division 1 locations are more hazardous than Division 2 locations but cannot state absolutely that Class I locations are more hazardous than Class II or III locations.

Since classification is so complex, industry often relies on examples seen in similar industries to decide whether a location is Division 1 or 2 or not of sufficient hazard to classify at all. Examples of some widely seen hazardous locations are as follows:

Description	Classification
Paint spray areas (flammable paint)	Class I, Division 1
Areas adjacent to but outside of paint spray booth	Class I, Division 2
Areas of open tanks or vats of volatile flammable solvents	Class I, Division 1
Storage areas for flammable liquids	Class I, Division 2
Inside refrigerators containing open or easily ruptured containers of volatile flammable liquids	Class I, Division 1
Gas generator rooms	Class I, Division 1
Grain mills or processors	Class II, Division 1
Grain storage areas	Class II, Division 2
Coal pulverizing areas	Class II, Division 1
Powdered magnesium mill	Class II, Division 1
Parts of cotton gin locations	Class III, Division 1
Excelsior storage areas	Class III, Division 2
Closed piping for flammable liquids (piping has no valves, checks, meters, or similar equipment)	Not classified

Equipment qualifying for hazardous locations usually qualifies for both Division 1 and 2. When in doubt, most industries want to install equipment approved for Division 1 in order to be prepared for the worst. Most violations of code are not for selection of Division 2 equipment when Division 1 should have been selected; most violations are from the use of thin-walled conduit and conventional electrical equipment in Division 1 or 2 locations.

"Approval" of electrical equipment for use in hazardous locations means that the manufacturer's design has been tested and approved by a recognized testing laboratory such as Underwriters' or Factory Mutual. A label must be on the equipment designating its classification if that equipment is to be used in hazardous locations.

Figure 13.11 shows several examples of "explosion-proof" equipment approved for Class I, Division 1 locations. Explosion-proof conduit looks more like pipe than conventional thin-walled conduit, which looks more like tubing. Explosion-proof junction boxes are castings, as opposed to conventional formed sheet metal boxes. The complicated structures for telephones and even light switches make

Figure 13.11. Explosion-proof electrical equipment approved for Class I, Division I hazardous locations. Note heavy-duty, machined components. (a) Electrical outlet plug and receptacle; (b) wall switches. (Courtesy Appleton Electric Co.)

obvious the fact that explosion-proof equipment costs several times as much as conventional equipment.

In Division 1 locations it is recognized that there is no way to ensure that the vapors will be kept out of the conduit and equipment. During maintenance, installation, or other open periods, vapors will enter the system. Therefore, the electrical equipment designer assumes this inescapable reality and designs the Division 1 equipment to withstand an internal explosion and cool the explosion gases as they escape before they can ignite the entire area in a devastating explosion.

By contrast, Division 2 electrical equipment enjoys some isolation from dangerous explosive vapors *most* of the time. Therefore, if the Division 2 equipment can be properly sealed with gaskets to make it vapor-tight, it will be safe. Class I, Division 2 equipment is characterized as "vapor-tight," whereas Class I, Division 1 equipment is *not* vapor-tight but is "explosion-proof." Of course, if equipment is classified as Class I, Division 1 (explosion-proof), it is also acceptable for use in Class I, Division 2 areas even though the equipment is not vapor-tight.

The foregoing comparison of Division 1 and Division 2 equipment and locations acknowledged the generalization that any equipment approved for Division 1 locations is also acceptable for Division 2 locations *of the same classification*. However, it should be noted here that equipment approved for Class I locations is not necessarily approved for Class II or III. The hazard mechanisms of ignitable Class II dusts or Class III fibers are somewhat different from the hazards of Class I vapors. Dusts and fibers can settle on warm equipment, insulating it from necessary heat dissipation during operation. Such insulation can cause a substantial heat buildup on the equipment which can result in a smoldering dust ignition and subsequent explosion.

A common error when selecting electrical receptacles for hazardous locations is to mistake weatherproof electrical outlets for approved equipment. Ordinary weatherproof outlets, as shown in Figure 13.12, are not approved for any type of hazardous locations in either Division 1 or 2. The spring-loaded cover protects the receptacle from weather when the receptacle is not in use, but when a plug is inserted into the receptacle, the receptacle is exposed as much as any conventional one.

The Safety and Health Manager may be baffled when reading equipment labels to find that equipment is classified and labeled by Class and *Group*, rather than Class and *Division*. Like the "Class" designation, the "Group" designation

Figure 13.12. Ordinary weatherproof outlets not approved for hazardous locations.

identifies the type of material present in the atmosphere, but the "Group" classification is more detailed. Four of the Groups belong to Class I and three to Class II and are summarized as follows:

Class	Group	Description	Examples	Industries
I	A		Acetylene	Welding fuel generators
I	B	Highly flammable gases and some liquids	Hydrogen	Chemicals and plastics
I	C	Highly flammable chemicals	Ethyl ether Hydrogen sulfide	Hospitals, chemical plants
I	D	Flammable fuels, chemicals	Gasoline	Refineries, chemical plants, paint spray areas
II	E	Metal dusts	Magnesium	Chemical plants
II	F	Carbon, coke, coal dust	Carbon black	Mines, steel mills, power plants
II	G	Grain dusts	Flour, starch	Grain mills and elevators

A typical classification label will state "approved for Class I, Groups A and B," omitting the Division designation. When the Division is omitted, the classification is invariably acceptable for both Division 1 and 2 locations. If the equipment is merely vapor-tight and approved only for Division 2, the Division designation should be displayed on the label. Class I equipment is usually either approved for Groups C and D or is approved for all four Class I Groups: A, B, C, and D.

TEST EQUIPMENT

Pursuant to general electrical safety there are a few inexpensive items of test equipment which the Safety and Health Manager should have access to for occasional in-house inspections. Brief descriptions of these items follow.

Circuit Tester

A circuit tester (Figure 13.13) simply has two wire leads connected by a small lamp bulb, usually neon. Whenever one of the leads touches a hot wire and the other lead touches a grounded conductor, completing the circuit, the bulb glows. The tester only works for a given voltage range, but most are able to handle both 110- and 220-

Figure 13.13.   Circuit tester.

volt circuits. The circuit tester is somewhat of a safety device in itself, enabling maintenance workers to be sure that power is turned off before touching hot wires. The circuit tester can also be used to determine whether various exposed parts of machines or conductors are "live" or "energized."

Receptacle Wiring Tester

One of the simplest and most widely used test devices is the receptacle wiring tester, which can be carried around in one's pocket (see Figure 13.14). No battery or cord is required. The user simply plugs the device into any standard 110-volt outlet and interprets the arrangement of indicator lights to determine whether the receptacle is wired improperly. Note carefully the use of the word "improperly" instead of the

Figure 13.14.   Receptacle wiring tester.

word "properly" in the preceding sentence. Only certain errors will be detected, and a "correct" indication of the lights means only that certain errors were not found. The receptacle could still be wired incorrectly.

Figure 13.15 reveals that the receptacle wiring tester is nothing more than three simple circuit testers combined into one tester. The lower portion of the figure shows types of wiring errors which can be interpreted from the indicator lights. One of the most common wiring errors—"ground jumped to neutral"—is shown by the tester to be wired correctly.

Continuity Tester

Sometimes it is useful to check a dead circuit simply to see if all the connections are complete or to see whether a break in a conductor has occurred. Figure 13.16 illustrates a simple continuity tester which is similar to a circuit tester except that a

Indicator lights

Condition description	1	2	3
Correct wiring	⊗	⊗	
Reversed polarity	⊗		⊗
Open ground	⊗		
Ground jumped to neutral or Neutral and ground reversed	Both of these errors go undetected by the tester because the tester will indicate:		

⊗　　⊗

(i.e., correct wiring!)

Figure 13.15. Schematic of receptacle wiring tester.

Figure 13.16.   Continuity tester.

small battery is included to provide power for the light. The continuity tester also has much less electrical resistance than the circuit tester because it is used on dead circuits, not live ones. If the continuity tester is used on a live 110-volt circuit, it will immediately blow a fuse or throw a breaker in the circuit.

An important application for a continuity tester is to check the path to ground. One terminal of the tester can be connected to the exposed equipment case of a machine in question and the other terminal to a known-grounded object. If the lamp lights, the machine is known to be grounded. If the lamp does not light, there is a break somewhere in the path to ground.

FREQUENT VIOLATIONS

Having been versed in the principles of electrical hazards, it should be apparent to the Safety and Health Manager what to look for when making an inspection. But for review, the remainder of this chapter will describe frequently cited violations of the *National Electrical Code®* .

Grounding of Portable Tools and Appliances

It should come as no surprise that the most frequently cited electrical code is for grounding. This is the code cited for ungrounded electric drills, sanders, saws, and other portable hand-held equipment. Nonportable equipment such as refrigerators, freezers, and air conditioners also must be grounded. Guarded motors and the metal frames of certain electrically heated appliances are not excepted. Special attention should be given to the use of portable electric tools used in damp or wet locations and, of course, to plugs that have the ground pin broken off.

Exposed Live Parts

Almost as frequent as ungrounded portable equipment is the existence of exposed live parts. If the actual equipment conductors cannot be insulated and terminals covered, locked rooms should be used to prevent exposure to workers. One of the most frequent observations of exposed live parts is sloppy electrical installations in which covers are left off junction boxes or receptacle cover plates are missing. A switchbox, fuse, or breaker box in which the door is open also constitutes "exposed live parts."

Improper Use of Flexible Cords

Makeshift or temporary electrical installations are prohibited as substitutions of flexible cords for fixed wiring. Obvious instances are flexible cords run through holes in the walls, ceilings, floors, doorways, or windows.

Marking of Disconnects

This is an easy thing to correct. The disconnect box or switch panel for motors and appliances must be identified so that equipment can be quickly and confidently disconnected. Also, the origins of branch circuits, such as in the breaker box, must be labeled to indicate their purposes. If the location or arrangement of the disconnect makes it obvious which machines or circuits it controls, marking may not be necessary. Very few of the disconnect marking violations are designated as "serious."

Connection of Plugs to Cords

Here is another extremely simple item. When repairing a plug to an extension cord, appliance cord, or any other flexible cord, be sure to tie a knot in the cord or do something that will prevent a pull on the cord from being transmitted directly to joints or terminal screws. This is just a basic principle of electrical maintenance.

SUMMARY

This chapter began with some sobering thoughts of what electricity, even in small amounts, can do to the human body. The greatest hazard is with 110-volt circuits, not 220-volt or higher circuits, because of the popularity of 110-volt circuits and the complacency of the people who use them. Besides electrocution hazards, electricity also presents fire hazards, not to speak of burns and other electrical exposure hazards.

Simple testers can demonstrate quickly some of the frequent wiring mistakes that might otherwise go undetected in normal operations. But some of these testers are too simple and overlook common errors such as "ground jumped to neutral."

Providing for electrical equipment in industrial processes that produce flammable vapors, dusts, or fibers is a difficult and expensive task. The definitions and codes for hazardous locations with explosive atmospheres are also complicated and tricky. The Safety and Health Manager is advised to look for required markings on electrical equipment installed in hazardous locations to be sure that the equipment is approved for the Class and Division of the location in which it has been installed.

Violations of electrical code are often for conditions that are easy to correct. The largest single item to remember is grounding—grounding both of portable and fixed equipment and the circuits that serve them. Of less frequency but of great seriousness are the OSHA citations for unapproved conduit and electrical equipment used in explosive atmospheres of flammable vapors or dusts.

EXERCISES AND STUDY QUESTIONS

13.1. What are the two principal hazards of electricity?

13.2. Approximately how many people each year die of electrocution in the United States?

13.3. Compare 110-, 220-, and 440-volt hazards of electricity.

13.4. At approximately what current levels through the heart and lungs does the sustained flow of electricity become fatal?

13.5. What part does grounding play in an electric circuit? Why do safety regulations require equipment to be grounded?

13.6. Compare the functions of hot, neutral, and ground wires.

13.7. What is a GFCI, and where is it used?

13.8. Explain the term "double insulation."

13.9. What are the hazards of reversed polarity?

13.10. Compare electrical "arcs" and "sparks."

13.11. Explain the difference between Class I, II, and III hazardous locations.

13.12. What do the terms "Division" and "Group" mean in the classification of hazardous locations?

13.13. Explain the difference between a continuity tester and a circuit tester.

13.14. Name some of the most frequent violations of electrical code.

13.15. A certain string of Christmas tree lights has eight bulbs, each rated at 5 watts. If there is no ground fault, how much current flows through the hot wire at the receptacle plug? How much flows through the neutral?

13.16. In Question 13.15, if all the current flowing through the hot wire would pass through a person's heart muscle, would the current likely be fatal?

13.17. A maintenance worker wires a 110-volt electric lamp without disconnecting the circuit. With no load energized in the circuit, the worker accidentally contacts the bare hot wire. What will probably happen? Under what conditions would the worker be likely to be killed?

13.18. In Question 13.17, suppose that the wire contacted were the neutral wire. What would probably happen? Are there conditions under which the worker would be killed?

13.19. In Question 13.17, suppose that the wire contacted were the equipment ground wire. What would probably happen?

13.20. Portable "trouble lights" used in automobile repair have been involved in many fatalities. Explain probable hazard mechanisms.

13.21. The following fatality actually occurred. A factory worker was electrocuted when he grasped the small chuck key (Figure 13.17) tied with a wire to the electrical cord of his portable electric drill. The twisted wire tie had worn through the cord insulation during constant use. Comment on how this fatality might have been prevented.

13.22. In the fatality case history of Question 13.21, suppose that the ground pin had been broken off the electric plug for the drill. How would this code violation have affected this fatality?

13.23. A worker was electrocuted when he drilled through a wall and the drill bit contacted an electric wire inside the wall and cut through the insulation. The ground pin of the electric plug for the drill was broken off. Describe several ways in which this fatality could have been prevented.

13.24. The ground pin broken off represented a code violation in the fatality case described in Question 13.23. How did this contribute to the hazard? What would probably have happened had the ground pin been intact?

Figure 13.17. Cause of electrocution fatality: Chuck key secured to cord by twisted wire tie. (a) Unsafe, improvised attachment of chuck key. Wire or tape will eventually damage cord insulation. (b) Safer method of attaching chuck key.

14

Construction

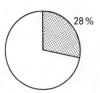

28%

Percent of all OSHA
citations addressing
construction

The Safety and Health Manager for a construction company will find this chapter of primary importance, but even the general industry Safety and Health Manager should find the chapter useful. Almost every industry has occasional remodeling or expansion projects that result in construction. Even if the company contracts out the construction project, what goes on is still important to the company because its own employees can become exposed to hazards created by the contract construction personnel. Such hazards can be both physical and legal.

GENERAL FACILITIES

Lighting

Curiously, there are standards for adequate lighting on construction sites, specifying minimum illumination intensities for various areas, whereas general industry has no such table of intensities. The reason perhaps has to do with trip hazards and pitfalls (literal sense of the word) common to construction sites, hazards that are intensified by poor illumination. The minimum illumination standard of 5 foot-candles is really quite low for general construction area lighting, and the standard drops even lower to 3 foot-candles for concrete placement, excavation and waste areas, accessways, active storage areas, loading platforms, refueling and field maintenance areas. Lighting requirements are higher for most construction *shops* and indoor areas.

Materials Handling and Storage

The real structural test of most buildings is during their own construction. The heaviest load a floor will probably ever support is the stacked materials used during its own construction. Planning is needed to prevent overloading and possible collapse. All nails should be withdrawn from used lumber before it is stacked.

Rigging equipment for material handling such as chain, rope, and wire rope is unfortunately often used until failure. If failure does occur on a construction site, the Safety and Health Manager must be prepared to explain why because rigging equipment is required to be inspected prior to use on *each shift*.

Disposal of scrap material requires vigilance throughout the construction phase. It is difficult to concentrate on scrap and waste removal when working on a tight project completion schedule, but haphazard scrap accumulation is not only unsafe, it slows progress. Enclosed chutes are needed for dropping materials from distances of greater than 20 feet.

PERSONAL PROTECTIVE EQUIPMENT

Construction requirements for personal protective equipment are similar to those for general industry, the principal difference being one of emphasis.

Hardhats

At the top of the list is head protection; indeed, the hardhat is a symbol of the construction industry. So obvious is the absence of a hardhat in a construction work crew that the hardhat rule can be a source of embarrassment to worker and manager alike. Only general conditions are laid out to describe *when* hardhats are needed. This lays the decision in the Safety and Health Manager's lap, and as was pointed out in Chapter 8, discretion is advised.

Hearing Protection

It may surprise some readers that hearing protection is a concern for *construction,* but construction work often involves damaging levels of noise. Consider the noise levels and durations of exposures of the compressed air "jackhammer," for example.

Eye and Face Protection

The greatest concern for construction workers' eyes is for mechanical injury, as from using structural steel riveters, grinders, powder-actuated tools, woodworking tools, concrete nozzles, and other spark- and chip-producing equipment. Surprisingly, construction workers can even be exposed to lasers used as tools for checking steel girder alignment and deflection in bridges and buildings.

Fall Protection

On the bases of both fatalities and injuries, falls are probably the greatest hazard in construction work. Where general industry has a permanent wall, construction may have only a guardrail. Where general industry has a permanent guardrail, construction may have a temporary guardrail or perhaps will have no protective structure at all. Where general industry has a permanent stairway, construction may have a temporary ladder. General industry fixed ladders may have cages or ladder safety devices; construction ladders often have no safety devices.

Personal protective equipment is the answer to many construction industry fall hazards, simply because protection by other means may be awkward or even impossible. Safety belts and lanyards tied to lifelines are essential to the safety of many construction workers.

One mistake made in selecting fall protection equipment is to improvise with ordinary leather belts and ropes. An ordinary leather belt and hardware will not satisfy the specified 4000-pound tensile test for safety belt hardware. No one weighs 4000 pounds, but what matters in a fall is the shock load, which can be several times as great as the ordinary dead weight. A falling 200-pound person can therefore result in ½ ton to 1 ton of force upon the belt and lifeline. Using a safety factor of approximately 4, it is easy to see why the standard specifies a 4000-pound tensile load limit.

The lanyard is that part of the fall protection system which attaches to the safety belt on one end and the lifeline or structure on the other. The lanyard must have a nominal breaking strength of 5400 pounds. The applicable standard specifies "½-inch nylon or equivalent." Beware of substituting materials of equivalent breaking strength to ½-inch nylon. Tensile or breaking strength is not the only consideration in selecting a lanyard. A certain resiliency or elasticity exists with artificial fiber ropes which lessens the shock load when arresting a fall.

An important point with safety belt lanyards is that they must not be too long. The standard specifies "a maximum length to provide for a fall of no greater than 6 feet." The rationale is that there is no point in breaking a fall with the lanyard if the worker has already fallen so far that the shock of the rope will be lethal. However, the standard is often misunderstood on this point. Note carefully that the wording quoted above does not limit the lanyard length to 6 feet. Figure 14.1 clarifies this point.

One difficulty is attempting to attach a lanyard to a vertical lifeline, especially when the lanyard must be adjusted upward or downward on the lifeline, such as in use with a scaffold. It is necessary for the attachment to slide easily when this is intended, but the attachment must lock and hold if the worker falls. There are mechanical devices, but an easy-to-tie knot for this purpose is the triple rolling hitch shown in Figure 14.2.

Fall protection is usually considered from the standpoint of height, but working over or near water presents a different hazard. Even the best of swimmers will have difficulty with a fall into water if the worker is fully dressed and perhaps burdened by tools, equipment, or materials such as rivets or bolts. If the water is

4'

Working level

(a) Lanyard length is 6 feet

10'

Working level

(b) Lanyard length is 12 feet

Figure 14.1. Maximum lanyard length must provide for a fall of no greater than 6 feet. Both diagrams show compliance with the OSHA standard. (a) Lanyard length is 6 feet; (b) lanyard length is 12 feet.

(a)

(b)

(c)

Figure 14.2. (a) Triple rolling hitch tied with free end of lanyard; (b) easily raised and lowered slipping on cable or lifeline; (c) when the lanyard is pulled tight as in a fall the triple rolling hitch will not slip on the lifeline.

329

rather cold, the danger of hypothermia increases the drowning hazard. However, some feel that the standard "goes overboard" in requiring:

1. Life jackets or buoyant work vests *and*
2. Ring buoys every 200 feet *and*
3. A lifesaving skiff

whenever employees work over or near water and a danger of drowning exists.

FIRE PROTECTION

From a property loss standpoint, fires are more dangerous after a building is completed, but to protect construction workers fire hazards must also be controlled *during* construction. Construction sites have somewhat more latitude in distributing fire extinguishers than do general industries. Even an ordinary garden hose may be used in place of fire extinguishers on construction sites. But there are so many restrictions on such use of garden hose that most Safety and Health Managers will regret having considered this alternative.

The biggest problem with fire prevention during construction is the handling of flammable liquids. For ordinary flammable liquids such as gasoline, quantities handled must be no more than 1 gallon unless approved metal safety cans are used. Approved metal safety cans must be used even for quantities up to 1 gallon unless the flammable liquid is used out of its *original* container. Keep the containers off stairways and away from exits and aisles.

TOOLS

It has been well publicized that mushroomed heads on chisels, wedges, and other impact tools are unsafe. The hazard is that a sliver of metal can break off and cause a severe eye injury, even total loss of sight. Another problem with hand tools is defective handles, in particular loose hammer heads.

Construction sites may employ pneumatic tools, such as jackhammers, staplers, or nailers. Pneumatic tools need to be secured to the hose by some positive means to prevent accidental disconnection. Hoses larger than ½ inch inside diameter need a pressure reducer device to prevent whip action in case of hose failure. Figure 14.3 is a diagram of a pressure reducer device for this purpose. It is a simple in-line device, usually placed between the hose and the compressor. Unfortunately, the device reduces the overall capacity of the system. Operating at full capacity, as from several tools operating at once, the pressure downstream from the device becomes so low that the device (a spring-loaded valve) begins to close as if the downstream line had ruptured. This shuts off or nearly shuts off the supply of air to the tools, making the system useless at that capacity. The result is that the worker removes the

Figure 14.3. In-line device for preventing whip action in event of pneumatic hose failure.

device from the line, and often as not it is soon lost. This is a sore point with many construction Safety and Health Managers.

Some construction applications are being found for hydraulically operated tools, especially in the public utility construction field. Hydraulic tools operate under the same principle as pneumatic tools but use liquids instead of air as the medium. Fluid pressures in these tools can reach 3000 psi gauge, approximately 20 times the maximum pressure achievable with pneumatic tools. Such tremendous pressures give hydraulic tools much greater power than pneumatic tools, but a hazard can exist if operating pressure limits of the equipment are exceeded. Under the criterion of occupational noise, however, hydraulic tools have a safety and health advantage over pneumatic tools. Additional hazards of hydraulic fluids are electrical conductivity and fire. These hazards tend to conflict in that the more fire resistant fluids are electrical conductors. When working in construction and alteration of electric utility transmission and distribution systems, the hazard of electrical conductivity is more serious than the hazard of fire. The hydraulic fluids used for the insulated sections of derrick trucks, aerial lifts, and hydraulic tools which are used on or around energized lines and equipment for power transmission and distribution are required to be of the insulating type. The fluids for hydraulic tools used in other applications are required to be fire resistant.

Powder-actuated tools carry an explosive charge to provide the driving force. The applications of these tools are increasing because they are both fast and effective. Driving fasteners into concrete, masonry, or steel demands large, accurately placed impact forces. Powder-actuated tools are able to provide these forces in a convenient way, speeding up production on construction projects. But together with this speed, force, and convenience come safety hazards.

A powder-actuated tool looks and operates very much like a handgun, as can be seen in Figure 14.4. Even the powder cartridges look like bullets for a gun. In powder-actuated tools, however, the projectile is separate from the cartridge, as shown in Figure 14.4.

In some ways powder-actuated tools are even more dangerous than handguns. Powder-actuated tools are capable of handling a variety of cartridges with a wide range of power ratings. These cartridges must be selected with care by a knowledgeable person. Insufficient power will fail to do the job, but too much power may drive the fastener completely through the material and kill a co-worker in another room (this has actually happened). To protect against such hazards and for convenience, cartridges are color coded for easy identification. Metal-cased

Fastener

Power load

Base material

Fastener driven by gas

Gas released against fastener

Figure 14.4. Powder-actuated fastening tool. (From NIOSH, Ref. 63.)

cartridges are available in a range of 12 different power ratings, as shown in Table 14.1. It may be necessary to back the material with a substance which will prevent the fastener from passing completely through.

Knowledge and judgment are also required in avoiding very hard or brittle materials, such as cast iron, glazed tile, surface-hardened steel, glass block, live rock, face brick, or hollow tile. If the material is spalled or cracked by an unsatisfactory previous fastening, the new fastening must be driven elsewhere. Fasteners driven too close to the edge of the material can cause explosive chipping of the material at the edge. Obviously, eye protection is always needed when using powder-actuated tools.

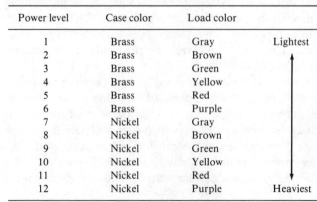

TABLE 14.1 COLOR IDENTIFICATION FOR CASED POWER LOADS FOR POWDER-ACTUATED TOOLS

Power level	Case color	Load color	
1	Brass	Gray	Lightest
2	Brass	Brown	
3	Brass	Green	
4	Brass	Yellow	
5	Brass	Red	
6	Brass	Purple	
7	Nickel	Gray	
8	Nickel	Brown	
9	Nickel	Green	
10	Nickel	Yellow	
11	Nickel	Red	
12	Nickel	Purple	Heaviest

ELECTRICAL

Construction workers are often in close contact with ground and frequently work in wet locations under adverse conditions. Electrocution ranks with falls near the top of the list of causes of fatalities among construction workers.

A principal requirement on construction sites is that all 15- and 20-ampere outlets have either ground fault circuit interruptor (GFCI) protection or a program of equipment ground conductor assurance, including inspection, testing, and record-keeping. The Safety and Health Manager in a construction company is faced with a decision between the two alternatives, and the following paragraphs are intended to assist in this decision.

The purpose and operating principle of the GFCI were discussed in Chapter 13. The hazards of electrical shock by faults to ground are greater on construction sites than in the general industrial workplace. Because of the increased hazards, the *National Electrical Code*® specified GFCIs for construction but not for general industry, although they would be of value in any electric circuit serving hand-held appliances or tools. Many individuals have installed them in their homes.

It seems that almost every safety device has its drawbacks, and the GFCI is no exception. The GFCI closely monitors any difference in current flow between the ground and neutral conductors, as low as fractions of a milliampere. But there are ways in which these currents can be unbalanced when no hazard exists. Tiny amounts of current leakage to ground occur for quite innocent reasons. Damp or weakened insulation might produce tiny currents in various locations. Even an extension cord that is too long can create a condition of capacitance between the conductor and the ground, resulting in a tiny leak. Although none of these conditions of itself is a significant hazard, the cumulative effect is one that may be great enough to trip the GFCI, shutting down the entire circuit. This is so-called "nuisance tripping" and has made the GFCI a controversial issue in the construction industry.

Chapter 13 mentioned a permissible alternative to the GFCI: the careful maintenance of the grounding conductors of electrical equipment. Such maintenance includes regular inspections and records of these inspections. The idea behind the assured equipment grounding conductor program is that if an electrical tool shorts to the case or handle, the third-wire grounding system will shunt the current to quickly throw the circuit breaker. A good grounding conductor can therefore provide protection similar to the GFCI.

The assured equipment grounding conductor program is appealing to many construction companies because they can avoid buying the GFCI equipment. They can also avoid the nuisance tripping of the GFCIs discussed earlier. But although the costs are less tangible, they nevertheless exist with the grounding assurance alternative. It takes instruments and time to test the grounding conductors, and the business of recordkeeping always involves intangible costs. An economic impact analysis of the two alternatives has been made which estimated a cost of compliance of $87.5 million for purchase, installation, and first-year maintenance of GFCIs.

The study estimated for the alternative assured equipment grounding conductor program a similar cost of $36 to $43.8 million. Inflation changes absolute annual cost estimates, but the relative difference between the costs of the two alternatives suggests that the assured equipment grounding conductor program is cheaper.

Temporary lighting is an electrical problem on construction sites more than in general industry. Often seen are ordinary incandescent bulbs suspended from electrical cords. When suspended in this way the cords and lights must actually be *designed* for this purpose; all electric cords for temporary lighting must be heavy duty, and insulation must be maintained in safe condition. To prevent accidental contact with bulbs, the bulbs must be guarded unless the construction of the reflector is such that the bulbs are deeply recessed.

A construction site is a profusion of temporary conditions, and electrical cords or extension cords strung about the area are a common sight. Unfortunately, the area is also visited by heavy-duty vehicles such as excavation equipment, heavily loaded trucks, and very heavy concrete delivery trucks. The situation is too hazardous to permit electrical cords to pass through work areas unless covered or elevated to protect from hazardous damage. No splices are permitted in flexible cord unless properly molded or vulcanized.

LADDERS AND SCAFFOLDS

The "care and use" provisions for ladders are most important for construction ladders, just as they are for ladders used in general industry covered in Chapter 5. Construction ladders have some differences, however, which have caused some problems.

Job-Made Ladders

Construction companies often make their own ladders, and such ladders are not illegal if made properly. The first question is to determine how many persons will be needing the ladder. If simultaneous two-way traffic is anticipated, a conventional ladder will not work, and a double-cleat ladder as shown in Figure 14.5 should be used. In fact, if the ladder is the only means of access or exit from a working area for 25 or more employees, the double-cleat ladder is *mandatory*. However, care must be taken not to use double-cleat ladders over 24 feet long (see Table 5.1). If greater length is needed, the solution is to provide landings with guardrails and toeboards on exposed sides. To be effective, each landing must offset two sections of ladders so that the length of a fall is minimized by the position of the landing.

The biggest mistake made in building job-made or homemade ladders is to fail to inset the cleats into the side rails (see Figure 14.6). It is so much more trouble to inset the cleats or use filler blocks than to simply nail the cleats in place. But the security and stability of the cleats are increased many times by this additional effort to make the ladder safe.

Most commercial ladders are flared top to bottom. It is permissible to flare

Figure 14.5. Double-cleat ladder for simultaneous two-way traffic.

Figure 14.6. Two acceptable ways of setting cleats in construction job-made ladders: (a) cleats inset into side rails; (b) cleats braced by filler blocks.

job-made ladders top to bottom, but it is also permissible to make the side rails parallel. The flare must be very small if used at all and must not be more than ¼ inch for each 2 feet in length. This determines limits for the width of the bottom cleats because top cleats must be at least 15 inches but no more than 20 inches wide. Thus the bottom cleat must be at least 15 inches wide but no more than given by the following formula:

$$W_B \;=\; W_T + \frac{1}{2}\left(\frac{L}{2}\right)$$

where

W_B = bottom cleat width in inches

W_T = top cleat width in inches

L = ladder length in feet

Scaffolds

The subject of scaffolds can become somewhat technical, and these technical details can be very important. The Safety and Health Manager will find it useful and in some cases imperative to obtain the services of a Registered Professional Engineer if he or she is not one him/herself. This is one area in which the credential as well as the knowledge can be quite useful.

One of the technical aspects of scaffolds is safety factor. Design safety factor for scaffolds and their components is a factor of 4. This factor increases to a factor of 6 for the suspension ropes supporting the suspended type of scaffolding. The application of counterbalances, tie-downs, footings, and all the allowance for wind loading all can be quite technical, and an engineering evaluation is advisable.

The Safety and Health Manager may be frustrated by the many bewildering names of scaffolds listed in applicable construction standards. But the majority of the scaffolds with unfamiliar names, such as "window-jack scaffolds," "outrigger scaffolds," and "chicken ladders," are rarely seen. The most popular scaffolds are the following types:

- Welded frame (or "bedstead" scaffolds)
- Manually propelled mobile scaffolds (on casters)
- Two-point suspension (or "swinging" scaffolds)
- Tube and coupler scaffolds

Some scaffolds, such as tube and coupler scaffolds and welded frame scaffolds, are supported by structure on the ground and must have sound footings. If the ground slopes, scaffold jacks may be necessary to assure that the footings are level. Certain engineered "cribbing" may be acceptable, but such unstable objects as barrels, boxes, loose brick, or concrete blocks are criticized. Concrete blocks are a real problem because most people feel that they are very strong and rigid. But scaffolds direct very highly concentrated loads upon their relatively tiny feet, and such loads can break through a molded concrete block. Once a scaffold support breaks through its footing, a major shift can occur aloft, and the consequences can be extremely serious. Such a mishap is most likely to occur at the worst time (i.e., when personnel are on the scaffold).

For scaffolds suspended from above such as from two-point suspension (swinging) scaffolds, the security of the attachment on the roof is of obvious importance. Since rooftops vary in structure and design, an engineer is very useful in assuring a safe anchor point. Cornice hooks are designed to hook over the edge, not into it. In addition, tiebacks are needed as a secondary means of support. Sometimes a sound structure for the tieback is not available on the rooftop, and the only solution is to cross the entire roof and go down to the ground on the other side of the building. Tying a scaffold to an ordinary vent pipe on the roof is asking for trouble.

Safety belts and lifelines for personnel on suspended swinging scaffolds must be tied to the building, not to the scaffold. Thus if the scaffold falls, the personnel can still be saved. For guidance on the attachment of the lanyard to the lifeline, refer back to the discussion of fall protection covered earlier in this chapter.

The floor of the scaffold is also important. Loose planks can be especially hazardous if the overhang beyond the support is insufficient for security. But too much overhang can be dangerous also because a worker might step beyond the support and cause the plank to tip like a seesaw. Figure 14.7 illustrates minimums and maximums for plank overhang. Figure 14.8 illustrates the minimum for overlap.

Scafflel Safety but 4
Payer 6

Figure 14.7. Scaffold plank overhang specifications.

Lightely   ≤ 4×4=2k
med   8'   =32
Length 10'   40

Figure 14.8. Scaffold plank overlap specifications.

FLOORS AND STAIRWAYS

As in general industry the standard for guarding of open-sided floors and platforms is one of the most frequently cited standards in construction. Curiously, though, instead of setting the vertical fall distance at 4 feet as in general industry, the specification for construction is *6 feet.* Thus an open-sided floor or platform 6 feet or more above the adjacent floor level is required to be guarded by a standard railing. Runways 4 feet high or more must be guarded.

During construction of buildings with stairways, construction employees use the new stairways during finishing operations for the building. If the building's stairways are designed properly (see Chapter 5), the Safety and Health Manager for the construction company has little to worry about.

A pitfall needs mentioning on the subject of use of new building stairways during construction. Many stairways and landings today are of steel construction with hollow pan-type treads which are filled on-site with concrete or other materials. Contractors often save until last the job of pouring the treads. Meanwhile during building construction workers are walking on the unfilled treads and are subject to the trip hazard created by the steel nosing along the leading edge of the riser. Of course, exposure to the hazard is unavoidable during the actual construction of the stairway itself. But after the stairways have been installed, lumber or other temporary material can be used to fill the hollow space and eliminate the trip hazard during completion of the building.

CRANES AND HOISTS

Cranes, hoists, and other material or personnel handling equipment are essential tools of the construction industry. Chapter 10 investigated the hazards of these machines in detail. This chapter discusses the subject from the viewpoint of the construction industry, the most important user of these machines.

On construction sites the simultaneous execution of many parts of the overall project means that personnel will often be working around or near a crane in operation. The crane operator will avoid moving the bucket or other load over personnel and will attempt to keep the cab from striking personnel, but it is impossible for the crane operator to watch all moving parts at all times. Particularly hazardous is the rear of the cab, which on many models swings outside the crawler or other substructure when the cab and boom rotate, as shown in Figure 14.9. This motion generally occurs on every cycle of a crane's operation and is a constant threat to personnel on the ground. The hazard is a very serious one, and accidents carry a high

Swing

Figure 14.9. Rear of crane cab is a hazardous area due to swing radius. (Courtesy of the Construction Safety Association of Ontario)

likelihood of fatality. The employee may be either struck or crushed between the cab and some other object, such as a building wall, stack of materials, or another vehicle.

Another serious hazard with construction cranes is the possibility of contact with exposed live overhead utility lines. Crane boom contact with high-voltage transmission lines results in fatalities every year. Figure 14.10 compares safety standards for general industry and construction regarding crane clearance from high-voltage electric lines.

Figure 14.10. Clearance for cranes from electric transmission lines.

A common worker practice is "riding the headache ball." Figure 14.11 identifies the headache ball as the ball-shaped weight used to keep a necessary tension on the wire rope when the hook is not loaded. It is possible for workers to stand on this ball and take a ride, using the crane as an elevator, a practice that usually horrifies the uninitiated passers-by. On crawler cranes used on construction, however, OSHA does not expressly prohibit the practice of riding the headache ball. What is needed is to make the ride safe for the passenger to attach a safety belt lanyard to the wire rope above the ball while riding.

Hammerhead tower cranes are large structures which take advantage of counterweights on the end of the jib opposite the work (see Figure 14.12). This crane is often used for large building construction. Sometimes construction workers will have duties *on* the horizontal jib of a hammerhead tower crane. This presents a fatal fall hazard, and guardrails or safety belts, lanyards, and lifelines are needed to protect the worker.

Helicopters are sometimes used in construction for such operations as the

Figure 14.11. Headache ball. These balls range in weight from less than 100 pounds to a ton or more. The weight of the ball overcomes the friction of the sheaves for the running rope of the crane. The headache ball should not be confused with the much heavier "wrecking ball."

Figure 14.12. Hammerhead tower crane.

placement of a steeple. Although use of helicopters in construction is very rare, the nature of the hazards of the operation is somewhat strange and therefore deserves some mention. Ordinary tag lines, used with all cranes to control the load from below, can be a hazard to helicopters because the lines can be drawn up into the rotors, resulting in a tragedy. Tag lines must be of a length that will not permit their being drawn up into the rotors.

Cargo hooks are another problem with helicopters used as cranes. With ordinary cranes the sole concern is that the hook will hold and not release at the wrong time. With helicopters there is this concern and the additional concern that the hook might *not* release at the right time. Cargo hooks for helicopter cranes need to have an emergency mechanical control to release the load in case the electrical release fails.

Another unusual effect that can represent a hazard is the generation of a static electrical charge on the load. This charge is developed by the friction of the air on the rotor and other moving parts. To deal with this hazard a grounding device can be used to dissipate the charge before ground personnel touch the load, or protective rubber gloves can be worn. Once the load actually touches down, the static charge is generally dissipated through the load itself directly into the ground.

An indirect hazard that can develop from the use of helicopters is fire on the ground. The rotating blades generate such a wind on the ground that it is considered unsafe to have open fires in the path of a low-flying helicopter.

Material and Personnel Hoists

Temporary external elevators are often used on construction sites to move workers and materials and must be designed, maintained, and used properly to avoid a serious hazard. The seriousness of this hazard is quite personal to me because my brother narrowly escaped a fatal accident on such a hoist when two of five members of his engineering inspection section were killed when a personnel hoist failed in an oil refinery. Two others were totally and permanently disabled, and the fifth, my brother, was uninjured because he had been asked to remain in the office that fateful day to complete an engineering drawing.

The design requirements for personnel hoists and material hoists are different, and one of the principal safety factors is maintaining the distinction between the two hoists during use. Material hoists must be conspicuously marked "No Riders Allowed." On the other hand, it is permissible to move material on a personnel hoist provided that rated capacities are not exceeded.

Latched gates are needed to guard the full width of the landing entrance for both material and personnel hoists. In the case of personnel hoists an electrical interlock must not allow movement of the hoist when the door or gate is open. Furthermore, on personnel hoists, the hoistway doors or gates must have mechanical locks which are accessible only to persons in the car.

Aerial Lifts

An alternative to scaffolds, ladders, and hoists is needed for high and awkward locations on a construction site. Increasing popularity is being seen in the use of vehicle-mounted boom platforms or aerial "buckets." The boom is usually "articulating" (capable of bending in the middle) or is "hydraulically extensible" (telescoping) or both.

The biggest problem with aerial lifts is not their construction but the way they are used. Anyone in an aerial bucket needs to recognize the difference between his or her perch and terra firma. The floor of the bucket is the only place to stand—not on a ladder or plank carried aloft. Sitting on the bucket's edge is also dangerous. Even if the worker *does* stand properly in the bucket, a body belt with lanyard tied to the boom or bucket is needed to protect against a hazard such as a surprise encounter with a tree limb.

HEAVY VEHICLES AND EQUIPMENT

Next to falls and electrocutions, more construction fatalities involve vehicles, tractors, and earthmoving equipment than any other hazard source. The fatality hazard is both to drivers of the equipment and to their co-workers. Vehicle *rollovers* are the principal cause of driver fatalities, whereas vehicle *runovers* are the problem for co-workers. Not to be excluded, however, is a significant number of fatalities from repairing tires for these vehicles.

ROPS

The acronym ROPS (rhymes with "hops") represents the term "rollover protective structures" and is a major change in construction vehicle design brought about by recent safety standards. The purpose of the ROPS is to protect the operator from serious injury or death in the event the vehicle rolls over. The following kinds of construction equipment require ROPS:

- Rubber-tired, self-propelled scrapers
- Rubber-tired, front-end loaders
- Rubber-tired dozers
- Wheel-type agricultural and industrial tractors
- Crawler tractors
- Crawler-type loaders
- Motor graders

Exempted are sideboom pipelaying tractors.

To be effective, the ROPS system must be able to withstand tremendous shock loads, which increase as the weight of the vehicle increases. The standards are quite specific regarding the structural tests to which ROPS systems must be subjected in order to qualify. For all wheel-type agricultural and industrial tractors used in construction either a laboratory test or a field test is required to determine whether performance requirements are met. The laboratory test may be either static or dynamic. In the static test the stationary tractor chassis is gradually loaded while strain is measured by deflection instruments. The required input energy is a function of tractor weight, which in turn is required to be a function of tractor horsepower. In other words, the tractor's gross weight cannot be lightened below rated horsepower limits in order to meet ROPS tests.

The dynamic test, an alternative to the static test, uses a 2-ton[1] pendulum which provides an impact load on the rear and side of the ROPS in successive tests. The height from which the pendulum is dropped is dependent on the calculated tractor weight based on horsepower, as in the static test described above. Deflection limits must not be exceeded.

If the field test is used, the tractor is actually rolled over, both rearward and sideways, as both types of accidents can easily occur. If the actual weight of the tractor is less than specified for its horsepower, ballast must be added for the tests.

For all ROPS tests it is a good idea to remove protective glass and weathershields, which would probably be destroyed during the test. If there is any question whether such shields may absorb some of the energy, thereby assisting ROPS to pass the test, the shields *must* be removed.

Having read thus far, the reader can see that retrofitting an old tractor to meet current requirements for ROPS is not an easy task. The Safety and Health Manager is cautioned against taking the tractor to a local welder and requesting a ROPS system to be fabricated. Unless the tractor is actually going to be subjected to the standard ROPS test, the frame should be of a design identical to a frame actually tested for the model tractor in question. Very old or rare model tractors are obviously a problem. If a qualified ROPS system is removed for any reason, it must be remounted with bolts or welding of equal or better quality than those required for the original. The ROPS must be permanently labeled with manufacturer's or fabricator's name and address and the machine make, model, or series number that the structure is designed to fit. This labeling requirement is a sobering thought for the welder or fabricator who is asked to retrofit an old tractor with a ROPS system. All told, it is easy to see why most companies dump the old equipment on the used equipment market, and many are shipped to foreign countries which do not require ROPS.

After the Safety and Health Manager has assured that all appropriate construction equipment has been equipped with ROPS systems, the next task is to assure that the equipment is used properly. Essential to the effectiveness of the ROPS system is that the operator wear a seat belt. If the operator is thrown out of the vehicle, the ROPS will afford no protection at all and may actually contribute to a fatality. Passengers are another hazard unless the vehicle is equipped with seat belts for passengers. Hitchhiking on heavy equipment at construction sites is a dangerous practice.

---

[1]More specifically, 4410 lb. (2000 kg).

Runover Protection

Most of the balance of fatalities with heavy construction equipment is due to personnel being run over by the equipment. Confrontation of this major fatality category has two main thrusts: operator visibility and pedestrian awareness.

Operator visibility as good as the visibility in a private automobile is simply not feasible for a huge piece of earthmoving machinery. It is no wonder that runovers occur frequently on construction sites. The operator needs all the help affordable, but ironically some of the poorest windshield conditions occur on construction equipment. In the morning the operator, foreman, and everyone concerned is anxious to get equipment rolling, but if the morning is cold, defrosting and defogging are essential. Often the defrosting or defogging equipment is ineffective. Dirty or cracked windshields are also a common sight in the harsh environment of the construction job.

The second link in the hazard prevention chain for runovers is the horn used by the operator to warn personnel when visibility *is* good enough to notice the endangered worker on the ground. Personnel are generally distributed all over a construction site, and the operator needs a good operable horn to warn them when they are dangerously close.

Besides the ordinary horn, many construction vehicles also need "backup alarms." Earthmoving equipment and construction vehicles that have an obstructed view to the rear need these backup alarms if they are used in reverse gear. The term "obstructed view" may be somewhat vague, but most safety and health professionals are taking the position that earthmoving machines of all types need these backup alarms. Too many people have been killed by machines backing over them to take this requirement lightly. This is said although it is acknowledged that the steady beep-beep-beep of the backup alarms can be very monotonous on a construction site and perhaps even lead to a certain complacency on the part of the personnel endangered. An alternative to the beeping backup alarms is the use of an observer standing behind the machine to alert others every time the machine backs up. This is expensive, though, and has the disadvantage of being an "administrative" or "work practice" control, instead of the preferred "engineering control" represented by the backup alarm.

Dump Trucks

One more hazard with construction vehicles and equipment needs emphasis. Dump trucks can cause a terrible accident if the raised dump body falls while the driver or some other worker has crawled into the exposed area for maintenance or inspection work. Sometimes all that holds the dump body aloft is the pressure in a hydraulic line, which can be suddenly lost due to any of a variety of failure modes. For this reason the safety of the maintenance or inspection worker inside the exposed area demands that the truck be equipped with some positive means of support, permanently attached and capable of being locked into position.

TRENCHING AND EXCAVATIONS

A major cause of construction fatalities is the sudden collapse of the wall of a trench or excavation. It is difficult to imagine the drama of digging for a co-worker who has been literally buried alive in such a cave-in. Before becoming involved in the field of occupational safety and health, the author coincidentally became an eyewitness to such a drama in Tempe, Arizona. The trench was located directly below a public stairway landing on which I happened to be standing, and thus my vantage point was directly over the scene of the cave-in. The impression was unforgettable, and its memory would motivate anyone to try to prevent such accidents in the future, a point supporting the principles of hazard avoidance set forth in Chapter 3.

Let it be understood without doubt that OSHA *has* had an impact on the reduction of fatalities due to cave-ins. Recognizing the seriousness of this hazard, OSHA undertook a comprehensive program of "Special Emphasis" on trenching and excavation cave-ins in the mid-1970s. The result was an impressive reduction in fatalities from this hazard source.

All trenches are excavations, but not all excavations are trenches. Trenches are narrow, deep excavations; the depth is greater than the width, but the width is no greater than 15 feet, according to the standard definition. A trench is more confined and generally more dangerous than other excavations, especially because both walls can collapse, trapping the worker. However, the walls of a trench are easier to shore than the walls of an excavation. Both are dangerous if over 5 feet deep and will easily snuff out the life of anyone who stands in the path of a collapsing wall. The hazard is not simply one of suffocation. A cave-in generally represents tons of falling earth which can crush the body and lungs of the worker even if the face and breathing passages are left clear.

The "angle of repose" is defined as the greatest angle above the horizontal plane at which a material will lie without sliding. The angle naturally varies with the material, and approximate angles are shown in Figure 14.13. The science of soil

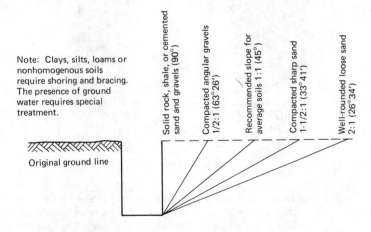

Figure 14.13. Approximate angle of repose for sloping the sides of excavations.

slides is not exact, and the uncertainty thwarts attempts to control the hazard. It is difficult to say whether a particular soil type is "typical" or "compacted angular gravels," or something between the two. The specifications for trench shoring are more detailed, as can be seen in Table 14.2.

Adding to the uncertainty of the cave-in hazard are certain hazard-increasing factors, such as:

- Rainstorms, which soften the earth and promote slides
- Vibrations from heavy equipment or street traffic nearby
- Previous disturbances of the soil, as from previous construction or other excavations
- Alternate freezing and thawing of the soil
- Large static loads, as from nearby building foundations or stacked material

Although judgment is required in deciding whether to employ shoring, the hazard is so serious that it is wise to adopt a conservative policy, well clear of the marginal area where a cave-in might or might not occur. One thing is certain: *after* the cave-in *does* occur, and there is a fatality, the OSHA officer will come to the scene, and everyone (including the OSHA officer) will conclude that the shoring or cave-in protection was insufficient.

A trench shoring system is shown in Figure 14.14. The trench jacks may be

(a)   (b)   (c)

Figure 14.14. Trench shoring system: (a) bracing used with two lengths of sheet piling; (b) bracing with screw jacks, hard soil; (c) screw jacks used with complete sheet piling. (Courtesy of the National Safety Council, Chicago, Illinois; used with permission.)

TABLE 14.2 TRENCH SHORING—MINIMUM REQUIREMENTS

Depth of trench (ft)	Kind or condition of earth	Uprights Minimum dimension (in.)	Uprights Maximum spacing (ft)	Stringers Minimum dimension (in.)	Stringers Maximum spacing (ft)	Cross braces[a] (in.) Width of trench (ft) Up to 3	3-6	6-9	9-12	12-15	Maximum spacing (ft.) Vertical	Horizontal
5-10	Hard, compact	3 × 4 or 2 × 6	6			2 × 6	4 × 4	4 × 6	6 × 6	6 × 8	4	6
	Likely to crack	3 × 4 or 2 × 6	3	4 × 6	4	2 × 6	4 × 4	4 × 6	6 × 6	6 × 8	4	6
	Soft, sandy, or filled	3 × 4 or 2 × 6	Close sheeting	4 × 6	4	4 × 4	4 × 6	6 × 6	6 × 8	8 × 8	4	6
	Hydrostatic pressure	3 × 4 or 2 × 6	Close sheeting	6 × 8	4	4 × 4	4 × 6	6 × 6	6 × 8	8 × 8	4	6
10-15	Hard	3 × 4 or 2 × 6	4	4 × 6	4	4 × 4	4 × 6	6 × 6	6 × 8	8 × 8	4	6
	Likely to crack	3 × 4 or 2 × 6	2	4 × 6	4	4 × 4	4 × 6	6 × 6	6 × 8	8 × 8	4	6
	Soft, sandy, or filled	3 × 4 or 2 × 6	Close sheeting	4 × 6	4	4 × 4	4 × 6	6 × 6	6 × 8	8 × 8	4	6
	Hydrostatic pressure	3 × 6	Close sheeting	8 × 10	4	4 × 6	6 × 6	6 × 8	8 × 8	8 × 10	4	6
15-20	All kinds or conditions	3 × 6	Close sheeting	4 × 12	4	4 × 6	6 × 6	6 × 8	8 × 8	8 × 10	4	6
Over 20	All kinds or conditions	3 × 6	Close sheeting	6 × 8	4	4 × 12	8 × 8	8 × 10	10 × 10	10 × 12	4	6

[a]Trench jacks may be used in lieu of, or in combination with, cross braces. Shoring is not required in solid rock, hard shale, or hard slag. Where desirable, steel sheet piling and bracing of equal strength may be substituted for wood.

Source: Code of Federal Regulations 29 CFR 1926.652.

347

either screw-type or hydraulically operated. They need to be secured to prevent falling or sliding if they loosen as the sidewalls adjust slightly. Care must be taken to level the jacks and also to be sure that they are below the plane of the surface of the surrounding earth. A trench jack placed too high can be subjected to bending stresses, as shown in Figure 14.15. This can damage or even ruin the trench jack. The shoring system should be removed slowly and carefully. It is sometimes necessary to remove braces or jacks by means of ropes from above after everyone has cleared the trench.

A cave-in is not the only hazard to working in trenches and excavations. All workers in the ditch are subject to the hazards of falling rocks, tools, timbers, or pipes. Head protection is needed for workers below, and good housekeeping is important along the edges of the excavation.

Even when personnel are not inside the excavation and machines do all the work, another hazard presents itself. Utility lines are often broken, resulting in fire, explosion, or inhalation hazards. Such accidents are not only dangerous but are always costly and require additional coordination with the utility company, coordination that would have been better handled in advance of the excavation. The Safety and Health Manager should install a procedure that assures that someone stops and checks with utility companies before proceeding "blind" into an excavation project.

Signs of an imminent cave-in, rupture of a utility line, dangerous accumulations of toxic gases in deep excavations, or other possible emergency situations dictate the need for quick and easy exit. A ladder, steps, or other adequate means of exit should be located so as to require no more than 25 feet of lateral travel for escape.

Figure 14.15.   Improperly placed trench jack.

CONCRETE WORK

One of the most and perhaps *the* most dramatic industrial accident in history was the partial collapse of a vertical wall of concrete. The year was 1978, and the location was Willow Island, West Virginia, where a huge cooling tower was under construction for a nuclear power plant. The poured concrete walls of the structure supported scaffolds for workers 170 feet above the ground. At the time of the accident the "green" concrete wall was insufficiently cured to accept the load. The wall failed, dropping the scaffold; 51 workers fell to their deaths.

The pressure to keep a construction project on schedule is antagonistic to the careful curing of concrete. But the consequences of rushing the job are serious even if no one is injured. The memory of the tragic Willow Island accident serves to remind concrete project managers of these consequences. Concrete curing time is a function of both time and temperature. About 70 °F is ideal, and temperatures either too hot or too cold can delay curing.

Even before the concrete is poured there are hazards to the placement of the reinforcing steel "rebars." For vertical structures rebars need guys or other support to prevent collapse. Another hazard is the protruding points of exposed vertical rebars. It may sound farfetched, but workers have actually been *impaled* by falling on these bars. The most serious hazards are for workers on ladders when these ladders are placed over protruding rebars. But even at floor or ground level a trip and fall on an exposed rebar can be fatal. The writer of this text knows of at least one individual case in which a worker stumbled at ground level and fell on an exposed rebar which impaled his neck and ruptured his jugular vein. The life of the worker was saved by the quick action of a trained first-aid person. The answer to the exposed rebar problem is to bend down the ends, cover them with plywood, or wrap with canvas until ready for pouring.

Concrete forms need carefully designed shoring systems to prevent collapse together with hazards reminiscent of the excavation cave-in hazards discussed earlier. The hydrostatic pressure of wet concrete can be very great just after it is poured. Then at a time when the forms' stress is greatest, vibrating equipment is often applied to assure even distribution, adding to the stresses on the forms. Overdesign is necessary to prevent the hazard of forms "kickout."

Concrete is often poured by buckets handled by cranes. Unfortunately, vibrator crews have to work with the freshly poured concrete in close proximity to the moving bucket. The pouring strategy should be to keep the vibrator crews away from the overhead path of the bucket. Riding the concrete bucket is prohibited.

STEEL ERECTION

Who has not marveled at the daring of the highrise steelworker "walking the beam" hundreds of feet up the steel superstructure of a new building under construction? Perhaps this work will always be dangerous, but the hazard has been mollified somewhat by requiring safety nets to be installed whenever the fall distance exceeds two stories or 25 feet. An alternative is to use scaffolds or temporary floors.

A safety railing around the perimeters of temporary floors is now required for tier buildings and other multifloored structures. However, during structural steel assembly the use of ½-inch wire rope approximately 42 inches high is permitted for the safety railing. To be effective the wire rope should be checked frequently to be sure that it remains taut.

To maintain structural integrity on the way up, the permanent floors should follow the structural steel as the work progresses. The general role is no more than eight stories between the erection floor and the uppermost permanent floor. No more than four floors or 48 feet of unfinished bolting or welding is permitted above the foundation or uppermost permanently secured floor.

Structural steel erection sites are subject to constant hazards of falling objects. Rivets, bolts, and drift pins are required to be stored in secured containers, and if all else fails, there is the hardhat to protect the worker below. For some objects at the steel erection site, the hardhat is no protection, however. The writer is reminded of an instance in Chicago where a falling steel beam had flattened a parked car from left taillight to right headlight. The car was parked next to a sidewalk along which this writer walked to and from a meeting. The accident occurred during the meeting; luckily, no one was injured.

DEMOLITION

Some would say that the subject of demolition does not belong in a chapter entitled "Construction," but demolition and construction are actually closely related. If the construction site is not clear, demolition of previous structures may be the first step in construction of a new building. Many of the tools and equipment are the same, such as cranes and bulldozers.

People identify skill, knowledge, and quality as important to construction jobs, but most people do not think of demolition as requiring skill and knowledge. But often the engineering expertise required of a demolition job far exceeds the engineering for the original construction. Buildings to be demolished have often been previously damaged by fire or may have been condemned for some serious reason such as structural damage. Required for every demolition operation is a written report of an engineering survey conducted in advance.

A demolition operation begins with manual operations such as disassembly of salvage items and then proceeds to material teardown and dumping to street level. Dangers exist in the debris dumping operation. The area below needs protection if it is outside the walls of the structure. Well-designed chutes capable of withstanding impact loads, together with substantial discharge gates, are needed to control the dropping material. One hazard is that personnel can fall down the chute while dumping debris. A substantial guardrail about 42 inches high is needed to protect against this hazard. A toeboard or bumper is also needed if wheelbarrows are used to prevent losing the wheelbarrow down the chute.

Once light teardown operations are completed, heavier demolition equipment such as cranes with wrecking balls are used. Most walls are unstable without lateral support, so they should not be allowed to stand alone at heights greater than one story. No unstable standing wall should be left at the end of a shift.

One sensational demolition technique that is gaining in popularity is controlled explosive demolition, illustrated in Figure 14.16. In this method carefully engineered explosive charges are detonated to precipitate a catastrophic failure of the building structure, resulting in an immediate and total collapse. The operation has been carried out successfully upon many downtown buildings in U.S. cities, typically triggered at the quiet time of dawn on Sunday mornings. Although the operation is dramatic and seems dangerous, it is really quite safe and avoids many of the hazards of a slow teardown process.

Figure 14.16. Building collapses under controlled explosive demolition. (Courtesy of Jim Wolfe Photo, Tulsa, Oklahoma.)

EXPLOSIVE BLASTING

Demolition is only one application for blasting; the construction industry has others. The preparation of roadway cuts is the most important. The chief concern with construction blasting is the safe handling, storage, and transportation of the explosives themselves. The reader might want to review some of the concepts of explosives handling covered in Chapter 7 of this text.

Most everyone has witnessed the familiar warning to "turn off two-way radio" when in a blasting area. The chance is remote, but an electrically fired blasting cap could be detonated by a small stray current induced by a radio transmitter. Lightning is even more of a hazard, and all blasting operations should cease when an electrical storm is near. Radar, nearby power lines, or even dust storms can also be sources of stray currents.

Good visibility reduces many hazards, and explosive blasting hazards are in this category. Aboveground blasting should be conducted in the daytime only. Black powder blasting has been replaced by safer modern methods, and black powder blasting is now prohibited in construction.

The transportation of explosive materials is subject to DOT (Department of Transportation) regulations familiar to suppliers and most construction operators who use explosives. Signs saying EXPLOSIVES in large (4 inch) red letters are required on all four sides of the vehicle. Blasting caps should be transported in a separate vehicle from the vehicle transporting other explosives, and both should be transported separate from other cargoes.

Vehicles for explosives need a good fire extinguisher rated at least 10-ABC on board. It would be foolhardy to attempt to control a fire in the cargo compartment of a vehicle transporting explosives. However, most vehicle fires begin in the engine compartment or outside but adjacent to the vehicle. Such fires can sometimes be controlled with a good fire extinguisher wielded by a trained operator, thereby averting a major explosives catastrophe.

ELECTRIC UTILITIES

A specialized type of construction is the erection and modification of electric transmission and distribution lines and equipment. The efficient transmission of usable levels of electrical power necessitates very high voltages. The rules for handling high voltages are quite different from the rules for handling ordinary household and industrial/commercial voltages. For instance, with ordinary voltages a danger is the contact with exposed live parts. For high voltages it can be dangerous even to approach the *vicinity* of live parts, as is reflected in Table 14.3 based on the OSHA standard. For voltages in the kilovolt range the atmosphere may not be an effective insulator, and arcing becomes a hazard. Therefore, safety distances must be maintained. Of course, the distances shown in Table 14.3 apply safety factors. The actual physical arcing distances are much smaller, but there is an element of

TABLE 14.3 MINIMUM CLEARANCE DISTANCES FOR LIVE-LINE BARE-HAND WORK (ALTERNATING CURRENT)

Voltage range (phase-to-phase) (kV)	Distance (feet and inches) for maximum voltage	
	Phase to ground	Phase to phase
2.1–15	2–0	2–0
15.1–35	2–4	2–4
35.1–46	2–6	2–6
46.1–72.5	3–0	3–0
72.6–121	3–4	4–6
138–145	3–6	5–0
161–169	3–8	5–6
230–242	5–0	8–4
345–362	7–0[a]	13–4[a]
500–552	11–0[a]	20–0[a]
700–765	15–0[a]	31–0[a]

[a]For 345–362, 500–552, and 700–765 kV, the minimum clearance distance may be reduced provided that the distances are not made less than the shortest distance between the energized part and a grounded surface.
Source: Code of Federal Regulations 29 CFR 1926.955.

uncertainty due to such factors as humidity and barometric pressure. Furthermore, the electric utility lineman may not be able conveniently to estimate precisely his distance from the high-voltage line or equipment, making safety factors essential.

Personal protective equipment for high-voltage work takes on a new dimension, that of *degree* of protection. Ordinary insulators on tools, protective gloves, and other insulating equipment which are effective insulators for ordinary applications might completely break down in high-voltage exposures. The whole business of working with energized high-voltage lines is a strange world to the uninitiated—a world fraught with curious physical effects. The electric utility industry offers a prime example of an industry in which training in awareness and understanding of hazards is the key to a safe workplace, echoing the principles set forth in Chapter 3.

SUMMARY

The construction industry deserves special consideration because it is so dangerous and also because OSHA has watched construction more closely than general industries. The Safety and Health Manager for construction jobs should remember that the principal task is to avoid *fatalities*. The top five categories of fatalities in the construction industry are

- Falls
- Electrocutions
- Vehicle rolling over
- Personnel runover by vehicle
- Excavation cave-ins

If the Safety and Health Manager will keep these fatality categories in mind, it will help to place overall efforts in proper perspective on the construction site.

Critics of OSHA should recognize one clear fact: OSHA has had a significant effect on reducing the number of fatalities from trenching and excavation cave-ins in the construction industry. Most trench workers are young males, and many of these are fathers with several dependents. The reduction in the number of deaths from this cause has a far-reaching value which is difficult to measure.

EXERCISES AND STUDY QUESTIONS

14.1. What is the minimum illumination level permitted for general construction areas? In what areas may lighting be reduced to 3 foot-candles?

14.2. How are lasers used in construction?

14.3. What tensile strength is specified for safety belt hardware? Give two reasons why the specification is so much higher than the weight of any human being.

14.4. What is a safety belt lanyard? What nominal breaking strength is specified for lanyards?

14.5. What is a triple rolling hitch?

14.6. Compare hydraulic versus pneumatic power tools from a safety and health standpoint.

14.7. How are construction cranes particularly hazardous to persons on the ground?

14.8. Why are helicopter hooks (for load attachment) more complicated than those for ordinary construction cranes?

14.9. What does the acronym ROPS represent?

14.10. What are the two principal strategies for preventing personnel from being run over by construction equipment?

14.11. What is the difference between a trench and an excavation?

14.12. How can trench jacks be damaged by improper placements?

14.13. What is a rebar? Why is it dangerous?

14.14. When must steel erection workers be protected against falls with safety nets?

14.15. What type of safety rail construction is permitted during steel erection?

14.16. Why is an engineering survey required prior to demolition of a building?

14.17. Is it a good idea to carry a fire extinguisher aboard a truck that transports explosives? Why or why not?

14.18. A crawler-type construction crane is operating near a 550-kilovolt power line. What is the minimum distance the boom should approach the line?

14.19. On the way to the worksite the crane passed under this same power line at a different location. What minimum distance is specified for this situation?

14.20. Suppose in Question 14.18 that the crawler crane had been employed in general industry instead of construction. What minimum distance would be permitted in general industry?

14.21. What minimum distance would be permitted for the crane in transit?

Bibliography

1. *Accident Facts*. Chicago: National Safety Council, 1981.

2. *Arc Welding and Gas Welding and Cutting,* Safety and Health in (NIOSH No. 78-138). Cincinnati, OH: U.S. Department of Health, Education and Welfare (NIOSH), 1978.

3. *Arkansas Workers' Compensation Laws and Rules of the Commission* (rev. ed.). Little Rock, AR: Workers' Compensation Commission, 1982.

4. *Auto and Home Supply Stores*, Health and Safety Guide for (NIOSH No. 76-113). Cincinnati, OH: U.S. Department of Health, Education and Welfare (NIOSH), 1975.

5. Bland, Jay, ed., *The Welding Environment*. Miami, FL: American Welding Society, 1973.

6. "Brake Monitoring," *Manufacturing Engineering,* February 1976.

7. *CGA Specification G-7.1* (ANSI 286.1-1972). New York: Compressed Gas Association, 1972.

8. Christensen, Herbert E., et al., eds., *Registry of Toxic Effects of Chemical Substances* (1975 ed.). Rockville, MD: U.S. Department of Health, Education and Welfare (NIOSH), 1975.

9. Christensen, Herbert E., et al., eds., *Suspected Carcinogens.* Rockville, MD: U.S. Department of Health, Education and Welfare, 1975.

10. *A Common Goal*. Little Rock, AR: Arkansas Department of Labor, 1975.

11. *Concepts and Techniques of Machine Safeguarding* (OSHA No. 3067). Washington, DC: U.S. Department of Labor, Occupational Safety and Health Administration, 1980.

12. *Concrete Products Industry*, Health and Safety Guide for (NIOSH No. 75-163). Cincinnati, OH: U.S. Department of Health, Education and Welfare (NIOSH), 1975.

13. *Construction and Related Machinery Manufacturers,* Health and Safety Guide for (NIOSH No. 78-103). Cincinnati, OH: U.S. Department of Health, Education and Welfare (NIOSH), 1977.

14. *Construction Industry OSHA Safety and Health Standards* (29 CFR 1926/1910; OSHA No. 2207) (rev.). Washington, DC: U.S. Department of Labor, Occupational Safety and Health Administration, 1979.

15. *Cost of Government Regulation Study.* Chicago: Arthur Andersen & Company, 1979.

16. Covan, John M., *Electrical Hazards Control Manual.* Little Rock, AR: Arkansas Department of Labor, 1977.

17. *Cranes,* A Guide to Good Work Practices for Operators (NIOSH No. 78-192). Cincinnati, OH: U.S. Department of Health, Education and Welfare (NIOSH), 1978.

18. Dickie, D. E., *Crane Handbook.* Toronto, Ontario: Construction Safety Association of Ontario, 1975.

19. *Engineering Control of Welding Fumes* (NIOSH No. 75-115). Cincinnati, OH: U.S. Department of Health, Education and Welfare (NIOSH), 1975.

20. *Fabricated Structural Metal Products Industry,* Health and Safety Guide for (NIOSH No. 78-100). Cincinnati, OH: U.S. Department of Health, Education and Welfare (NIOSH), 1977.

21. *Fiberglass Layup and Sprayup* (NIOSH No. 76-158). Cincinnati, OH: U.S. Department of Health, Education and Welfare (NIOSH), 1976.

22. *Fire Protection Handbook* (15th ed.). Quincy, MA: National Fire Protection Association, 1981.

23. *Flash Point Index of Trade Name Liquids* (9th ed.). Boston: National Fire Protection Association, 1978.

24. *Foundries,* Health and Safety Guide for (NIOSH No. 76-124). Cincinnati, OH: U.S. Department of Health, Education and Welfare, 1976.

25. Fox, Stephen, "Manhole Cover Industry Hurt by Industry Imports," *Northwest Arkansas Times,* 1981. (An Associated Press interview with Mr. Jim Pinkerton, Pinkerton Foundry, Lodi, CA.)

26. *Fumes and Gases in the Welding Environment.* Miami, FL: American Welding Society, 1979.

27. *General Industry OSHA Safety and Health Standards* (29 CFR 1910; OSHA No. 2206) (rev.). Washington, D.C.: U.S. Department of Labor, Occupational Safety and Health Administration, 1981.

28. Garg, A., D. B. Chaffin, and G. D. Herrin, "Prediction of Metabolic Rates for Manual Materials Handling Jobs," *American Industrial Hygiene Association Journal,* 39 (1978), pp. 661–674.

29. *Grain Mills,* Health and Safety Guide for (NIOSH No. 75-144). Cincinnati, OH: U.S. Department of Health, Education and Welfare (NIOSH), 1975.

30. Grimaldi, John V., and Rollin H. Simonds, *Safety Management* (3rd ed.). Homewood, IL: Richard D. Irwin, Inc., 1975.

31. Hamilton, Robert W., "The Role of Nongovernmental Standards in the Development of Mandatory Federal Standards Affecting Safety or Health," *Texas Law Review,* 56, No. 8 (November 1977).

32. Hammer, Willie, *Occupational Safety Management and Engineering* (2nd ed.). Englewood Cliffs, NJ: Prentice-Hall, Inc., 1981.

33. *Handbook of Organic Industrial Solvents*, Technical Guide No. 6 (5th ed.). Chicago: Alliance of American Insurers, 1980.

34. *Health and Safety Guide for Public Warehousing*. Cincinnati, OH: U.S. Department of Health, Education and Welfare (NIOSH), 1978.

35. Heinrich, H. W., *Industrial Accident Prevention* (4th ed.). New York: McGraw-Hill Book Company, 1959.

36. *Hotels and Motels*, Health and Safety Guide for (NIOSH No. 76-112) Cincinnati, OH: U.S. Department of Health, Education and Welfare (NIOSH), 1975.

37. Hyatt, E. C., et al., "Effect of Facial Hair on Worker Performance," *American Industrial Hygiene Association Journal,* April 1973.

38. Imre, John, Unpublished doctoral dissertation, Michigan State University, East Lansing, MI, 1974.

39. *The Industrial Environment—Its Evaluation and Control*. Cincinnati, OH: U.S. Department of Health, Education and Welfare (NIOSH), 1973.

40. *Industrial Noise Control Manual* (NIOSH 79-117) (rev. ed.). Cincinnati, OH: U.S. Department of Health, Education and Welfare, 1978.

41. *Industrial Ventilation* (15th ed.). Lansing, MI: Committee on Industrial Ventilation, American Conference of Governmental Industrial Hygienists (ACGIH), 1978.

42. *Lithographic Printing Industry,* Employee Health and Safety (NIOSH No. 77-223). Cincinnati, OH: U.S. Department of Health, Education and Welfare (NIOSH), 1977.

43. Lovett, John N., *Industrial Noise Control Manual*. Little Rock, AR: Arkansas Department of Labor, 1976.

44. *Machine Guarding: Assessment of Need* (NIOSH No. 75-173). Cincinnati, OH: U.S. Department of Health, Education and Welfare (NIOSH), 1975.

45. *Manual of Respiratory Protection against Radioactive Materials* (NUREG-0041). U.S. Atomic Energy Commission, Directorate of Regulatory Standards, 1974.

46. *Manufacturers of Paints and Allied Products,* Health and Safety Guide for (NIOSH No. 75-179). Cincinnati, OH: U.S. Department of Health, Education and Welfare (NIOSH), 1975.

47. (a) McElroy, Frank E., ed., "Administration and Programs," *Accident Prevention Manual for Industrial Operations* (8th ed.). Chicago: National Safety Council, 1981.

47. (b) _____ , "Engineering and Technology," *Accident Prevention Manual for Industrial Operations* (8th ed.). Chicago: National Safety Council, 1981.

48. *Metal Stamping Operations,* Health and Safety Guide for (NIOSH No. 75-174). Cincinnati, OH: U.S. Department of Health, Education and Welfare (NIOSH), 1975.

49. *Method of Recording and Measuring Work Injury Experience* (ANSI Z16.1-1967). New York: American National Standards Institute, 1967.

50. *NFPA Inspection Manual* (3rd ed.). Boston: National Fire Protection Association, 1970.

51. *NIOSH Certified Equipment,* Cumulative Supplement (NIOSH No. 77-195). Morgantown, WV: U.S. Department of Health, Education and Welfare (NIOSH), 1977.

52. *National Electrical Code®* (NEC). Boston: National Fire Protection Association, 1971 and 1981.

53. *Occupational Diseases: A Guide to Their Recognition*. Cincinnati, OH: U.S. Department of Health, Education and Welfare, 1951.

54. *Occupational Exposure Sampling Manual* (NIOSH No. 77-173). Cincinnati, OH: U.S. Department of Health, Education and Welfare (NIOSH), 1977.

55. *Occupational Safety and Health Cases*, Vols. 1–9. Washington, DC: Bureau of National Affairs, Inc., 1974–1982.

56. *Occupational Safety and Health Directory* (NIOSH No. 80-124). Cincinnati, OH: U.S. Department of Health and Human Services (NIOSH), 1980.

57. *Occupational Safety and Health in Vocational Education* (NIOSH No. 79-125). Cincinnati, OH: U.S. Department of Health and Human Services (NIOSH), 1979.

58. *Occupational Safety and Health Reporter* (weekly periodical), Vols. 1–11. Washington, DC: Bureau of National Affairs, Inc., 1971–1982.

59. Olishifski, Julian B., and Frank E. McElroy, eds., *Fundamentals of Industrial Hygiene*. Chicago: National Safety Council, 1971.

60. *Overhead and Gantry Cranes* (ANSI B.30.2.0-1976). New York: American National Standards Institute, 1976.

61. Peterson, Donald R., and David B. Thomas, *Fundamentals of Epidemiology*. Lexington, MA: Lexington Books, 1978.

62. *Plumbing, Heating and Air Conditioning Contractors,* Health and Safety Guide for (NIOSH No. 76-127). Cincinnati, OH: U.S. Department of Health, Education and Welfare (NIOSH), 1975.

63. *Powder Actuated Fastening Tools* (NIOSH No. 78-178A). Cincinnati, OH: U.S. Department of Health, Education and Welfare (NIOSH), 1978.

64. *A Prescription for Battery Workers* (NIOSH No. 76-153). Cincinnati, OH: U.S. Department of Health, Education and Welfare (NIOSH), 1976.

65. *Public Law 91-596* (The Williams-Steiger Occupational Safety and Health Act of 1970), U.S. Congress, December 19, 1970.

66. *Recordkeeping Requirements under the Occupational Safety and Health Act of 1970*. Washington, DC: U.S. Department of Labor, Occupational Safety and Health Administration, 1978.

67. *Respiratory Protection,* A Guide for the Employee (NIOSH No. 78-193B). Cincinnati, OH: U.S. Department of Health, Education and Welfare (NIOSH), 1978.

68. *Respiratory Protection,* An Employer's Manual (NIOSH No. 78-193A). Cincinnati, OH: U.S. Department of Health, Education and Welfare (NIOSH), 1978.

69. *Rigging Manual*. Toronto, Ontario: Construction Safety Association of Ontario, 1975.

70. *Safety Requirements for Woodworking Machinery* (ANSI 01.1-1975). New York: American National Standards Institute, 1975.

71. *Safety Standard for Monorail Systems and Underhung Cranes* (ANSI B30.11-1973). New York: American National Standards Institute, 1973.

72. Sax, N. Irving, *Dangerous Properties of Industrial Materials* (5th ed.). New York: Van Nostrand Reinhold Company, 1975.

73. *Sign and Advertising Display Manufacturers,* Health and Safety Guide for (NIOSH No. 76-126). Cincinnati, OH: U.S. Department of Health, Education and Welfare (NIOSH), 1976.

74. Smith, Harry C., "BLEVE Can Blow You to Oblivion," *Ohio Monitor,* 52, No. 10, October 1979 p. 16.

75. *Threshold Limit Values* (TLV Booklet). Cincinnati, OH: American Conference of Governmental Industrial Hygienists.

76. *The Welding Environment.* Miami, FL: American Welding Society, 1973.

77. *Wooden Furniture Manufacturing,* Health and Safety Guide for (NIOSH No. 75-167). Cincinnati, OH: U.S. Department of Health, Education and Welfare (NIOSH), 1975.

78. *Work Injury and Illness Rates.* Chicago: National Safety Council, 1981.

79. *Work Practices Guide for Manual Lifting* (NIOSH No. 81-122). Cincinnati, OH: U.S. Department of Health and Human Services (NIOSH), 1981.

Appendix

OSHA Permissible Exposure Levels

TABLE A.1 GENERAL TABLE OF PELS

Substance	p/m^a	mg/m$^{3\ b}$		Substance	p/m^a	mg/m$^{3\ b}$
Acetaldehyde	200	360		Benzoyl peroxide		5
Acetic acid	10	25		Benzyl chloride	1	5
Acetic anhydride	5	20		Biphenyl, (*see* Diphenyl)		
Acetone	1000	2400		Boron oxide		15
Acetonitrile	40	70	C	Boron trifluoride	1	3
Acetylene dichloride (*see* 1,2-				Bromine	0.1	0.7
Dichoroethylene)				Bromoform—skin	0.5	5
Acetylene tetrabromide	1	14		Butadiene (1,3-butadiene)	1000	2200
Acrolein	0.1	0.25		Butanethiol (*see* Butyl mercaptan)		
Acrylamide—skin		0.3		2-Butanone	200	590
Aldrin—skin		0.25		2-Butoxy ethanol (butyl		
Allyl alcohol—skin	2	5		cellosolve)—skin	50	240
Allyl chloride	1	3		Butyl acetate (n-butyl acetate)	150	710
C Allylglycidyl ether (AGE)	10	45		sec-Butyl acetate	200	950
Allyl propyl disulfide	2	12		tert-Butyl acetate	200	950
2-Aminoethanol (*see* Ethanolamine)				Butyl alcohol	100	300
2-Aminopyridine	0.5	2		sec-Butyl alcohol	150	450
Ammonia	50	35		tert-Butyl alcohol	100	300
Ammonium sulfamate (ammate)		15	C	Butylamine—skin	5	15
n-Amyl acetate	100	525	C	tert-Butyl chromate (as CrO$_3$)—		
sec-Amyl acetate	125	650		skin		0.1
Aniline—skin	5	19		n-Butyl glycidyl ether (BGE)	50	270
Anisidine (o, p-isomers)—skin		0.5		Butyl mercaptan	10	35
Antimony and compounds (as Sb)		0.5		p-tert-Butyltoluene	10	60
ANTU (alpha naphthyl thiourea)		0.3.		Calcium oxide		5
Arsenic organic compounds (as As)		0.5		Camphor		2
Arsine	0.05	0.2		Carbaryl (Sevin)		5
Azinphos-methyl—skin		0.2		Carbon black		3.5
Barium (soluble compounds)		0.5		Carbon dioxide	5000	9000
p-Benzoquinone (*see* Quinone)						*(continued)*

361

TABLE A.1 (CONTINUED)

	Substance	p/m^a	mg/m^3 b		Substance	p/m^a	mg/m^3 b
	Carbon monoxide	50	55		Dichloromonofluoromethane	1000	4200
	Chlordane—skin		0.5		1,1-Dichloro-1-nitroethane	10	60
	Chlorinated camphene—skin		0.5	C	1,2-Dichloropropane (*see* Propylenedichloride)		
	Chlorinated diphenyl oxide		0.5		Dichlorotetrafluoroethane	1000	7000
C	Chlorine	1	3		Dieldrin—skin	0.1	0.25
	Chlorine dioxide	0.1	0.3		Diethylamine	25	75
C	Chlorine trifluoride	0.1	0.4		Diethylamino ethanol—skin	10	50
C	Chloroacetaldehyde	1	3		Diethyl ether (*see* Ethyl ether)		
	α-Chloroacetophenone (phenacylchloride)	0.05	0.3		Difluorodibromomethane	100	860
	Chlorobenzene (monochlorobenzene)	75	350	C	Diglycidyl ether (DGE)	0.5	2.8
	o-Chlorobenzylidene malononitrile (OCBM)	0.05	0.4		Dihydroxybenzene (*see* Hydroquinone)		
	Chlorobromomethane	200	1050		Diisobutyl ketone	50	290
	2-Chloro-1,3-butadiene (*see* Chloroprene)				Diisopropylamine—skin	5	20
	Chlorodiphenyl (42 percent chlorine)—skin		1		Dimethoxymethane (*see* Methylal)		
					Dimethyl acetamide—skin	10	35
	Chlorodiphenyl (54 percent chlorine)—skin		0.5		Dimethylamine	10	18
					Dimethylaminobenzene (*see* Xylidene)		
	1-Chloro, 2,3-epoxypropane (*see* Epichlorhydrin)				Dimethylaniline (n-dimethyl-aniline)—skin	5	25
	2-Chloroethanol (*see* Ethylene chlorohydrin)				Dimethylbenzene (*see* Xylene)		
	Chloroethylene (*see* Vinyl chloride)				Dimethyl 1,2-dibromo-2,2-dichloroethyl phosphate (dibrom)		3
C	Chloroform (trichloromethane)	50	240		Dimethylformamide—skin	10	30
	1-Chloro-1-nitropropane	20	100		2,6-Dimethylheptanone (*see* Diisobutyl ketone)		
	Chloropicrin	0.1	0.7		1,1-Dimethylhydrazine—skin	0.5	1
	Chloroprene (2-chloro-1,3-butadiene—skin	25	90		Dimethylphthalate		5
	Chromium				Dimethylsulfate—skin	1	5
	Soluble chromic, chromous salts as Cr		0.5		Dinitrobenzene (all isomers)—skin		1
	Metal and insoluble salts		1		Dinitro-o-cresol—skin		0.2
					Dinitrotoluene—skin		1.5
					Di-sec-octyl phthalate (di-2-ethylhexylphthalate)		5

	Substance		
	Coal tar pitch volatiles (benzene soluble fraction) anthracene, BaP, phenanthrene, acridine, chrysene, pyrene		0.2
	Cobalt, metal fume and dust		0.1
	Copper		
	Fume		0.1
	Dusts and mists		1
	Cotton dust (raw)		1[d]
	Crag herbicide		15
	Cresol (all isomers)—skin	5	22
	Crotonaldehyde	2	6
	Cumens—skin	50	245
	Cyanide (as CN)—skin		5
	Cyclohexane	300	1050
	Cyclohexanol	50	200
	Cyclohexanone	50	200
	Cyclohexene	300	1015
	Cyclopentadiene	75	200
	2,4-D		10
	DDT—skin		1
	DDVP—skin		1
	Decaborane—skin	0.05	0.3
	Demeton[R]—skin		0.1
	Diacetone alcohol (4-hydroxy-4-methyl-2-pentanone)	50	240
	1,2-Diaminoethane (see Ethylenediamine)		
	Diazomethane	0.2	0.4
	Diborane	0.1	0.1
	Dibutyl phosphate	1	5
	Dibutylphthalate		5
C	o-Dichlorobenzene	50	300
	p-Dichlorobenzene	75	450
	Dichlorodifluoromethane	1000	4950
	1,3-Dichloro-5,5-dimethyl hydantoin		0.2
	1,1-Dichloroethane	100	400
	1,2-Dichloroethylene	200	790
C	Dichloroethyl ether—skin	15	90
	Dichloromethane (see Methylenechloride)		
	Dioxane (diethylene dioxide)—skin	100	360
	Diphenyl	0.2	1
	Diphenylmethane diisocyanate [see Methylene bisphenyl isocyanate (MDI)]		
	Dipropylene glycol methyl ether—skin	100	600
	Endrin—skin		0.1
	Epichlorhydrin—skin	5	19
	EPN—skin		0.5
	1,2-Epoxypropane (see Propyleneoxide)		
	2,3-Epoxy-1-propanol (see Glycidol)		
	Ethanethiol (see Ethylmercaptan)		
	Ethanolamine	3	6
	2-Ethoxyethanol—skin	200	740
	2-Ethoxyethylacetate (Cellosolve acetate)—skin	100	540
	Ethyl acetate	400	1400
	Ethyl acrylate—skin	25	100
	Ethyl alcohol (ethanol)	1000	1900
	Ethylamine	10	18
	Ethyl sec-amyl ketone (5-methyl-3-heptanone)	25	130
	Ethyl benzene	100	435
	Ethyl bromide	200	890
	Ethyl butyl ketone (3-heptanone)	50	230
	Ethyl chloride	1000	2600
	Ethyl ether	400	1200
	Ethyl formate	100	300
C	Ethyl mercaptan	10	25
	Ethyl silicate	100	850
	Ethylene chlorohydrin—skin	5	16
	Ethylenediamine	10	25
C	Ethylene glycol dinitrate and/or Nitroglycerin—skin	0.2[e]	1
	Ethylene glycol monomethyl ether acetate (see Methyl Cellosolve acetate)		

(continued)

C	Substance	p/m^a	mg/m$^{3\ b}$
	Ethylene imine—skin	0.5	1
	Ethylene oxide	50	90
	Ethylidine chloride (see 1,1-Dichloroethane)		
	N-Ethylmorpholine—skin	20	94
	Ferbam		15
	Ferrovanadium dust		1
	Fluoride (as F)		2.5
	Fluorine	0.1	0.2
	Fluorotrichloromethane	1000	5600
	Formic acid	5	9
	Furfural—skin	5	20
	Furfuryl alcohol	50	200
	Glycidol (2,3-epoxy-1-propanol)	50	150
	Glycol monoethyl ether (see 2-Ethoxyethanol)		
	Guthion (see Azinphosmethyl)		
	Hafnium		0.5
	Heptachlor—skin		0.5
	Heptane (n-heptane)	500	2000
	Hexachloroethane—skin	1	10
	Hexachloronaphthalene—skin		0.2
	Hexane (n-hexane)	500	1800
	2-Hexanone	100	410
	Hexone (methyl isobutyl ketone)	100	410
	sec-Hexyl acetate	50	300
	Hydrazine—skin	1	1.3
	Hydrogen bromide	3	10
	Hydrogen chloride	5	7
C	Hydrogen cyanide—skin	10	11
	Hydrogen peroxide (90%)	1	1.4
	Hydrogen selenide	0.05	0.2
	Hydroquinone		2
C	Iodine	0.1	1
	Iron oxide fume		10
	Isoamyl acetate	100	525

C	Substance	p/m^a	mg/m$^{3\ b}$
	Methyl chloroform	350	1900
	Methylcyclohexane	500	2000
	Methylcyclohexanol	100	470
	o-Methylcyclohexanone—skin	100	460
	Methyl ethyl ketone (MEK) (see 2-Butanone)		
	Methyl formate	100	250
	Methyl iodide—skin	5	28
	Methyl isobutyl carbinol—skin	25	100
	Methyl isobutyl ketone (see Hexone)		
	Methyl isocyanate—skin	0.02	0.05
C	Methyl mercaptan	10	20
	Methyl methacrylate	100	410
	Methyl propyl ketone (see 2-Pentanone)		
C	Methyl styrene	100	480
C	Methylene bisphenyl isocyanate (MDI)	0.02	0.2
	Molybdenum		
	Soluble compounds		5
	Insoluble compounds		15
	Monomethyl aniline—skin	2	9
C	Monomethyl hydrazine—skin	0.2	0.35
	Morpholine—skin	20	70
	Naphtha (coaltar)	100	400
	Naphthalene	10	50
	Nickel carbonyl	0.001	0.007
	Nickel, metal and soluble compounds, as Ni		1
	Nicotine—skin		0.5
	Nitric acid	2	5
	Nitric oxide	25	30
	p-Nitroaniline—skin	1	6
	Nitrobenzene—skin	1	5
	p-Nitrochlorobenzene—skin		1
	Nitroethane	100	310

	Substance		
	Isoamyl alcohol	100	360
	Isobutyl acetate	150	700
	Isobutyl alcohol	100	300
	Isophorone	25	140
	Isopropyl acetate	250	950
	Isopropyl alcohol	400	980
	Isopropylamine	5	12
	Isopropylether	500	2100
	Isopropyl glycidyl ether (IGE)	50	240
	Ketene	0.5	0.9
	Lindane—skin		0.5
	Lithium hydride		0.025
	LPG (liquified petroleum gas)	1000	1800
	Magnesium oxide fume		15
	Malathion—skin		15
	Maleic anhydride	0.25	1
C	Manganese		5
	Mesityl oxide	25	100
	Methanethiol (see Methyl mercaptan)		
	Methoxychlor		15
	2-Methoxyethanol (see Methyl cellosolve)		
	Methyl acetate	200	610
	Methyl acetylene (propyne)	1000	1650
	Methyl acetylene-propadiene mixture (MAPP)	1000	1800
	Methyl acrylate—skin	10	35
	Methylal (dimethoxymethane)	1000	3100
	Methyl alcohol (methanol)	200	260
	Methylamine	10	12
	Methyl amyl alcohol (see Methyl isobutyl carbinol)		
	Methyl (n-amyl) ketone (2-heptanone)	100	465
	Methyl bromide—skin	20	80
	Methyl butyl ketone (see 2-Hexanone)		
C	Methyl Cellosolve—skin	25	80
	Methyl Cellosolve acetate—skin	25	120

	Substance		
C	Nitrogen dioxide	5	9
	Nitrogen trifluoride	10	29
C	Nitroglycerin—skin	0.2	2
	Nitromethane	100	250
	1-Nitropropane	25	90
	2-Nitropropane	25	90
	Nitrotoluene—skin	5	30
	Nitrotrichloromethane (see Chloropicrin)		
	Octachloronaphthalene—skin		0.1
	Octane	500	2350
	Oil mist, mineral		5
	Osmium tetroxide		0.002
	Oxalic acid		1
	Oxygen difluoride	0.05	0.1
	Ozone	0.1	0.2
	Paraquat—skin		0.5
	Parathion—skin		0.1
	Pentaborane	0.005	0.01
	Pentachloronaphthalene—skin		0.5
	Pentachlorophenol—skin		0.5
	Pentane	1000	2950
	2-Pentanone	200	700
	Perchloromethyl mercaptan	0.1	0.8
	Perchloryl fluoride	3	13.5
	Petroleum distillates (naphtha)	500	2000
	Phenol—skin	5	19
	p-Phenylene diamine—skin		0.1
	Phenyl ether (vapor)	1	7
	Phenyl ether-biphenyl mixture (vapor)	1	7
	Phenylethylene (see Styrene)		
	Phenyl glycidyl ether (PGE)	10	60
	Phenylhydrazine—skin	5	22
	Phosdrin (Mevinphos)—skin	0.1	0.1
	Phosgene (carbonyl chloride)	0.1	0.4
	Phosphine	0.3	0.4
	Phosphoric acid		1
	Phosphorus (yellow)		0.1

(continued)

TABLE A.1 (CONTINUED)

Substance	p/m[a]	mg/m[3] [b]
Phosphorus pentachloride		1
Phosphorus pentasulfide		1
Phosphorus trichloride	0.5	3
Phthalic anhydride	2	12
Picric acid—skin		0.1
Pival (2-pivalyl-1,3-indandione)		0.1
Platinum (soluble salts) as Pt		0.002
Propane	1000	1800
n-Propyl acetate	200	840
Propyl alcohol	200	500
Propylene dichloride	75	350
Propylene imine—skin	2	5
Propylene oxide	100	240
n-Propyl nitrate	25	110
Propyne (see Methylacetylene)		
Pyrethrum		5
Pyridine	5	15
Quinone	0.1	0.4
Rhodium		
Metal fume and dusts, as Rh		0.1
Soluble salts		0.001
Ronnel		15
Rotenone (commercial)		5
Selenium compounds (as Se)		0.2
Selenium hexafluoride	0.05	0.4
Silver, metal and soluble compounds		0.01
Sodium fluoroacetate (1080)—skin		0.05
Sodium hydroxide		2
Stibine	0.1	0.5
Stoddard solvent	500	2900
Strychnine		0.15
Sulfur dioxide	5	13
Sulfur hexafluoride	1000	6000
Sulfuric acid		1

Substance	p/m[a]	mg/m[3] [b]
Tetrachloronaphthalene—skin		2
Tetraethyl lead (as Pb)—skin		0.075
Tetrahydrofuran	200	590
Tetramethyl lead (as Pb)—skin		0.075
Tetramethyl succinonitrile—skin	0.5	3
Tetranitromethane	1	8
Tetryl (2,4,6-trinitrophenyl-methylnitramine)—skin		1.5
Thallium (soluble compounds) as Tl—skin		0.1
Thiram		5
Tin		
Inorganic compounds, except oxides		2
Organic compounds		0.1
C Toluene-2,4-diisocyanate	0.02	0.14
o-Toluidine—skin	5	22
Toxaphene (see Chlorinated camphene)		
Tributyl phosphate		5
1,1,1-Trichloroethane (see Methyl chloroform)		
1,1,2-Trichloroethane—skin	10	45
Titaniumdioxide		15
Trichloromethane (see Chloroform)		
Trichloronaphthalene—skin		5
1,2,3-Trichloropropane	50	300
1,1,2-Trichloro 1,2,2-trifluoroethane	1000	7600
Triethylamine	25	100
Trifluoromonobromomethane	1000	6100
2,4,6-Trinitrophenol (see Picric acid)		
2,4,6-Trinitrophenylmethyl-nitramine (see Tetryl)		
Trinitrotoluene—skin		1.5
Triorthocresyl phosphate		0.1
Triphenyl phosphate		3

Substance		
Sulfur monochloride	1	6
Sulfur pentafluoride	0.025	0.25
Sulfuryl fluoride	5	20
Systox (see Demeton)		
2,4,5-T		10
Tantalum		5
TEDP—skin		0.2
Tellurium		0.1
Tellurium hexafluoride	0.02	0.2
TEPP—skin		0.05
C Terphenyls	1	9
1,1,1,2-Tetrachloro-2,2-difluoroethane	500	4170
1,1,2,2-Tetrachloro-1,2-difluoroethane	500	4170
1,1,2,2-Tetrachloroethane—skin	5	35
Tetrachloromethane (see Carbon tetrachloride)		
Turpentine	100	560
Uranium		
Soluble compounds		0.05
Insoluble compounds		0.25
C Vanadium		
V_2O_5 dust		0.5
V_2O_5 fume		0.1
Vinyl benzene (see Styrene)		
Vinylcyanide (see Acrylonitrile)		
Vinyl toluene	100	480
Warfarin		0.1
Xylene (xylol)	100	435
Xylidine—skin	5	25
Yttrium		1
Zinc chloride fume		1
Zinc oxide fume		5
Zirconium compounds (as Zr)		5

Note: A "C" preceding an entry indicates the ceiling value.

[a] Parts of vapor or gas per million parts of contaminated air by volume at 25 °C and 760 mm Hg pressure.

[b] Approximate milligrams of particulate per cubic meter of air.

(No footnote "c" is used to avoid confusion with ceiling value notations.)

[d] This standard applies in cotton yarn manufacturing until compliance with § 1910.1043 (c) and (e) is achieved.

[e] An atmospheric concentration of not more than 0.02 ppm, or personal protection may be necessary to avoid headache.

Source: Code of Federal Regulations 29 CFR 1910.1000.

TABLE A.2 PELS AND STELS FOR SOME SPECIALIZED MATERIALS

Material	8-hour time weighted average	Acceptable ceiling concentration	Acceptable maximum peak above the acceptable ceiling concentration for an 8-hour shift	
			Concentration	Maximum duration
Benzene (Z37.40—1969)[a]	10 ppm	25 ppm	50 ppm	10 minutes
Beryllium and beryllium compounds (Z37.29—1970)	2 $\mu g/m^3$	5 $\mu g/m^3$	25 $\mu g/m^3$	30 minutes
Cadmium dust (Z37.5—1970)	0.2 mg/m^3	0.6 mg/m^3		
Cadmium fume (Z37.5—1970)	0.1 mg/m^3	0.3 mg/m^3		
Carbon disulfide (Z37.3—1968)	20 ppm	30 ppm	100 ppm	30 minutes
Carbon tetrachloride (Z37.17—1967)	10 ppm	25 ppm	200 ppm	5 minutes in any 4 hours
Chronic acid and chromates (Z37.7—1971)		1 $mg/10\ m^3$		
Ethylene dibromide (Z37.31—1970)	20 ppm	30 ppm	50 ppm	5 minutes
Ethylene dichloride (Z37.21—1969)	50 ppm	100 ppm	200 ppm	5 minutes in any 3 hours
Fluoride as dust (Z37.28—1969)	2.5 mg/m^3			
Formaldehyde (Z37.16—1967)	3 ppm	5 ppm	10 ppm	30 minutes
Hydrogen fluoride (Z37.28—1969)	3 ppm			
Hydrogen sulfide (Z37.2—1966)		20 ppm	50 ppm	10 minutes once only if no other measurable exposure occurs
Mercury (Z37.8—1971)		1 $mg/10\ m^3$		
Methyl chloride (Z37.18—1969)	100 ppm	200 ppm	300 ppm	5 minutes in any 3 hours
Methylene chloride (Z37.23—1969)	500 ppm	1000 ppm	2000 ppm	5 minutes in any 2 hours
Organo(alkyl)mercury (Z37.30—1969)	0.01 mg/m^3	0.04 mg/m^3		
Styrene (Z37.15—1969)	100 ppm	200 ppm	600 ppm	5 minutes in any 3 hours
Tetrachloroethylene (Z37.22—1967)	100 ppm	200 ppm	300 ppm	5 minutes in any 3 hours
Toluene (Z37.12—1967)	200 ppm	300 ppm	500 ppm	10 minutes
Trichloroethylene (Z37.19—1967)	100 ppm	200 ppm	300 ppm	5 minutes in any 2 hours

[a]Occupational exposures to benzene are subject to the requirements of § 1910.1028 except as specifically exempted by §1910.1028(a)(2). Exposures exempted by § 1910.1028(a)(2) are covered by this § 1910.1000.

Source: Code of Federal Regulations 29 CFR 1910.1000.

TABLE A.3 PELS FOR MINERAL DUSTS

Substance	Mppcf[a]	Mg/m^3
Silica		
Crystalline		
Quartz (respirable)	$\dfrac{250^b}{\%SiO_2 + 5}$	$\dfrac{10 \text{ mg/m}^{3c}}{\%SiO_2 + 2}$
Quartz (total dust)		$\dfrac{30 \text{ mg/m}^3}{\%SiO_2 + 2}$
Cristobalite: Use ½ the value calculated from the count or mass formulas for quartz.		
Tridymite: Use ½ the value calculated from the formulas for quartz.		
Amorphous, including natural diatomaceous earth	20	$\dfrac{80 \text{ mg/m}^3}{\%SiO_2}$
Silicates (less than 1% crystalline silica)		
Mica	20	
Soapstone	20	
Talc (nonasbestos form)	20[d]	
Talc (fibrous). Use asbestos limit		
Tremolite (*see* talc, fibrous)		
Portland cement	50	
Graphite (natural)	15	
Coal dust (respirable fraction less than 5% SiO$_2$)		2.4 mg/m^3
		or
For more than 5% SiO$_2$		$\dfrac{10 \text{ mg/m}^3}{\%SiO_2 + 2}$
Inert or nuisance dust		
Respirable fraction	15	5 mg/m^3
Total dust	50	15 mg/m^3

Note: Conversion factors:

$$\text{mppcf} \times 35.3 = \text{million particles per cubic meter}$$
$$= \text{particles per cubic centimeter}$$

[a] Millions of particles per cubic foot of air, based on impinger samples counted by light-field technics.

[b] The percentage of crystalline silica in the formula is the amount determined from airborne samples, except in those instances in which other methods have been shown to be applicable.

[c] Both concentration and percent quartz for the application of this limit are to be determined from the fraction passing a size selector with the following characteristics:

[d] Containing <1% quartz; if 1% quartz, use quartz limit.

Aerodynamic diameter (unit density sphere)	Percent passing selector
2	90
2.5	75
3.5	50
5.0	25
10	0

The measurements under this note refer to the use of an AEC instrument. The respirable fraction of coal dust is determined with a MRE; the figure corresponding to that of 2.4 mg/m^3 in the table for coal dust is 4.5 mg/m^3.

Source: Code of Federal Regulations 29 CFR 1910.1000.

Index

D